A2 Geography
for Edexcel

Digby ▸ Hurst ▸ Ch........ ▸ ......ng ▸ Dale

OXFORD

OXFORD
UNIVERSITY PRESS

Great Clarendon Street, Oxford OX2 6DP

Oxford University Press is a department of the University of Oxford.
It furthers the University's objective of excellence in research,
scholarship, and education by publishing worldwide in

Oxford   New York

Auckland   Cape Town   Dar es Salaam   Hong Kong   Karachi
Kuala Lumpur   Madrid   Melbourne   Mexico City   Nairobi
New Delhi   Shanghai   Taipei   Toronto

With offices in

Argentina   Austria   Brazil   Chile   Czech Republic   France   Greece
Guatemala   Hungary   Italy   Japan   Poland   Portugal   Singapore
South Korea   Switzerland   Thailand   Turkey   Ukraine   Vietnam

Oxford is a registered trade mark of Oxford University Press
in the UK and in certain other countries

© Oxford University Press 2009

Authors: Bob Digby, Catherine Hurst, Russell Chapman, Dan Cowling,
Tony Dale

The moral rights of the authors have been asserted

Database right Oxford University Press (maker)

First published 2009

British Library Cataloguing in Publication Data

Data available
ISBN 978-0-19-913483-0

10 9 8

Printed in Malaysia by Vivar Printing Sdn Bhd.

Acknowledgements

The publisher and authors would like to thank the following for permission to use
photographs and other copyright material:

p6 Reed Saxon/Associated Press; p8 David McNew/Staff/Getty Images News/Getty Images
North America/Getty Images; p9 Str Old/Reuters; p10t Hank Morgan – Rainbow/Science
Faction/Getty Images; p10b Greg Smith/Corbis; p11 Ian Bracegirdle/Shutterstock; p13t
Copyright 2008 3TIER, Inc.; p13b Albuisson, M., Lefevre M. and Wald L./ Mines ParisTech;
p15 Bob Sacha/Corbis; p16t Caroline Penn/Photolibrary; p16b Practical Action/Karen
Robinson; p17 www.CartoonStock.com; p18 Natalie Behring-Chisholm/Stringer/Getty
Images News/Getty Images AsiaPac/Getty Images; p19m David Hume Kennerly/Reportage/
Getty Images; p19b Steve Vidler/Photolibrary; p22 Lynn Stone/Photolibrary; p24 Handout/
Getty Images News/Getty Images Europe/Getty Images; p25t Misha Japaridze/Associated
Press; p25b Jeremy Nicholl/Alamy; p26t Jeremy Nicholl/Alamy p26b Sergei Chuzavkov/
Associated Press; p28(t) Barbara Cushman Rowell/Mountain Light/Alamy; p30t Todd Korol/
Dinodia; p30b Eamon Mac Mahon/Associated Press; p32 Bullit Marquez/Associated Press;
p35 Jupiter Images; p39 Christian Charisius/Reuters; p41 Erik S. Lesser/EPA; p42 Gen-Gerolf
Niebner/A1PIX; p44t Ian Waldie/Staff/Getty Images News/Getty Images North America/Getty
Images; p44b Petros Giannakouris/Associated Press; p49 Max Earey/Shutterstock; p52t David
Paul Morris/Stringer/Getty Images News/Getty Images North America/Getty Images; p52b
MODIS Rapid Response/NASA; p53 Alamy; p57t John C Dohrenwend/Southwest Satellite
Imaging; p57m John C Dohrenwend/Southwest Satellite Imaging; p57b Mark & Audrey
Gibson/Dinodia; p58 Doc Searls; p59/David McNew/Staff/Getty Images News/Getty Images
North America/Getty Images; p62 US Geological Survey; p63t John Javellana/Reuters; p63b
The United Nations; p65 Dadang Tri/Reuters; p66 Andrew Mercer; p70 Alamy; p71 Ali Kabas/
Alamy; p72t Yaniv Nadav Photography/Photographers Direct; p72b Gil Hadani Photography/
Photographers Direct; p74 Ammar Awad/Reuters; p75 Gideon Lichfield; p79 Yves Gellie/
Corbis; p80 WaterAid (www.wateraid.org); p81l WaterAid (www.wateraid.org); p81r
WaterAid (www.wateraid.org); p85 Vickie Burt/Photographers Direct; p88 China Photos/
Stringer/Getty Images News/Getty Images AsiaPac/Getty Images; p90 Julie Plasencia/
Associated Press; p94 Ted Kerasote/Photolibrary; p96 Andy Soloman/Reuters; p97t Steven J.
Kazlowski/Alamy; p97b Subhankar Banerjee/Gwich'in Steering Committee; p98 James P.
Blair/National Geographic/Getty Images; p99 John Gaps III/Associated Press; p103 F Hart/The
Bridgeman Art Library/Getty Images; p104 Konstantin Mikhailov/Nature Picture Library/
p106l Gerry Pearce/Alamy; p106r Bob Digby; p107 Stock Connection Blue/Alamy; p107bl
Alison Thompson/Photographers Direct; p107br Getty Images; p110t Port Douglas Daintree
Tourism Limited; p110b Jeff Hunter/ Photographer's Choice/Getty Images; p111t Holger
Leue/Alamy; p111b Lincoln Fowler/Icons International; p112t Holger Leue/Lonely Planet
Images; p112b Bob Digby; p115 Jon Arnold Images Ltd /Alamy; p118t Ray Dennis/Alamy;
p118b Aristidis Vafeiadakis/Alamy; p119t Jesse Allen/Earth Observatory/NASA; P119b Getty
Images; p120t Getty Images; p120b Tim Graham/Alamy; p122t Mark Edwards/Still Pictures;
p122b Yann Arthus-Bertrand/Corbis; p123 Jeremy Horner/Alamy; p124 G P Bowater/Alamy;

p125t Roger Cracknell 01/Classic/Alamy; p125b Getty Images; p127 Alamy; p128 Binsar
Bakkara/Associated Press; p129 Dirk Enters/Imagebroker/Alamy; p130t Jorgen Schytte/Still
Pictures; p130b Miroslava Vilimova/Alamy; p131 Aditya Singh/Dinodia Photo Library; p136
Wolfgang Rattay/Reuters; p137 Spectrum Photofile Inc./Photographers Direct; p139 SASI
Group (University of Sheffield) and Mark Newman (University of Michigan); p140 SASI
Group (University of Sheffield) and Mark Newman (University of Michigan); p142 Bob Digby;
p143 Aldona Sabalis/Science Photo Library; p145 Dmitri Kessel/Stringer/Time & Life Pictures/
Getty Image; p148 Vipin chandren/The Hindu Images; p151t Bob Digby; 151b Associated
Press; p152 Olivier Asselin/Associated Press; p155t Victor Englebert/Photographers Direct;
p155b Victor Englebert/Photographers Direct; p156 Renee Morris / Alamy; p157 Paul
Brehem/Photographers Direct; p159t Alamy; 159b Bert Wiklund/Photographers Direct; p161
BMW; p162 China Photos/Stringer/Getty Images News/Getty Images AsiaPac/Getty Images;
p163 Ng Han Guan/Associated Press; p164 Dima Korotayev/AFP; p165t Sergei Karpukhin/
Associated Press; p165b Iain Masterton /Alamy; p166 Feije Riemersma Photography/
Photographers Direct; p172 Bob Digby; p174t SASI Group (University of Sheffield) and Mark
Newman (University of Michigan); p174b Urbanmyth/Alamy; p178 Amy Sancetta/Associated
Press; p179t Scott Peterson/Contributor/Getty Images News/Getty Images North America/
Getty Images; p179b Cheryl Ravelo/Reuters; p180b John Henry Claude Wilson/Robert
Harding World Imagery/John Henry Claude Wilson/Getty Images; p184t Euan Denholm/
Reuters; p184b Sue O'Connor/Alamy; p185t Wandering Spirit Travel Images/Photographers
Direct; p185b Glenna Gordon; p188t Eduardo Munoz/Reuters; p188b Arnd Wiegmann/
Reuters; p191 Data: AVHRR, NDVI, Seawifs, MODIS, NCEP, DMSP and Sky2000 star catalog;
AVHRR and Seawifs texture: Reto Stockli; Visualization: Marit Jentoft-Nilsen; p194t Walter
Bibikow/Getty Images; p194b Jerome Delay/Associated Press; p195 Alamy; p196t Vickie Burt
Photography/Photographers Direct; p196b Jon Hrusa/Associated Press; p198 Matthias Geel;
p199 Richard Hammerton/Art Directors & Trip Photo Library; p200 Robert Wallis/Corbis;
p201 Manor Photography/Alamy; p202 Wandering Spirit Travel Images/Photographers
Direct; p203 Glenna Gordon; p204 Simon Rawles/Alamy; p205 Paul Joynson-hicks/Art
Directors & Trip Photo Library; p206t Marcia Levy; p206b Karel Prinsloo/Associated Press;
p207 Brian Cruickshank/Lonely Planet Images; p208 Martin Barlow/Art Directors & Trip
Photo Library; p209 Caroline Penn/Corbis; p211t James Akena; p211b James Akena; p212
Hugh Scudder; p216t Medical-on-Line/Alamy; p216b Karen Kasmauski/Corbis; p220 Mike
Booth/Alamy; p222t Sean Sprague/Sprague Photo Stock; p222b Theo Allofs/zefa/Corbis; p225
Greg Balfour Evans/Alamy; p226 Michael Dunning/Photographer's Choice/Getty Images;
p228 Skyscan/Corbis; p230 James Marko; p231 Pavel Rahman/Associated Press; p232 Mark
Edwards/Still Pictures; p233 Cavallini James/Photolibrary; p234 Joe Raedle/Staff/Getty
Images News/Getty Images North America/Getty Images; p235 Kiersten Israel-Ballard; p237
Bruno Vincent/Staff/Getty Images News/Getty Images Europe/Getty Images; p238 Wakil
Ehsass/VOA; p239 Brent Stirton/Staff/Getty Images News/Getty Images Europe/Getty Images;
p240t Alternative Energy Development Board, Pakistan; p240b Alternative Energy
Development Board, Pakistan; p241 Barefoot; p242 Robert Pratta/Reuters; p243 Paul A.
Souders/Corbis; p244 Mike Jackson/Still Pictures; p246 Tata Motors Limited; p247 Alamy;
p249 Mark Edwards/Still Pictures; p250 Spectrum Photofile Inc./Photographers Direct; p251
Sean Sprague/Sprague Photo Stock; p252 Bruno Vincent/Staff/Getty Images News/Getty
Images Europe/Getty Images; p253t Roger Angel/University of Arizona; p253b David Harlow/
Stringer/Time & Life Pictures/Getty Images; p254 Dr Klaus Lackner/Columbia University;
p255 Columbia University; p256 Isifa/Contributor/Getty Images News/Getty Images Europe/
Getty Images; p261m D.W. Peterson/US Geological Survey; p261bl Bob Digby; p261br Marri
Nogueira/Associated Press; p262t Pablo Bartholomew/Contributor/Getty Images News/Getty
Images North America/Getty Images; p262m Jochen Tack/Alamy; p262b Istockphoto; p264
Kevin West/Contributor/Getty Images News/Getty Images North America/Getty Images;
p265t Andrew Woodley/Alamy; p265b Montserrat Tourist Board; p268 Liu Jin/AFP; p269t
Kyodo/Reuters; 269b Kimimasa Mayama/Reuters; p273t Radius Images/Jupiter Images; p273b
Photolibrary; p274t Caro Photo Agency; p274b Oddvar Midtkandal; p276t Karen Mills;
p276b Patrick Ward/Alamy; p277tr Robert Estall photo agency/Alamy; p277m Norfolk
Museums and Archaeology Service; p277br David Moore/Alamy; p277bl Norfolk Wildlife
Trust; p280 Peter Wiles/Photographers Direct; p281 Prashant Bhoot/Associated Press; p284
Eduardo Munoz/Reuters; p285 Robert Frerck/Odyssey Productions, Inc.; p288 Rob Walls/
Alamy; p289t Sipa Press/Ref Features; p289b Jayawardene Travel Photo Library/
Photographers Direct; p290t Puah Sze Ning/Sze Ning Photography; p290b Billion Lim/AFP;
p291m Nils Jorgensen/Rex Features; p291b '©2008 The Financial Times Limited. All rights
reserved'; p293t www.yellowbookltd.com; p293b Photofusion/Photographers Direct; p296
State Library of Western Australia, The Battye Library; p301bl Joerg Boethling/Still Pictures;
p301br Philip Wolmuth/Alamy; p304 Juan Manuel Borrero/Nature Picture Library; p306t
Mariana Bazo/Reuters; p306b Peter Jordan/Alamy; p309 David McNew/Staff/Getty Images
News/Getty Images North America/Getty Images; p311 Associated Press; p312t Christine
Osborne Pictures/Alamy; p312b Jose Fuste Raga/Corbis; p314 Kamran Jebreili/Kamran
Jebreili/Associated Press; p315 Nakheel PJSC; p316bl Christine Osborne/Corbis; p316br
Nakheel PJSC; p317 Anwar Mirza/Reuters; p319t Stock Connection Blue/Alamy; p319b
Robert Pratta/Reuters.

Cover photo: Oxford University Press

Illustrations are by Barking Dog Art.

**Bob Digby would like to thank Joe Powell, Tom Youngman, and Ben White for
their help in research.**
Dan Cowling would like to thank James Stock for his help in research.
**Russell Chapman would like to thank Hannah Chapman and Ben Gibson for their
help in research.**

Every effort has been made to contact copyright holders of material reproduced in this
book. Any omissions will be rectified in subsequent printings if notice is given to the
publisher.

Contents

1 Energy security — 5

1.1	Energy security in the USA	6
1.2	The energy mix	11
1.3	Access to energy	14
1.4	China: the new economic giant	17
1.5	Energy connections and geopolitics	22
1.6	The race for new resources	28
1.7	Energy supply players	32
1.8	Is there enough energy?	36
1.9	Business as usual	40
1.10	Meeting future energy needs (s)	46

2 Water conflicts — 51

2.1	California calling	52
2.2	California – environment at risk	58
2.3	Global imbalance	60
2.4	Global water crisis	62
2.5	Water insecurity	66
2.6	Managing water insecurity	70
2.7	Water, wealth and poverty – 1 (s)	76
2.8	Water, wealth and poverty – 2	82
2.9	The price of water	88

3 Biodiversity under threat — 93

3.1	The great Alaskan wilderness	94
3.2	Global distribution of biodiversity	100
3.3	Biodiversity under threat	103
3.4	The Daintree rainforest	106
3.5	Threats to the Daintree	110
3.6	Can the threats to biodiversity be successfully managed?	115
3.7	Mangroves	118
3.8	Threats to mangroves	122
3.9	Managing the threats to mangroves	127
3.10	Biodiversity – what is the future? (s)	130

4 Superpower geographies — 135

4.1	Changing world order	136
4.2	Who are the superpowers?	138
4.3	The earliest superpowers	142
4.4	Cold War and superpower rivalries	146
4.5	Colonialism – gone but not forgotten?	152
4.6	China – economic superpower?	158
4.7	Russia – the re-emerging superpower?	164
4.8	Superpower influences over nation states	168
4.9	Cultural superpowers (s)	172

5 Bridging the development gap — 177

5.1	Food in crisis	178
5.2	Identifying the development gap	180
5.3	Living on the wrong side of the gap	184
5.4	How we see the world	190
5.5	South Africa – the widening gap	194
5.6	Booming Bangalore	198
5.7	Ways forward 1: the importance of trade	202
5.8	Ways forward 2: aid or investment? (s)	206
5.9	Ways forward 3: the Millennium Development Goals	214

6 The technological fix? — 219

6.1	The geography of technology	220
6.2	Distribution of technology	222
6.3	Access to technology – 1	228
6.4	Access to technology – 2	232
6.5	The technological fix?	236
6.6	Technological leapfrogging	238
6.7	Impacts of technology	242
6.8	The effects of technology	244
6.9	Contrasts in technology	248
6.10	The global fix	252
6.11	Technology and the future (s)	256

(s) Each of these symbols indicates a synoptic section.

● ● ● **Unit 4: Geographical research introduction**   **261**

7   **Tectonic activity and hazards**   **263**

7.1   Abandoning paradise   264
7.2   Earthquakes in China and Japan   268

8   **Cold environments: landscapes and change**   **271**

8.1   The 'Big Thin' and the new 'Cold War'   272
8.2   Uncovering the past   276

9   **Life on the margins: the food supply problem**   **279**

9.1   The Kalahandi Syndrome   280
9.2   Global food crisis   284

10   **The world of cultural diversity**   **287**

10.1   Culture shock!   288
10.2   The attack of the clones   292

11   **Pollution and human health at risk**   **295**

11.1   The invisible hazard   296
11.2   Asbestos closer to home   300

12   **Consuming the rural landscape: leisure and tourism**   **303**

12.1   Save the rainforests!   304
12.2   Postcards from Paradise?   308

13   **Contested planet – synoptic investigation**   **311**

13.1   The Vision – economic success without oil   312
13.2   What has been going on in Dubai?   314
13.3   Consequences of Dubai's development   316

EXAMS: HOW TO BE SUCCESSFUL   **319**

Glossary   330
Index   334

About the questions in this book

Over to you

'Over to you' questions mostly provide you with opportunities for collaboration, for pair or group work.

On your own

'On your own' questions mostly provide you with opportunities for independent work.

Exam question: There are 'exam-style' questions, with marks allocated. These are clearly identified. They are included in certain 'On your own' questions, and on the final page for each chapter.

Synoptic question: There are 'exam-style' synoptic questions, with marks allocated. These are clearly identified in the synoptic sections.

You will encounter questions like these in your exam, so they give you the chance of valuable exam practice.

The exam questions and synoptic questions are based on Edexcel examination questions; they have not been taken from past exam papers, and they have not been provided by Edexcel.

There are also 'What do you think?' boxes. These ask questions about controversial issues, and will challenge your critical thinking. Questions of this type will not appear in exam papers.

Should energy be used as an economic or political weapon?

What do you think?

Answers

The *A2 Geography for Edexcel Activities and Planning OxBox CD-ROM* provides:

• answer guidance for the 'Over to you', 'On your own', and 'What do you think?' questions
• answer information and mark schemes for the exam questions and synoptic questions.

Edexcel accepts no responsibility whatsoever for the accuracy or method of working in the answers given.

CONTENTS

What do I have to know?

This chapter is about tensions between those with energy supplies and those without. Global supplies are not evenly distributed. Some areas have energy surpluses and are energy secure, while others have too little and are insecure. Physical factors help to determine supplies, while human factors determine consumption. Growing energy demands do not match supply. Resources are often located far from where they are consumed, and the pathways between them can be insecure. The potential for conflict is high. Increased demand is leading to exploitation of energy reserves in sensitive areas. The Specification has three parts, shown below, with the examples used in this book.

1 Energy supply, demand and security

What you need to learn	Examples used in this book
• Energy sources are classified in different ways, and each has environmental costs. • Access to and consumption of energy is not evenly distributed. • Energy supply and consumption depends on physical factors, cost, technology. Some areas are energy poor, while others have a surplus. • Global demand for energy is growing. • Energy security depends on its availability and security of supply, and is affected by geopolitics.	• Different energy types; energy classification and costs. • Global energy reserves; global imbalance in supply and demand, e.g. Europe versus Russian and Middle Eastern supplies, USA. • Energy sources – their distribution globally and in California / USA; energy poverty in Kenya. • Availability of energy in the UK. • China's increasing demand for energy. • Energy insecurity – UK, China, Japan; securing supplies (e.g. in the Middle East).

2 The impacts of energy insecurity

What you need to learn	Examples used in this book
• Energy pathways are complex and show increasing levels of risk. • There are economic and political risks if energy supplies are disrupted	• Connections – the East Siberia-Pacific Ocean oil pipeline; the rise of the Russian energy oligarchies; Iraqi oil and conflict in the Middle East. • Threats to European supplies from Russia.
• Increasing energy insecurity has stimulated exploration of difficult and sensitive areas.	The search for new resources: • The Arctic and geopolitics (Russia's scramble for control). • Environmental costs (Canadian tar sands).
• Energy TNCs and large producers are powerful players in global energy supply.	• TNCs and energy (Exxon); Gazprom in Russia. • Influences in US politics. • International energy controls, e.g. OPEC.

3 Energy security and the future

What you need to learn	Examples used in this book
• Global energy supplies, reserves, and demand are uncertain.	• Future energy growth – 'peak oil and gas'. • Projections for energy production and consumption.
• Different responses to increasing energy demands. • Energy insecurity can lead to geopolitical tension and conflict. • Meeting future energy needs may require radical new approaches.	Future scenarios: • Business as usual (reliance on fossil fuels). • Alternative sources (nuclear, renewables, conservation, recycling). • Conserving resources and using less. • Global insecurity, e.g. terrorism. • OPEC and the future. • The Middle East – Iraq, Iran and oil security.

Energy security in the USA

In this unit you'll look at energy security issues in the USA as a whole and California in particular.

> *A barrel of oil now costs four times more than it did when President Bush took office in 2001. There are two oilmen in the White House [George W. Bush and Dick Cheney] and petrol now costs $4 a gallon.*
> **Nancy Pelosi, Democrat Speaker of the US House of Representatives, June 2008**

> *The price of oil is just scaring the American people to death.*
> **Republican Senator Pete Domenici of New Mexico, June 2008**

Why is the USA in energy crisis?

The USA is an energy hungry country. Oil is used in every sector of its economy, so any price rise threatens the economy by making everything more expensive. The fact that the USA obtains most of its oil from overseas (particularly from the Middle East) also threatens its **energy security**.

Consumption

The USA consumes vast quantities of energy, as the two tables show. In 2007, it consumed 21.3% of all global primary energy supplies and 23.8% of the world's oil. In terms of electricity consumption, an average American consumes over 100 times more electricity than a citizen of Kiribati – the country with the world's lowest consumption of electricity.

Reliance on imports

The USA's three main needs for energy are: electricity generation, transport and heating. Oil and natural gas are the major sources of fuel for these needs. Between 1960 and 2003, the USA's reliance on imported oil and gas increased from 18% to 58%. The figure for oil on its own is even higher, at 70%.

The terrorist attack on the World Trade Centre in New York on 11 September 2001 (9/11), and its links to the Middle East, heightened the USA's concern over its dependence on imports for its main energy sources – and also the source of those imports.

In 2006, President Bush said: 'America must end its dependence on oil. When you're hooked on oil from the Middle East, it means you've got an economic security issue and a national security issue.'

▲ Oil price inflation is pushing up the price of fuel and energy bills, as well as contributing to the increasing price of food and other essentials (because of higher transport costs)

> ● **Energy security** means access to reliable and affordable sources of energy. Areas or countries which have surplus energy (such as Russia) are said to be **energy secure**, whereas those with an energy deficit (such as the USA) suffer **energy insecurity**.

	USA	World
Oil	943.1 (23.8%)	3962.8
Natural gas	595.7 (22.6%)	2637.7
Coal	573.7 (18.1%)	3177.6
Nuclear	192.1 (30.9%)	622.0
HEP	56.8 (8.0%)	709.2
Primary energy	2361.4 (21.3%)	11099.3

▲ The USA's energy consumption in 2007, in million tonnes or equivalent per source (plus the USA's percentage of total world consumption in brackets)

Rank	Country	Consumption (GWh/year)	Date of information	Consumption per capita (KWh/year)
	World total	**16 790 000**	**2005**	**2215**
1	USA	3 816 900	2006	12 187
2	China	2 859 800	2006	2140
3	Russia	985 200	2007	5679
4	Japan	974 228	2006	7424
193	Nauru	21.39	2003	1945
194	Comoros	16.74	2003	22
195	Sao Tome and Principe	13.95	2003	100
196	Kiribati	11.16	2003	116

▲ World electricity consumption – the four largest and the four smallest consumers

Price

When President Bush made his 2006 statement, the price of a barrel of oil had risen steadily from $20 to $60 over the previous ten years (see the graph). However, since 2004, the price of a barrel has accelerated rapidly, so that by mid-2008, oil was $140 a barrel – with some people predicting price increases to $200 a barrel within two years. The USA has had to change from a nation that expects low-cost energy into one that accepts that those days are over.

A global problem

The problem for the USA and the rest of the world is that:

- reserves of fossil fuels are starting to run out (see pages 36-39). Oil and gas reserves should meet current levels of production for about 40 and 65 years respectively. Coal reserves are far more abundant, but coal is a heavily polluting fuel when used without expensive clean technology.
- global sources of energy are unevenly distributed. Many energy sources are concentrated in politically unstable parts of the world, such as the Middle East, which can lead to potential disruption of supplies.

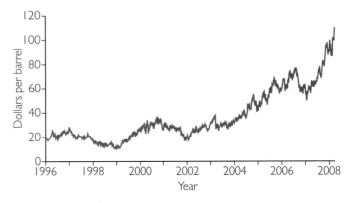

▲ *The rising price of oil*

- demand for energy is increasing. The map and graph below show how energy consumption is likely to rise around the world by 2030. By then, rapid economic growth in Asia, particularly in India and China, is likely to mean that the USA and Europe will have been overtaken as the world's largest energy consumers.

These three factors affect energy security – and the world is becoming more energy insecure as a result.

▼ *Predictions for world energy consumption by 2030 (in quadrillion Btu)*

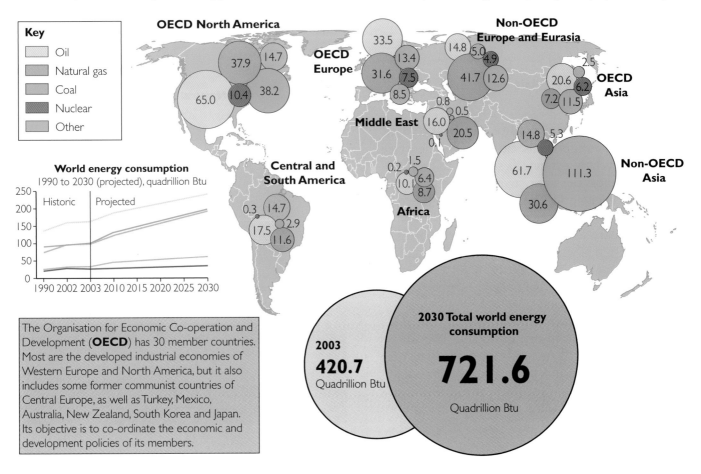

The Organisation for Economic Co-operation and Development (**OECD**) has 30 member countries. Most are the developed industrial economies of Western Europe and North America, but it also includes some former communist countries of Central Europe, as well as Turkey, Mexico, Australia, New Zealand, South Korea and Japan. Its objective is to co-ordinate the economic and development policies of its members.

Energy security in California

Achieving energy security

The following factors are important for energy security:

- Control over supplies
- Control over prices
- A variety of energy sources
- Political stability (both in the consuming country and the producing country)

Energy security can be threatened by:

- rapid price increases, e.g. of oil since 2004
- the political instability of suppliers, e.g. Georgia in August 2008
- the manipulation of supplies and prices (see page 9)
- attacks on infrastructure, e.g. by terrorists
- competition from other expanding economies, e.g. China
- environmental legislation which adds to the costs of finding, transporting and processing the resource

Energy security can be improved by:

- greater energy efficiency
- greater energy self-sufficiency
- decentralization of energy production
- short-term stockpiles (90 days)

California's energy crisis

California orders statewide blackouts

Californian power officials, citing a severe shortage of electricity, have extended a series of rolling blackouts across the state – to relieve the strain on the electricity grid. The blackouts have lasted longer and affected more of the state than power cuts in January, and many consumers fear that they are a sign of what to expect over the summer. The blackouts have been called in a desperate last-minute bid to avoid a widespread collapse of the entire electricity system.

President George W. Bush has said that the country faces a serious energy problem. 'The energy crunch we are in is a supply and demand issue – we need to reduce demand and increase supply' said President Bush.

Adapted from a BBC news article, March 2001

California – facts

California:

- has the largest state economy in the USA, and is richer than most countries
- has the lowest per capita energy consumption rate in the USA – partly due to the mild weather, which tends to reduce energy demands for heating and cooling
- has 16% of the USA's oil reserves, but only 3% of its gas reserves
- produces 5% of the USA's total electricity, and imports more electricity than any other state
- has more motor vehicles registered than any other state, with commuting times among the longest in the country

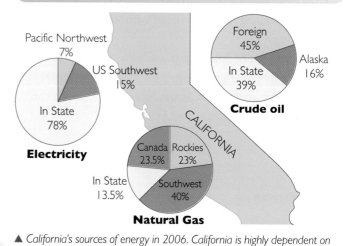

▲ California's sources of energy in 2006. California is highly dependent on imported natural gas for electricity generation

▲ A Californian ice cream parlour forced to close in March 2001 because a power cut melted the stock

The reasons for the Californian blackouts

The US energy market is entirely privatised. Instead of providing a service for its own sake, the market is driven by the desire to maximise profits, cut costs, and pay dividends to shareholders. The provision of energy infrastructure, such as power lines and power stations, is expensive. Therefore, US energy companies have been unwilling to invest the money needed to provide an energy infrastructure in the USA that can cope with the increasing energy demands being placed on it.

California suffered an energy crisis between June 2000 and May 2001 – characterised by price instability and major blackouts affecting millions of customers. A number of different factors influenced the Californian blackouts:

- The weather in 2000/2001:
  - Oregon and Washington states supply hydro-electricity to California. 2000 was the third year of a drought and they had less surplus electricity to export.
  - The summer of 2000 was unusually hot, which triggered increased use of electricity for air-conditioning.
  - The winter of 2000-01 was unusually cold, which increased the need for heating. Natural gas was in short supply, so prices soared overnight.
- Insufficient generating capacity. California introduced strong anti-pollution energy laws in the 1970s, which meant that new power stations were expensive to build. Energy companies were unwilling to invest in the new energy infrastructure required.
- The limited capacity of interstate power lines to import more electricity to make up the shortfall.
- The deregulation of the power industry in 1996. Before deregulation, 80% of supply was vertically integrated, which meant that the electricity supply companies also owned the power plants. After deregulation, power plants were bought by out-of-state companies – and some neglected to maintain a sufficient reserve capacity (infrastructure that could deliver 20% above any expected peak demand).
- Manipulation of the Californian energy market. Enron was a large and wealthy American energy company, based in Houston, Texas. In California, Enron used the supply and demand mechanism to ensure that energy prices remained high – even when supply was plentiful.

They did this by:

- obscuring the origin of electricity, in order to obtain higher prices (known as 'megawatt laundering')
- deliberately reserving more power line usage than was actually required (known as 'overscheduling')
- moving energy around the Californian electricity grid in order to receive payments from the state for 'relieving congestion'
- encouraging suppliers to shut down plants to perform unnecessary maintenance – thus reducing the supply to keep prices high

All of these factors meant that, in 2000-2001, the two major power companies in California – Pacific Gas and Electric, and Southern California Edison – were forced to shut off electricity supplies in order to conserve their limited stocks.

▲ Directing traffic by hand in San Francisco in March 2001, because a rolling blackout knocked out the traffic lights

California moves on

In 2003, Arnold Schwarzenegger (previously better known as an actor) became Governor of California. Under Governor Schwarzenegger, California began to lead the way in efforts to conserve energy supplies and seek alternative energy sources. The state declared that, by 2010, 20% of its electricity must come from renewable sources, such as solar power. It also introduced laws to make big cuts in greenhouse gases, and laid the foundations for a carbon market that would reward firms for cutting their emissions.

In 2007, in an attempt to reduce greenhouse gas emissions, Governor Schwarzenegger issued a directive to establish the world's first **low-carbon standard** for transportation fuels. He said: 'California's petroleum dependency contributes to climate change and leaves workers, businesses and consumers vulnerable to price shocks from an unstable global energy market. As a world leader in energy efficiency, alternative energy and reducing greenhouse gases, California's new low-carbon standard is an innovative action that will diversify our fuel supplies and establish a vibrant market for cleaner-burning fuels.'

Governor Schwarzenegger continued his low-carbon drive. In 2008, he signed agreements to cut the use of high-carbon petroleum sources, such as tar sands from Canada and elsewhere (see pages 30-31). Other US states were expected to join California's low-carbon standard.

> *Is the control of energy generation and supply too important to be left to the free market?*
>
> **What do you think ?**

▲ *Concentrating solar power, or solar thermal electricity, harnesses enough of the Sun's energy to provide large-scale domestically secure and environmentally friendly electricity. The world's largest solar-power facility – near Kramer Junction, California – consists of five solar electric generating stations, which produce enough power for about 150 000 homes*

▲ *Extracting oil from Canada's tar sands, which not only damages the environment but has a high carbon footprint*

Over to you

1 a In pairs, draw a spider diagram to identify the factors threatening the USA's energy security.
   b Which of the factors are **(i)** internal to the USA, **(ii)** external from overseas?
   c Using these factors, identify how the USA's energy security could be improved.
2 In pairs, decide how far you agree with the statement: 'Electricity is really different from everything else – it is a public good that must be protected from private abuse.'

On your own

3 Explain how Enron's involvement in the energy supply market caused the flow of energy to collapse in California.
4 Explain in 300 words the energy problems which the USA is facing, and why its energy insecurity is growing.

1.2 ## The energy mix

In this unit you'll investigate the environmental impacts of different energy sources, and the global distribution of the main energy sources.

Energy – classification and costs

There are many different sources of energy, which can be classified as:

- non-renewable (finite), e.g. oil and gas – their exploitation and use will eventually lead to their exhaustion
- renewable, e.g. solar, wind and wave power – flows of nature which are continuous and can be constantly re-used
- recyclable, e.g. reprocessed uranium and plutonium from nuclear power plants, and heat recovery systems (as outlined on the right)

Different sources of energy have different environmental costs – or impacts – associated with their production and use, as the table shows.

▶ *In many industries, heat goes straight up the chimney (up to 65% in conventional electricity power stations)*

Recycling energy

West Virginia Alloys (WVA) – of Charleston, West Virginia, USA – has entered an agreement with Recycled Energy Development (RED) to recycle energy, improve efficiency, and slash greenhouse gas emissions and other pollutants from its plant.

WVA uses electric arc furnaces to produce silicon. Waste heat from the furnaces is usually vented into the atmosphere. RED will install waste heat recovery boilers to convert the exhaust heat into steam. The steam will drive a generator to produce electricity – offsetting roughly a third of WVA's electricity consumption.

Adapted from an article by Reuters, 6 February 2008

▼ *The environmental costs of different energy sources*

Energy source	Environmental costs
Coal	It is produced by deep or opencast mining, which creates environmental damage. Burning it releases greenhouse gases, which contribute to global warming. It is heavy and bulky to transport, which creates further greenhouse gas emissions.
Oil and natural gas	New fields are often in environmentally sensitive areas, such as the Arctic. Their terminals and refineries take up large areas of land. There is a danger of oil spills and gas leaks causing major damage to vegetation and wildlife. Burning them releases greenhouse gases.
Nuclear energy	Safety is a major issue – an explosion at Chernobyl in the Ukraine spread radiation across Europe in 1986. Nuclear waste remains radioactive for many thousands of years, and is difficult to process and store.
Hydro-electric power	Hydro-electric plants require large areas of useful land to be flooded behind their dams – to create lakes to supply the water the plants need to generate electricity. The large amounts of vegetation drowned by the new lakes decays and releases methane and carbon dioxide (greenhouse gases). Silt from upstream areas is deposited in the new lakes, rather than enriching agricultural land downstream, which may mean that farmers have to use damaging chemical fertilizers instead. There is a danger of dam collapse, e.g. at Sichuan in China following the May 2008 earthquake, which weakened many HEP dams.
Geothermal energy	Its large-scale use is generally limited to volcanic areas, such as Iceland, where there is always a risk of volcanic eruptions, earthquakes, and emissions of sulfuric gases.
Wind power	Wind turbines can be visually unappealing in the landscape. They can also affect wildlife, particularly birds, and can cause noise pollution for local residents.
Solar energy	To make good use of solar energy, large areas of land need to be covered with solar panels or photovoltaic cells.
Tidal power	Barrages built to harness tidal power destroy wildlife habitats, both upstream and downstream of the barrage.
Biomass	This can release greenhouse gases – but if it is used as a methane gas fuel (e.g. in rural India) there is very little greenhouse gas emission.
Combined Heat and Power (CHP)	The environmental impact depends on the type of fuel used. Although greenhouse gases are released, it is a far more efficient method of energy production than, for example, conventional power stations (see pages 56-7 of the AS textbook).

The distribution of energy sources

The distribution of energy reserves

The world's energy sources are distributed unevenly. The map and tables below show the top ten countries (in 2007), in terms of their reserves of oil, natural gas, coal and uranium (for nuclear energy).

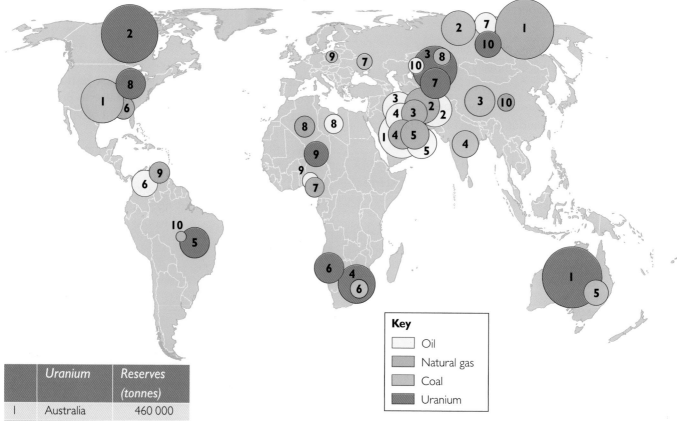

Key
- Oil
- Natural gas
- Coal
- Uranium

	Uranium	Reserves (tonnes)
1	Australia	460 000
2	Canada	426 000
3	Kazakhstan	254 000
4	South Africa	186 000
5	Brazil	112 000
6	Namibia	110 000
7	Uzbekistan	109 000
8	USA	102 000
9	Niger	94 000
10	Russia	75 000

	Coal	Reserves (million tonnes)
1	USA	267 554
2	Russia	173 074
3	China	126 215
4	India	101 903
5	Australia	86 531
6	South Africa	53 738
7	Ukraine	37 647
8	Kazakhstan	34 479
9	Poland	15 432
10	Brazil	11 148

	Oil	Reserves (billion barrels)
1	Saudi Arabia	262.3
2	Iran	136.3
3	Iraq	115.0
4	Kuwait	101.5
5	United Arab Emirates	97.8
6	Venezuela	80.0
7	Russia	60.0
8	Libya	41.5
9	Nigeria	36.2
10	Kazakhstan	30.0

(Note: Canada, which would be second in the above table – with reserves of 179.2 billion barrels – has been excluded because tar sand reserves are not included in these figures (see pages 30-31).)

	Natural gas	Reserves (billion cubic metres)
1	Russia	47 820
2	Iran	27 680
3	Qatar	25 783
4	Saudi Arabia	6847
5	United Arab Emirates	6065
6	USA	5913
7	Nigeria	5152
8	Algeria	4615
9	Venezuela	4343
10	China	3526

Renewable energy distribution

The distribution of renewable energy sources also varies globally. The first map below shows the variation in wind speeds around the world – on land and at sea. However, while some places may not be able to take advantage of wind power, they may be able to harness other sources of renewable energy instead – such as solar power (see the second map below).

The potential of solar energy is enormous – particularly for developing countries lying within the tropics, where sunshine is abundant and the cost of developing solar power may be less than the cost of extending electricity grids and transmission lines into rural areas.

One of the advantages of renewable energy is that the 'mix' of types is so varied that many parts of the world should be able to adopt at least one renewable energy source.

15km Global Wind Map at 80m
Mean Wind Speed for a single year
© Copyright 2008 3TIER, Inc.

developed by 3TIER

Wind speed over water
5 10 15 20 m/s

Wind speed over land
3   6   9 m/s

▲ Global mean wind speeds over land and sea at 80 metres above the surface

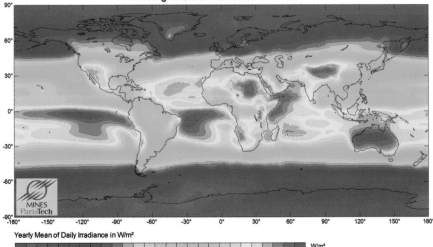

Averaged Solar Radiation 1990-2004

Yearly Mean of Daily Irradiance in W/m²
20 30 40 50 60 70 80 90 100 110 120 130 140 150 160 170 180 190 200 210 220 230 240 250 260 270 280 290   W/m²

▲ The average amounts of solar energy reaching the Earth, 1990-2004. The highest levels of solar energy are shown in red, followed by orange, yellow, green, blue, purple and finally pink.

Over to you

1 Compare the energy types listed in the table on page 11. In pairs, decide which source has **a** the greatest and **b** the least environmental impact.

On your own

2 Research energy reserves in two countries in different continents, e.g. Australia and Nigeria.
   a Where are the main reserves found?
   b How large are the reserves?
   c Are these countries energy secure?
   d What are the energy related issues in your chosen countries?
3 Find out how much oil, natural gas and coal the world consumes each day. Use the tables of energy reserves to work out:
   a how soon new energy sources need to be discovered, or alternatives found
   b which 'energy superpowers' currently control the world's energy resources.

Is a world free from dependence on fossil fuels an unattainable dream?

What do you think?

In this unit you'll learn how a range of factors affect the availability of, and access to, energy.

The UK and energy

For the majority of people in the UK, energy is delivered straight to the door, via gas pipes, electricity cables, or – in the case of oil – tankers. Electricity is available at the flick of a switch and is generated from a range of sources, as the pie chart shows.

However, electricity generation can also generate fierce debate. The energy company E.ON plans to build two new coal-fired power stations to replace the existing one at Kingsnorth in Kent, which is due to close in 2015. E.ON claims that the replacement power stations will be 20% cleaner than the existing one. If the plan is approved, it could pave the way for a new generation of coal-fired power stations in the UK. However, many people are unhappy with this prospect and have protested about E.ON's plans to replace the existing coal-fired, pollution-generating, power station with what the protestors perceive as more of the same.

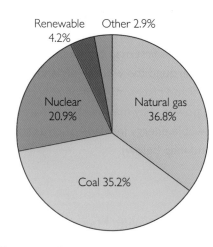

▲ The sources of the UK's electricity, 2005-2006

The new power station planned for Kingsnorth will output more CO_2 each year than the whole of Ghana.
World Development Movement

Nuclear power and public perceptions

All but one of the UK's current nuclear power stations are due to close by 2023 (leaving just Sizewell B in Suffolk). In 2008, the Labour government announced plans to build a new generation of nuclear power stations to replace the existing ones and improve Britain's energy security (Britain is becoming increasingly reliant on imported oil and gas from countries such as Russia, and this loss of control over future energy supplies worries many UK politicians).

However, many British people are concerned about nuclear power – particularly about its safety – and agree with the views of Friends of the Earth, an environmental pressure group. In 2008, Friends of the Earth opposed the British Government's plans by arguing that:

- Britain can meet its energy needs (and tackle climate change) through a programme of renewable energy, CHP, energy efficiency, and cleaner carbon technology
- nuclear power is expensive, dangerous, and leaves radioactive waste for thousands of years
- building new nuclear power stations will undermine the development of more-efficient renewable energy technology by steering investment away from it

Physical factors

Physical factors determine the distribution of energy resources (see pages 12-13), their access, and their generation.

Geology – Worldwide, the shallower reserves of fossil fuels have almost been exhausted, and known alternative reserves include some at great depth and others in environmentally sensitive areas – which will be difficult and expensive to extract. The UK's oil, gas and coal reserves have all peaked, and it now has to import energy to meet demand.

Location – The UK is located between the Atlantic Ocean and the North Sea, and has some of the best wind (see the top map on page 13), wave and tidal power resources in the world. The River Severn has the second highest tidal range in the world (up to 14 metres), and for years there have been suggestions about building a barrage across the Severn to harness its tidal power to generate electricity. The barrage would cost £15 billion to build, be 10 miles long, and could meet around 5% of England and Wales' current electricity demand. But it is also argued that the environmental effects of such a barrage could be disastrous.

Technology and cost

These go hand in hand as far as access to renewable energy is concerned.

- Although the cost of installing wind turbines has risen – due to the increasing prices of materials – wind power still remains viable, because the price of other energy sources, such as oil, looks likely to remain high.
- China has become a low-cost producer of solar photovoltaic (PV) cells, which generate electricity from sunlight. As more and more PV cells are produced, the price of each cell will fall – and potential access for poorer countries will increase. PV technology can now be incorporated into roof tiles, windows and walls.
- In developing countries, renewable energy can provide power more cheaply and quickly than extending transmission lines and constructing new power plants.

Unequal shares

As well as being distributed unevenly throughout the world, energy sources are also consumed unequally. The table shows the enormous differences in oil consumption for a range of countries – as well as the dependence of most of those countries on expensive oil imports to cover their consumption (even big oil producers like the USA). Those with an energy deficit suffer energy insecurity, even if they have the money to purchase the energy to make up that deficit.

However, oil is not the only energy source to be shared unequally. The graph below shows how the daily consumption of gas and coal varies for a range of countries. The USA heads the league for gas consumption by far, and – as with oil – is dependent on imports to meet demand. However, China consumes the largest volume of coal – without having to rely on any imports.

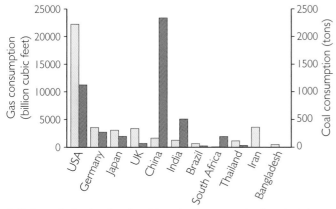

▲ Daily gas (yellow bars) and coal (grey bars) consumption for selected countries

Country	Oil (barrels a day) (A) = Production (B) = Consumption (C) = Net result	Population (millions)
USA	(A) 8 330 000 (B) 20 687 000 (C) -12 357 000	304
Japan	(A) 128 000 (B) 5 159 000 (C) -5 031 000	127
UK	(A) 1 689 000 (B) 1 830 000 (C) -141 000	61
China	(A) 3 845 000 (B) 7 201 000 (C) -3 356 000	1330
India	(A) 854 000 (B) 2 572 000 (C) -1 718 000	1148
Brazil	(A) 2 167 000 (B) 2 217 000 (C) -50 000	192
South Africa	(A) 204 000 (B) 504 000 (C) -300 000	44
Iran	(A) 4 148 000 (B) 1 686 000 (C) +2 462 000	65
Bangladesh	(A) 7000 (B) 90 000 (C) -83 000	154
D.R. Congo	(A) 20 000 (B) 10 000 (C) +10 000	67

▲ The global use of solar heating systems is growing at 15-20% a year. Rooftop solar collectors now provide hot water for 50 million households worldwide – 80% of them in China (as pictured here). Solar systems pay for themselves in a few years.

Access to energy – 2

Energy poverty

While some countries (mostly in the Middle East) have an **energy surplus**, the unequal distribution of energy – and lack of access to energy resources – means that some countries or areas suffer from **energy poverty**.

- 1.6 billion people – a quarter of the world's population – have no access to electricity, with 80% of these living in rural areas.
- 2.4 billion people rely on traditional biomass for cooking and heating.
- In rural sub-Saharan Africa, many women carry 20 kg of fuel wood an average of 5 km *every day*.

Not only that, but:

- the use of traditional biomass is killing people – 2.5 million women and children die each year from lung conditions caused by smoke from traditional cooking stoves.
- energy poverty keeps people poor by limiting women's ability to engage in education and income-generating activities.

People suffering from energy poverty need access to clean, sustainable and renewable energy, such as: solar power, small-scale hydropower, modern biomass, and wind power. Access to these can play a key role in reducing the burden of overall poverty and environmental degradation around the world.

▲ *Women in Ethiopia carrying dried animal dung and wood to use as fuel. In sub-Saharan Africa, women can spend hours every day collecting and carrying fuel sources such as these*

Micro-hydro

96% of Kenyans have no access to grid electricity. In rural homes, families spend at least a third of their income on kerosene for lighting and diesel for milling grain. Women spend hours collecting wood and dung for cooking.

Mbuiru village is a typical Kenyan village – 200 km north of Nairobi. The Tungu-Kabri Micro-hydro Power Project (funded by the United Nations Development Programme and developed by Practical Action East Africa and the Kenyan Ministry of Energy) harnesses the energy of falling water to create electricity. The project is cheap, sustainable and small scale. It generates enough electricity to benefit about 1000 people – providing light, saving time, and allowing people to run small businesses.

▲ *Local villagers were involved in building the micro-hydro power project*

Over to you

1 Look at the oil consumption table on page 15. Identify three key issues which come out of the table and explain these to your neighbour.
2 How far does access to energy depend on: physical factors, cost, technology, and public perception? Work in pairs to rank these factors and then justify your rankings.

In the UK, people object to nuclear power, coal-fired power stations, tidal barrages, wind farms and dams (for HEP production). Where should we get our energy from?

What do you think?

On your own

3 The world's largest consumer of oil is the USA, but China's consumption is growing rapidly. Research the potential for renewable energy in these two countries.
4 a Research the physical factors needed in the UK for **two** of the following: HEP, a nuclear power station, a wind farm, a coal-fired power station, a tidal barrage.
   b Generally speaking, what is the British public's opinion about your two chosen examples.
   c How is this likely to affect the UK's energy security?

In this unit you'll investigate how China is coping with its enormous demand for energy.

China's energy hunger

Since the early 1980s, China has undergone massive economic changes – its economy has doubled in size every eight years. It now has the largest sustained GDP growth in history, and is no longer considered a low-income country. However, while China's economic growth is impressive, this growth has required more and more energy supplies to sustain it. In 2001, China accounted for 10% of global energy demand. By 2007, this figure had grown to over 15%. China is now the world's second largest energy consumer (and is set to become the world's leading source of greenhouse gas emissions).

However, China's per capita energy demand remains relatively small, because of its huge population. In 2005, China consumed less than 7 million barrels of oil a day, which was a third of the USA's consumption and slightly more than Japan (with only a tenth of China's population). What would happen to the world if China's per capita energy demand were to reach that of the USA?

China controls 3% of the world's oil reserves. It was self-sufficient in oil until 1993, but since then has needed to import oil to fuel its rapid economic growth. The graphs opposite show Chinese oil production in 2005 (compared with other major oil producers), and also demonstrate how China's rapidly increasing oil consumption has outstripped production since 1993.

▶ *China's oil production and consumption, 1986-2006*

How did it happen?

After it took over in 1949, China's communist government kept the country separate and disconnected from the rest of the world. The economy was planned centrally, goods were produced for the consumption of China's own people, and no private wealth was permitted.

However, from 1986, China's government developed an 'Open-Door Policy' to overseas investment. In the 1990s, it transformed into a more capitalist economy – allowing individuals to accumulate wealth by producing goods and services without State interference. It is still not a pure free-market economy – most of China's largest companies are either totally or in part State-owned, so any profits are reinvested or ploughed back into State spending on health and education.

For more on China's rise as an economic superpower, see Unit 4.6 (pages 158-163).

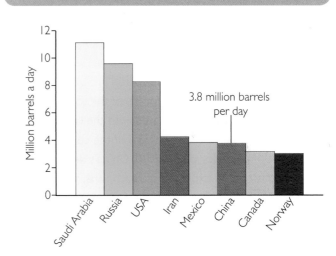

▲ *The world's main oil producers in 2005*

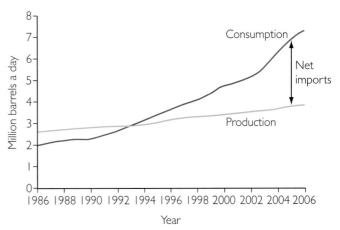

China: the new economic giant - 2

The graph shows how China's energy demand is projected to increase up to 2030. However, this steep rise in energy consumption is not just fuelled by basic economic growth and the demands of new industry but also by rapid urbanization and growing car ownership.

China's rate of urbanization is staggering. Rural-to-urban migration in China is the highest ever recorded – 8.5 million people a year. 45 million Chinese are expected to move to cities by 2012, and most will head for the industrial centres by the coast – where energy consumption is significantly higher than in rural areas.

China's roads used to be filled with bicycles, but now cyclists battle with increasing numbers of cars as the fruits of China's rapid economic growth lead to greater levels of car ownership.

- Car ownership is expected to jump from 16 cars per 1000 people in 2002, to 267 cars per 1000 people by 2030.
- This anticipated increase should account for 25% of the global demand for cars by 2030.
- By 2020, China is expected to have 140 million private cars on the road – even more than the USA. Currently, 1000 new cars arrive on Beijing's streets every day.
- China only uses 10% of its energy for transport at the moment (as the graph below shows), but it will need much more oil to fuel the anticipated growth in car use. Altogether, China accounts for 8% of the total global demand for crude oil (compared with the USA's 25%), BUT it also now accounts for 33% of the current growth in global demand for oil. By 2010, China's dependence on imported oil is expected to reach 60%, which raises issues about China's long-term energy security (see pages 20-21).

▲ China's projected growth in energy demand

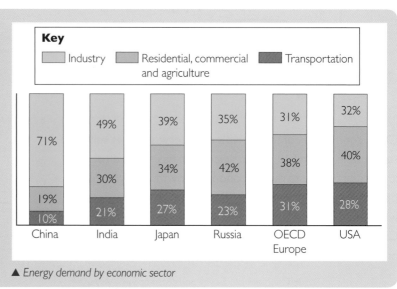

▲ Compared to the 1990s, Chinese cyclists in Beijing today have to compete with increasing numbers of cars

Background

Energy demand by economic sector

China is still largely an agricultural society. 60% of the population lives in rural communities, and 45% of the workforce is employed in agriculture. However, agriculture contributes only 13% of China's GDP, and only uses a fraction of its energy compared with industry (which accounts for 71% of China's total energy use). The graph on the right shows how China's energy use by sector compares with a number of other countries.

Key
Industry   Residential, commercial and agriculture   Transportation

	China	India	Japan	Russia	OECD Europe	USA
Industry	71%	49%	39%	35%	31%	32%
Residential, commercial and agriculture	19%	30%	34%	42%	38%	40%
Transportation	10%	21%	27%	23%	31%	28%

▲ Energy demand by economic sector

Where does China's electricity come from?

Coal

China is by far the world's biggest producer and consumer of coal. It relies on coal for 70% of its electricity generation, and will continue to rely heavily on coal for the foreseeable future – as the graph shows.

China's huge demand for electricity means that it is building an average of three new coal-fired power stations a week. In 2006, China added 102 gigawatts of generating capacity to its grid – as much as the entire capacity of France.

Coal is a cheap source of fuel, but it is also a dirty form of energy and, with its increasing pollution problems, China now needs to build new, cleaner, coal-fired power stations. It has also given the go-ahead for the construction of several huge new projects to turn 'dirty' coal into 'clean' gas. This is an expensive but necessary route to take if China is to overcome its serious pollution problems.

HEP

Hydro-electricity (HEP) accounts for 16% of China's energy production, and major HEP projects are part of China's long-term energy strategy.

On the middle reaches of the Yangtze, work is nearing completion on the Three Gorges Dam. When it is finished, the world's biggest turbines – deep inside the world's largest dam – will generate 25 gigawatts of electricity at its maximum output. That is equivalent to a third of the UK's total energy capacity.

But that is only the start. China wants to build HEP dams on all of its major rivers, which has caused concerns for environmental groups and neighbouring countries (see pages 66-67). The disastrous earthquake in Sichuan province, in May 2008, damaged a number of large dams there – leaving many dangerously close to collapse, and calling into question China's 'big dam' policy.

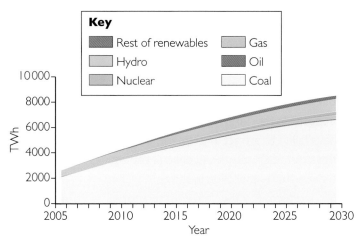

▲ The sources of electricity generation in China, 2005-2030 (projected)

Pollution is a major problem which China is beginning to address

▶ The Three Gorges Dam

China's energy insecurity

Oil

Production at China's largest oil fields has now peaked, and exploration has begun for new fields in the far west and offshore. But there are problems with this. Large oil deposits in the Tarim Basin in the far west have failed to attract investment, because of their remote location and difficult geology. Deepwater exploration in the South China Sea is affected by the danger of territorial disputes with neighbouring countries like Vietnam and the Philippines. So China is now importing more and more of its oil.

Coal

China's coal reserves are largely located in the north and far west, while the industry representing 71% of China's demand for energy is mainly located much further east and south. Having coal in the ground is one thing. However, it is quite another to be able to mine it, move it, and burn it quickly, cleanly, and in sufficient quantity to meet China's escalating demand. A surge in electricity demand in 2002 resulted in coal shortages, power cuts, spikes in oil demand, and a rapid deterioration in air quality due to pollution from the coal-fired power stations.

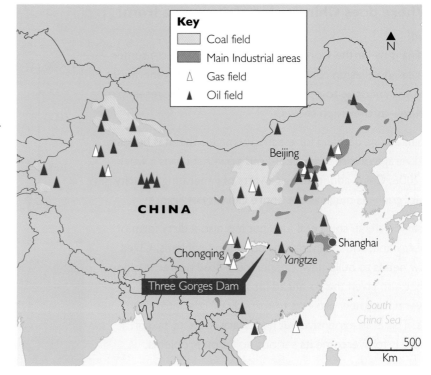

▲ China's energy resources are often located away from the major industrial and urban centres

Natural gas

Plans have slowed down to expand China's use of natural gas, which burns more cleanly and produces less greenhouse gases than coal or oil, to generate energy. It has proved costly and difficult to build pipelines from the gas fields in western China, while liquefied natural gas for transport in ships is in short supply.

The Strait of Malacca

Most of China and Japan's oil imports pass through the Strait of Malacca, between Malaysia and Sumatra (Indonesia), which connects the Pacific and Indian Oceans. It is 500 miles (800 km) long and only 1.5 miles (2.8 km) wide at its narrowest point, and is in constant danger from pirates (150 attacks in 2003) and from the haze of bush fires (caused by deforestation in the Indonesian rainforests), which can lead to collisions. Nevertheless, over 50 000 ships pass through the Strait every year.

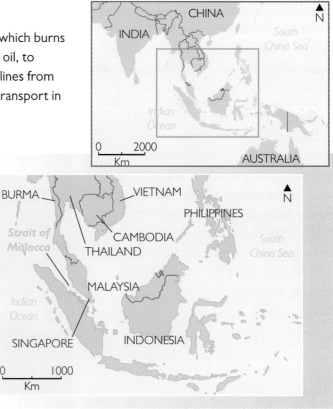

Background

Japan's energy insecurity

China's energy security problems matter to the rest of the world because of its sheer size – and also the impact that its search for greater energy supplies is having on the global economy (its increasing demand means increasing prices for everyone). However, its **energy dependency** (12%) is minor compared with the USA's (40%) and Japan's (80%).

Japan is the largest economy in Asia, yet it has the smallest energy reserves of any major economy. 99% of its oil and gas is imported, and energy security has long been a core goal of its domestic and foreign policy (it is one of the reasons why it went to war in 1941). The table compares China and Japan in terms of their energy security.

Since the collapse of the former Soviet Union, Japan has been seeking to reduce its dependence on the Middle East by engaging with the new Central Asian countries, such as ex-Soviet Kazakhstan. China also hopes to satisfy some of its future energy needs from places such as Kazakhstan – and this could be a cause of future tension between the two countries.

● **Energy dependency** is measured by the level of energy imports as a proportion of total energy consumption – the higher the proportion of energy imports, the more energy dependent the country is on others.

Energy security	Japan	China
Energy efficiency	High levels	Low levels
Diversification of energy sources	High diversity, e.g. 35% nuclear	Low diversity; it relies on coal for 70% of electricity generation
Diversification of energy suppliers	It is 80% dependent on Middle East suppliers	It has a wide range of suppliers from the Middle East, Africa, and Central Asia
Control over the means of supply	It has high control, due to the large Japanese merchant fleet	It has low control, with 90% foreign suppliers
Development of military capability to protect its energy supplies	It has been dependent on the US Navy since losing the Second World War	It has developed a powerful navy

The search for security

In its search for oil (and other raw materials – see pages 158-159), China is trawling the world. China's foreign policy could now be said to be 'resource based', and energy has become a national security issue. Examples of this include:

- Chinese companies getting Government support and guidance over which countries and sectors to invest in
- China protecting its oil imports from the Middle East by:
  - investing in the port of Gwadar in Pakistan
  - upgrading its military airstrip on Hainan in the South China Sea
  - closer ties with Burma
  - expanding naval access in Bangladesh
- China treating Central Asia as the most secure source of its future energy supplies, but also investing in – and importing energy from – countries such as Iran, Sudan, Canada and Australia
- China building up a powerful navy to protect its sea-lanes

Over to you

1 a Working in pairs, create a spider diagram to show the reasons for China's energy insecurity.

  b Now create a second spider diagram to show the steps which China is taking to tackle this energy insecurity.

On your own

2 Compare the map opposite, showing the distribution of China's coal, oil and gas fields, with an atlas map showing China's population distribution. Describe the differences in terms of imbalances, energy poverty and energy surplus.

3 What problems has China's increasing demand for energy created in the short term, and what are the long-term problems likely to be?

Exam question: Discuss how far economic development can be affected by energy security. **(15)**

In this unit you'll learn that energy pathways can be complex, and also about the economic and political risks if energy supplies are disrupted.

The East Siberia-Pacific Ocean oil pipeline

The Sri Lankan government pays fishermen to protect rare sea turtles. The Dominican Republic has a team dedicated to saving the Sisserou parrot. All over the world, governments are striving to preserve and protect 'flagship species'. So why is Russia building a $12 billion oil pipeline straight through the habitat of the world's rarest big cat [the Amur leopard]?

Adapted from *The Guardian*, 2 June 2005

The East Siberia-Pacific Ocean (ESPO) oil pipeline will be 2600 miles long when it is completed. Construction began in 2006 and is likely to take ten years. Originally Russia planned to route the pipeline through the area where the last remaining wild Amur leopards live (on the Chinese border, near Vladivostok) – on its way from Siberia to the Pacific coast. However, in July 2005, it was announced that the route would be changed because of logistical, safety and environmental objections. So, for now, the few remaining Amur leopards are relatively safe.

Apart from the issue of protecting the Amur leopards, the ESPO pipeline has run into other problems:

● The initial route of the ESPO also ran too close to the northern tip of Lake Baikal (the world's largest freshwater lake and a UNESCO protected site), and there were fears that oil spills might cause an environmental disaster. Therefore, the pipeline had to be re-routed further north, away from the lake, which massively increased the project's cost.

● The cost has also ballooned because of rising steel prices, and the challenge of building in soils affected by differing permafrost conditions. When completed, the ESPO will be the longest and most expensive pipeline ever built.

Constructing this pipeline from Siberia to the Pacific coast means that Russia will eventually have a new **energy pathway**, and will be able to export its oil to countries in Asia – as well as possibly North America.

▲ *The endangered Amur leopard – there are only 30-40 left in the wild*

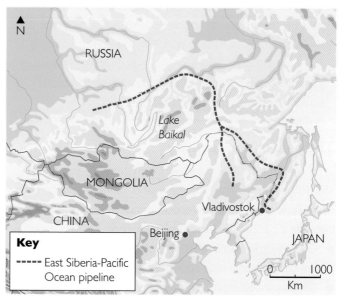

▲ *The route of the East Siberia-Pacific Ocean (ESPO) oil pipeline*

● **Energy pathways** refer to the flow of energy from a producer and the means by which it reaches the consumer. Energy pathways can be contentious, since routes may be complicated by:

  ▶ natural or environmental issues, e.g. crossing mountains or an earthquake zone

  ▶ human or political issues, e.g. conflicts between neighbouring countries

Geopolitics and the ESPO pipeline

The Russian, Chinese and Japanese governments are all interested in oil for a whole host of **geopolitical** reasons. China and Japan have been competing for access to Russia's oil and the ESPO pipeline project.

> ● **Geopolitics** is the study of the ways in which political decisions and processes affect the way space and resources are used. It is the relationship between geography, economics and politics.

China's needs v Japan's needs

China

China's needs for energy and its relationship with Russia are complex:

- China needs increasing amounts of energy to fuel its rapid economic growth. Access to foreign energy sources, and energy security, are vital for maintaining the control of the Chinese Communist Party and the security of the state.

- The main energy pathway for China's existing external oil supplies is vulnerable – 80% of Chinese oil imports currently pass through the Strait of Malacca (see page 20), and China wants to widen its supply options – both sources and routes. The graph shows the main sources of China's oil imports in 2005 and 2006.

- China and Russia have a joint political interest over the issue of the US military presence in Central Asia and the Middle East, and also US policies relating to the promotion of democracy and some regime change there. Beijing is concerned that the overall American strategy aims to block or prevent China's expanding global economic, political and energy ties.

Japan

Japan also wants a share of Russia's oil:

- Japan has almost no oil reserves of its own.
- It is the world's third largest oil consumer, after the USA and China.
- In 2007, Japan imported 76% of its oil from the Middle East – as the graph shows. The East Siberia-Pacific Ocean oil pipeline could reduce Japan's oil dependence on the Middle East by 10-15%.
- Japan also wants to engage with Russia and increase its economic and political influence, which declined in the 1980s and 1990s. During this period, China asserted and deepened its own relationship with Russia at the expense of Japan.

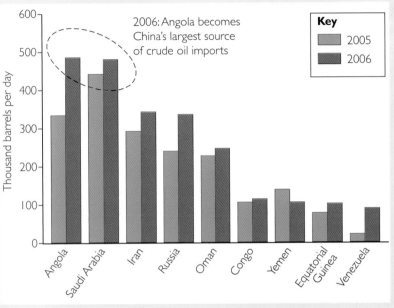

▲ The main sources of China's oil imports in 2005 and 2006

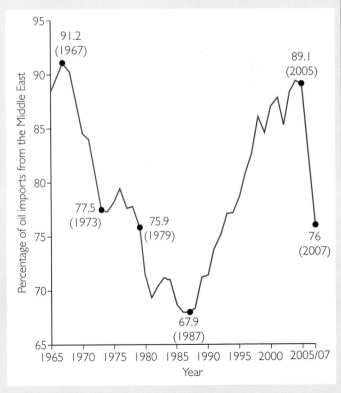

▲ Japan's relience on Middle Eastern oil imports, which it would like to reduce in order to increase its energy security

Energy connections and geopolitics – 2

Russia and China share a 4300 km long border, so a good relationship between them is vital. However, they are suspicious of each other, with Russia seeing China as a rival power and potential threat. Therefore, Russia has been reluctant to commit itself too heavily in terms of energy supply to China.

For the reasons outlined on the previous page, the Japanese government was keen for the ESPO pipeline not to end in China (as originally proposed), but for it to be extended to the Pacific coast (bringing the oil closer to Japan). Therefore, it offered to finance a large proportion of the ESPO project ($7 billion).

The Japanese financing will enable the Russians to build the most expensive pipeline in history, while restricting Chinese access to Russia's oil and helping to rebuild the relationship between Russia and Japan. Russia will also be able to export its oil more widely to other countries in Asia, and possibly to North America. However, a spur will still be built off the main ESPO pipeline to run into China and, as a 'sweetener', Russia has promised to increase oil exports to China by rail to 300 000 barrels a day.

Background

Energy pathways

Energy pathways refer to the flow of energy from producer to consumer. The ESPO oil pipeline is an energy pathway that will link Russia with China, Japan and any other Pacific countries which import Russian oil.

However, energy pathways can be complex, and they face risk and disruption as a result of both physical and human causes:

- In 1991, Iraqi troops retreating from Kuwait at the end of the first Gulf War, set fire to over 600 Kuwaiti oil wells. The fires burned for over eight months and consumed an estimated 6 million barrels of oil a day (and 70-100 million cubic metres of natural gas). This had a major impact on Kuwait's oil production in 1991-2, as the graph shows.
- In 2005, Hurricane Katrina affected oil production and refining in the Gulf of Mexico (together with the production of natural gas and the importation of both fuels). Oil and petrol prices rose as a direct result. Ten days after the hurricane, oil production was still only at 42% of its normal level.
- In 2005, the explosions and subsequent fire at the Buncefield Oil Storage Depot in Hertfordshire destroyed fuel worth £10 million. Buncefield supplied 30% of Heathrow Airport's fuel and, as a result of the fire, Heathrow had to ration fuel and some long-haul flights had to make unscheduled stops elsewhere to top up their fuel tanks. The fuel restrictions lasted for weeks.
- In 2006 and 2008, disputes between Russia and Ukraine disrupted gas supplies to Western Europe (see pages 26-7).

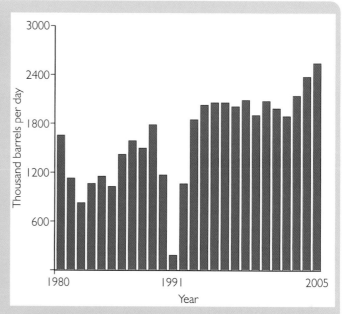

▲ Kuwaiti oil production, 1980-2005

▼ The Buncefield oil stage depot on fire in 2005

Energy control and disruption

Gazprom – Russia

Gazprom. The soaring concrete and glass office block thrusts towards the sky in downtown Moscow, like a rocket with boosters strapped to its side. After passing through security and handing over your passport to a stern-looking official, you cross a bleak windswept plaza to the tower of power. Once inside, your papers are scrutinised again and your bags put through X-ray machines. It is hard not to think of Fritz Lang's science-fiction film *Metropolis*, with its giant towers filled with toiling workers.

Adapted from *The Independent*, 3 September 2007

Gazprom – facts

Gazprom has rapidly become one of the world's most powerful companies. Based in Moscow, it is the world's largest gas supply company. Gazprom:

- controls about a third of the world's gas reserves.
- accounts for 92% of Russia's gas production.
- provides 25% of the EU's natural gas – in 2007, it provided 4% of Britain's gas supplies (expected to rise to 10% within five years). Over 80% of gas exports to Western Europe cross the Ukraine (a former Soviet state and **transit state**).
- is the sole gas supplier to Bosnia-Herzegovina, Estonia, Finland, Latvia, Lithuania, Macedonia, Moldova, and Slovakia.
- is the world's third largest corporation (after Exxon Mobil and General Electric). Gazprom's annual earnings (in 2006) were £31.55 billion.
- began in 1992, when the former Ministry of the Gas Industry was re-organised (see page 165 for more information about the history of Gazprom).The government of the Russian Federation still owns 50.002% of the shares in Gazprom, and Gazprom retains very close links to the State (see page 167).
- employed 432 000 people in 2006.

▲ *Gazprom is a towering economic presence in Russia*

> ● A **transit state** is a country or state through which energy flows on its way from producer to consumer.

Energy connections and geopolitics – 3

Russia's energy strategy – politics or economics?

Russia is re-emerging as a global player (see pages 164-167 and 170-171). Its economic power lies in its key natural resources – particularly oil and gas – with energy becoming a political tool. Energy has:

- helped to re-assert Russia's power and influence over former Soviet states (particularly in the South Caucasus and Central Asia)
- increased Russia's power within the region
- given Russia a way to restore its international position and regain geopolitical importance

Critics say that Russia has been using the supply of gas as a weapon:

- In November/December 2004, Ukraine (a former Soviet republic) had what became known as its Orange Revolution, which got rid of the pro-Russian government in favour of pro-Western reformers. A year later, in January 2006, Ukraine found its gas (supplied by Gazprom) cut off, after Russia decided to quadruple the price and Ukraine's new pro-Western government refused to pay.
- In March 2008, Gazprom again cut gas supplies to Ukraine (by 50%) over a dispute that Gazprom claimed was about debt. However, Ukraine was seeking to join NATO and the EU at the time, which angered Russia (see page 170 for more information about these Ukrainian and Russian disputes).

Alexei Pushkov, a professor of international relations and Russian TV presenter, says that it is a misconception if people think that Russia is using gas as an economic weapon. 'Gazprom is an instrument of foreign policy, like American oil companies are instruments of American foreign policy' he says. 'If Russia was using gas to blackmail foreign countries, we wouldn't have special deals with European nations who are eager to have long-term deals with Gazprom.'

Europe's energy security

The cutting off of gas supplies to Ukraine by Gazprom alarmed many European countries, because they knew how dependent they were on Russian gas. Was Europe right to be worried about its energy security? Yes and no.

Yes – because of the amount of gas Russia supplies to Europe (and the fact that most of it is currently piped through Ukraine). When Gazprom shut down the pipeline in 2006, the flow of gas to the rest of Europe fell by 40% in some areas.

▲ *Protestors in the Orange Revolution in Ukraine in November 2004. Colour or flower revolutions are the collective names for a series of related movements that developed in post-communist societies in Central and Eastern Europe and Central Asia*

▼ *Europe depends on this – a gas pipeline running through Ukraine from Russia, which was turned off in March 2008 in a dispute between Russia and Ukraine*

No – for the following reasons:

- Gazprom cannot cut off supplies within Russia (where demand is growing), and its export markets to Western Europe are too valuable to lose. So it is former Soviet states, such as Ukraine and Belarus, which are likely to lose out. Cliff Kupchan, an American analyst, said: 'Russia will be a quirky seller, but not unreliable. However, if you happen to live in a former Soviet satellite state, you'd better start stocking up on firewood!'
- Even during the Cold War (see page 146), the supply of Russian gas was stable.
- Gazprom is now helping to secure Europe's energy supplies, with the construction of new pipelines bypassing Ukraine and Belarus (see the map):
  - ▶ The Nord Stream pipeline will be a new energy pathway for Russian gas to Europe. It will run for 1200 km along the bed of the Baltic Sea – with no transit countries involved, thus reducing any possible political interference with energy supplies.
  - ▶ The South Stream pipeline will run under the Black Sea from the Russian coast to the Bulgarian coast.

- In an effort to enhance its energy security, the EU is planning its own pipeline. The Nabucco pipeline will bring gas from Central Asia and the Caspian Sea across Turkey into the EU. But its capacity will be small and it may only be able to supply about 5% of Europe's needs.
- The South Caucasus pipeline (opened in 2006 – see page 171) will bring gas from Azerbaijan to Europe via Turkey.
- The EU is also looking at alternative energy sources. However, developing serious alternatives to Russia's gas, possibly including a greater reliance on nuclear power, could take years.

Should energy be used as an economic or political weapon?

What do you think ?

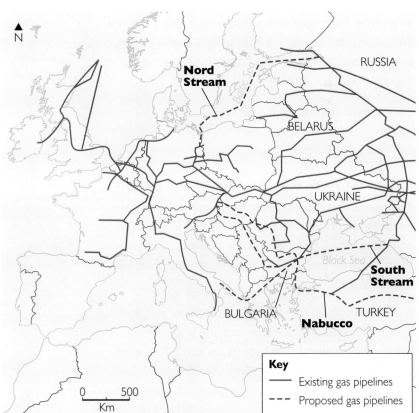

▼ *The main gas pipelines in Europe*

Key
— Existing gas pipelines
- - - Proposed gas pipelines

Over to you

1 Annotate a copy of the map on page 22, showing the route of the East Siberia-Pacific Ocean pipeline.
   a Explain the physical and human problems that this energy pathway faces.
   b Add the reasons why China and Japan are so keen to have access to Russia's oil.
2 Hold a class debate on the following topic: 'Russia uses its gas as an economic weapon'.

On your own

3 Write 300 words to explain why Gazprom's close links to the Russian Government could create problems for those countries dependant on Russian gas.
4 Conduct research into Gazprom's new Nord Stream and South Stream pipeline projects, using www.gazprom.com. Find out about their routes and why the pipelines are important.

In this unit you'll investigate how increasing energy insecurity is leading to the exploration of technically difficult and environmentally sensitive areas in the search for new resources.

A new, and cold, war

A new Cold War is brewing under the Arctic Ocean. It is all down to a dispute about who owns the Arctic Region and its resources – and it has been simmering for years.

On 2 August 2007, two Russian mini-submarines dived to the bottom of the Arctic Ocean under the North Pole and planted a titanium Russian flag on the seabed. In doing so, Russia symbolically claimed the rights to that seabed and its resources. Russia claims that the Lomonosov Ridge, an underwater mountain range located under the North Pole, is an extension of the Russian land mass – geologically linked to the Siberian **continental shelf** (see the background box opposite).

● The **continental shelf** is the relatively shallow submarine platform that borders continents. It is a continuation of the continental land mass.

Why everyone wants a slice of the Arctic

The United States Geological Survey estimates that the Arctic Region contains 25% of the world's unexploited oil and gas reserves (see right) – on a scale that could match Saudi Arabia's reserves – as well as diamonds, coal and other minerals.

As China and India demand more and more oil to fuel their rapid economic development, the price per barrel has spiralled upwards. Oil experts say that oil prices of around $70 a barrel make drilling in the Arctic a viable economic proposition. In 2007, prices reached nearly $100 a barrel, and, in mid-2008, the price soared to $147 before dropping back slightly. Small wonder, then, that Russia is keen to stake its claim to as big a slice of the Arctic's energy reserves as possible.

However, Russia is not alone. Eight countries form part of the Arctic Region (see the map opposite), and many of them have their eyes set on the vast energy and mineral deposits located there. They considered Russia's flag planting to be an old-fashioned land grab. Peter Mackay, Canada's Foreign Minister, said: 'This is not the fifteenth century. You cannot go around the world and just plant flags and say: "We're claiming this territory."'.

▲ A Russian nuclear ice-breaker at the North Pole. Under international law, no country owns the North Pole.

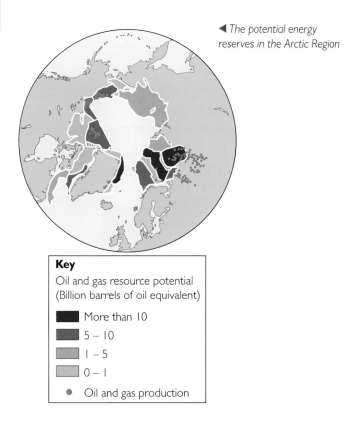

◄ The potential energy reserves in the Arctic Region

Key

Oil and gas resource potential (Billion barrels of oil equivalent)

■ More than 10

■ 5 – 10

□ 1 – 5

□ 0 – 1

● Oil and gas production

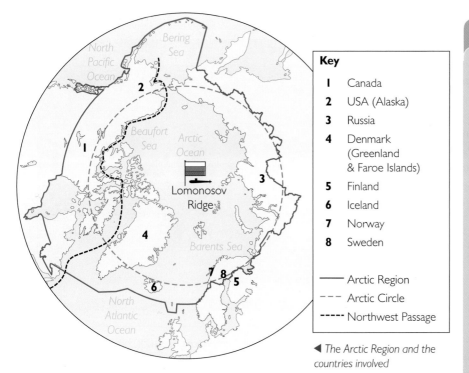

Key

1	Canada
2	USA (Alaska)
3	Russia
4	Denmark (Greenland & Faroe Islands)
5	Finland
6	Iceland
7	Norway
8	Sweden

———— Arctic Region

– – – – Arctic Circle

- - - - - Northwest Passage

◄ *The Arctic Region and the countries involved*

The environmental view

Environmental groups feel that oil companies have already wreaked havoc on large parts of Alaska and Siberia, and that they should keep out of the Arctic. But it is not just oil development at stake. Ben Stewart of Greenpeace said: 'Any country rushing for oil at a time of global warming is being deeply irresponsible. Aside from the issue that they might ruin the pristine environment of the Arctic, countries should be competing in a rush towards renewable energy, not in a rush for fossil fuels.' Add to that the fact that the new oil rush is only possible because of the accelerated shrinking of the polar ice cap due to global warming, and somehow things do not quite stack up (also see pages 272-275).

The race goes on

The Arctic states are battling on. Canada plans to open an army training centre at Resolute Bay and a deep water port on Baffin Island. In 2007, it announced plans to build a fleet of patrol boats to guard the Northwest Passage, which has now opened up as a result of increased ice melt due to global warming (see map above). In 2008:

- Russian ice-breakers patrolled the Arctic for months on end (see the top photo opposite), and Russian aircraft flew over the North Pole simulating strikes on bases and shipping.
- after joining forces to scientifically research the Arctic seabed over a number of years, Denmark and Canada claimed in August that the Lomonosov Ridge is actually connected to North America and Greenland, instead of Russia.
- Russia, Canada, the USA, Norway, and Denmark signed an agreement that they would all abide by the UN Law of the Sea Convention. A UN panel will decide about control of the Arctic by 2020.

Background

UN Law of the Sea Convention

Countries, such as the eight Arctic states (see left), are allowed to exploit offshore resources within 200 nautical miles of their territory. This limit can be extended if a country can prove that the continental shelf beneath the ocean is connected to their land – and is similar geologically. Russia's first attempt to push its maritime borders beyond the existing 200-mile zone was rejected by the UN in 2001.

The Lomonosov Ridge

The Lomonosov Ridge is about 100 miles wide and 1240 miles long, and it zig-zags through the ocean between Russia and Greenland – under the North Pole. Above the ridge, the ocean is just 1000 metres deep in places. Beyond the ridge, it is about 5000 metres deep. When the Russian submarines planted their flag on the ridge in August 2007, they also took away geological samples to try to prove that the ridge is part of the Russian continental shelf.

Alberta, Canada

The sky is blue and cloudless, but you can barely see it from John Martin's cab. John sits three stories above a sea of mud, dirt and sand – both hands working video-game-like toggles to manoeuvre an enormous shovel. It takes him 25 seconds to scoop out about 80 tons of sticky brown sand and release it into the world's largest dumper truck.

Adapted from an article on www.popularmechanics.com in March 2007

▶ *The Bucyrus 495 shovel can fill the world's biggest dumper truck with five shovel loads (in just over two minutes). It works 24/7 at the Muskeg River Mine in Alberta. Each dumper truck costs $5 million but can earn its price back with just eight days' work*

Where are tar sands located?

Tar sands are found in many countries, but the largest reserves are in Canada and in Venezuela in South America. Most of Canada's tar sands are found in three major deposits in northern Alberta – the Athabasca, the Cold Lake and the Peace River deposits (see the map). Together they cover an area larger than England.

The Muskeg River Mine, where John Martin works, is located 47 miles north of Fort McMurray in the Athabasca deposit. It is not an easy environment to work in – blazing hot in summer and -40°C in winter.

Why turn to tar?

Between 2000 and 2005, as oil prices and demand continued to rise, the oil industry spent more than $86 billion on difficult-to-exploit **frontier hydrocarbons**. One of the centres of this spending was Alberta in Canada. Below Alberta's forests lie oil reserves, in the form of tar sands, which could be as high as 180 billion barrels (second only to Saudi Arabia's reserves of conventional oil).

Extracting oil from tar sands is neither easy nor cheap. Although commercial production began in Alberta in 1967, it is only recently – as oil prices have risen – that the production of oil from tar sands has attracted the interest of big companies such as Exxon Mobil, Shell and BP.

Alberta's tar sands produced a million barrels of oil a day in 2003, and are expected to reach 3.5 million barrels a day by 2011. By 2030, they plan to produce at least 5 million barrels a day and export more than Nigeria and Venezuela.

● **Tar sands** are also known as oil sands, bituminous sands and extra heavy oil in Venezuela. They are naturally occurring mixtures of sand, or clay, and water – and a very dense viscous form of petroleum called bitumen.

● **Frontier hydrocarbons** are those grades of oil, such as tar sands, which are inferior to conventional sources of oil.

▼ *The landscape around Fort McMurray is scarred with toxic waste ponds, mines, hundreds of miles of pipes and burgeoning petrochemical works*

How is the oil extracted?

Conventional crude oil is normally extracted by drilling into a petroleum reservoir, but tar sands are mostly mined by opencast mining. The extracted material is then crushed, mixed with water and the bitumen separated out before it can be refined for use. Tar sands are also pumped out. High-pressure steam is injected underground to separate the bitumen from the sand.

Costs of exploiting tar sands

- Producing useable oil from tar sands is expensive; it is only viable when oil costs over $30 a barrel. It costs $15 a barrel to extract bitumen from tar sands, compared with $2 a barrel for conventional crude oil.
- Producing useable oil from tar sands is also very energy intensive. It takes the equivalent of one barrel of oil to produce three barrels of crude oil from tar sands. Conventional oil production requires much less energy.
- The Alberta Energy Research Institute says that processed tar sands are a large source of greenhouse gas emissions (due to their energy intensive production).
- Two tonnes of mined tar sands are required to produce one barrel of oil, which leaves huge quantities of waste sand and environmental destruction (see opposite).
- It also takes between two and five barrels of water to produce every barrel of oil.
- The environmental impact of mining Canada's tar sands includes the removal of trees, shrubs and soil. 470 km^2 of forest have been removed, and lakes of toxic wastewater cover 130 km^2.

Mike Hudema, a Greenpeace activist, has said: 'The tar sands are the greatest climate crime. Not only will their development produce 100 million tonnes of greenhouse gases a year by 2012, but it will kill off 147 000 square miles of forest which represents the greatest carbon sink in the world.'

Benefits of exploiting tar sands

- Tar sands provide an alternative source of oil when conventional sources are unavailable for political or access reasons – or existing fields cannot produce more.
- By 2030, tar sands could meet 16% of North America's demand for oil – providing a secure, sizable source of oil for Canada and the USA.
- Tar sands will provide an additional source of oil until more renewable sources and cleaner fuels can be developed.
- Mining companies are required to reclaim (i.e. replant) land disturbed by mining (but the reclamation lags behind the disturbance).
- Oil is vital to the Canadian economy. In 2007, the oil industry accounted for nearly 20% of the total value of Canadian exports.

The Canadian Association of Petroleum Producers has said: 'Far from being a huge global source of greenhouse gases, the tar sands produce only 4% of Canada's greenhouse gas emissions. Only 20% of deposits are close enough to the surface to be mined – the rest will be produced by technologies that have far less surface impact.'

Should North America rely on Canada's tar sands to provide an alternative source of oil, or look for alternatives to fossil fuels?

What do you think?

Over to you

1. a. Identify all the players involved in mining tar sands in Canada.
   b. Draw up and complete a conflict matrix to identify which players would agree or disagree with each other.
   c. Describe what the matrix shows about exploiting the tar sands.
2. Complete a table to assess the costs and benefits of drilling for oil in the Arctic (economic, social, and environmental).

On your own

3. Research another example where people are searching for oil in either extreme or environmentally sensitive areas, e.g. prospecting for oil off the north-west coast of Australia, or developing Colorado's oil shales. Use the following framework for your enquiry:
   a. How is the oil formed?
   b. What issues are involved in extracting the oil?
   c. Who is involved and how?
   d. What are the conflicts?
   e. How far are the conflicts being resolved?

In this unit you'll find out about the economic and political power of OPEC and energy companies in the USA.

Oil and economics

'Oil price soars as US woes mount' ran a BBC headline in June 2008. Oil prices surged upwards, the dollar slumped, share prices on Wall Street plummeted, and US unemployment had its biggest rise in 20 years. The price of oil was pushed even higher when Israel threatened to attack Iran (OPEC's third largest oil producer) over its suspected nuclear weapons programme – and Nigeria's oil production dropped by about 25% due to action by local militants.

As oil prices rose during the first half of 2008 (see the graph) the USA, UK and other major oil consumers urged producers like Saudi Arabia to increase supply in order to bring down the price.

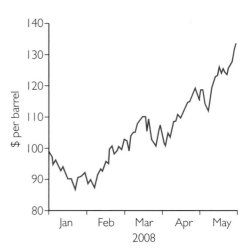

▲ The price of oil rose rapidly in the early part of 2008

Saudi oil output to rise

At the request of consumer nations, Saudi Arabia has agreed to pump an extra half a million barrels of oil a day from July 2008. The price of oil has risen to $135 a barrel – sparking protests around the world. Saudi Arabia has been under pressure from the USA to increase production – as American petrol prices hit record highs – but the Saudis have argued that the high oil prices are not just caused by excess demand and not enough supply, but also by 'speculators' in the oil markets.

Saudi Arabia is worried that the current high oil price will dampen growth in the industrialised West and lower demand – hurting Saudi Arabia's economy.

Adapted from an article in *The Independent*, 16 June 2008

▲ In the summer of 2008, protests related to the high price of oil swept across Europe and Asia, like these protests in the Philippines

OPEC

Saudi Arabia is a member of the Organisation of the Petroleum Exporting Countries (OPEC). OPEC is a permanent intergovernmental organisation, consisting of oil producing and exporting countries. For nearly all of them, oil is their main – or only – export, and is vital for their development and social and economic well-being. The table opposite shows OPEC members in 2008 (Ecuador is also a member).

OPEC's objective is to 'co-ordinate petroleum policies amongst its members, to ensure fair and stable prices for producers, an efficient, economic and regular supply to consumer nations, and a fair return on capital for those investing in the industry.'

OPEC was formed in 1960 to protect the interests of oil-producing countries. At that time, the Arab oil fields were controlled by multinational companies – who raised and lowered prices as they saw fit. In 1960, faced with a glut of oil, they lowered prices – thus reducing the amount that the producers received in taxes and royalties. Five of the producers (Iran, Iraq, Kuwait, Saudi Arabia and Venezuela) came together to create OPEC and demand higher prices – thus forming what some view as a **cartel**.

● A **cartel** is an association of producers or suppliers, formed to monopolise the production and distribution of a product or service to control prices, etc. In politics, a cartel is an alliance of parties or interests created to further common aims.

OPEC's role in energy supply

OPEC is a powerful player in the global energy supply business. It sets oil production quotas for member countries in response to economic growth rates and demand-and-supply conditions. If demand suddenly rises, OPEC can increase oil production to prevent a sharp price rise; similarly, oil production can be reduced to maintain the price if demand falls. OPEC's aim of ensuring fair and stable prices for its members, means that it aims to get the best possible price without either swamping the market (bringing the price down), or restricting it (forcing the price up).

At the end of 2006, the OPEC members had proven reserves of over 900 000 million barrels of crude oil (nearly 78% of the world's total reserves). They produce about 45% of the world's crude oil and 18% of its natural gas – and their oil is sold all over the world, as the table shows.

▼ Where OPEC's oil went in 2006 (barrels per day). Ecuador is missing from the table because it suspended its membership of OPEC between 1992 and 2007.
* Indonesia left OPEC in 2008, because it was no longer a net exporter of oil

OPEC member	Europe	North America	Asia and Pacific	Latin America	Africa	Middle East	Total world
Middle East							
Iran	1 084 000	0	1 605 000	0	149 000	0	**2 839 000**
Iraq	371 000	664 000	446 000	0	0	0	**1 481 000**
Kuwait	317 000	154 000	1 956 000	0	46 000	0	**2 473 000**
Qatar	0	3000	699 000	0	0	0	**701 000**
Saudi Arabia	1 163 000	1 501 000	4 721 000	85 000	340 000	497 000	**8 307 000**
United Arab Emirates	78 000	9000	2 741 000	0	44 000	0	**2 873 000**
Africa							
Algeria	488 000	729 000	59 000	88 000	2000	17 000	**1 382 000**
Libya	1 266 000	160 000	122 000	0	61 000	0	**1 609 000**
Nigeria	638 000	1 597 000	64 000	0	0	0	**2 299 000**
Angola	140 000	824 000	60 000	0	0	0	**1 025 000**
Asia							
Indonesia*	0	30 000	497 000	0	0	0	**527 000**
Latin America							
Venezuela	258 000	996 000	173 000	835 000	3000	0	**2 265 000**

Background

The price of oil

The final price of oil is decided by two factors – the price on the free market (i.e. demand and supply conditions) and the quality of the oil.

Crude oil is now the most actively traded commodity in the world. It can be sold as a 'futures contract' – where the buyer agrees to take delivery and the seller agrees to provide a fixed amount of oil at a pre-arranged price at a specified location. The aim of this 'futures trading' is to try to smooth out price fluctuations by assuring a future demand for a future supply for those industries that rely on it, such as airlines.

Crude oil comes in different varieties and qualities, depending on where it is pumped up from. Buyers and sellers refer to a limited number of reference, or benchmark, crude oils. Brent crude from the North Sea is accepted as the world benchmark and other varieties of oil are priced in relation to Brent crude.

In the first half of 2008, oil prices continued to rise rapidly. The supply of oil did not appear to be keeping pace with increasing demand – especially from China and India – and speculators were investing in oil, gambling that the steep price rise would continue. (Speculators are a problem, because – by betting on price rises and falls – they disturb the balance that normal 'futures trading' maintains.) OPEC was worried about increasing the supply of oil, in case investors then felt that there was too much oil on the market, stopped investing and caused a collapse in the price. A fall in oil price would hit the economies of OPEC members badly.

USA – oil, politics and economics

Oil, politics and the economy are inextricably linked – and nowhere more so than in the USA. The USA is the world's biggest consumer of oil – taking a 24% share of the world's oil in 2007. Oil matters.

Oil and gas – contributions to politics

The total financial contributions from the oil and gas industry to federal candidates and political parties for the 2008 US election cycle (up to the end of April 2008) stood at over $14 million. The pie chart and table show how this was split between Republicans and Democrats, and the top ten contributing companies.

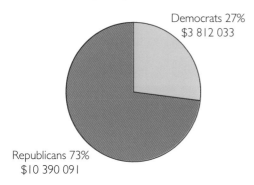

Democrats 27%
$3 812 033

Republicans 73%
$10 390 091

Background

Political funding and the USA

The oil and gas industry (which includes TNCs, independent oil and gas producers and refiners, gas pipeline companies, service stations, and oil dealers) has a long history of influence on the US government. Since the 1990 US election cycle, the oil and gas industry has contributed $182 million to politicians – and 75% of this has gone to the Republicans.

There are two main political parties in the USA – **Republicans** (right wing) and **Democrats** (liberal centre). Many individuals and organisations contribute funding to them – US elections are an expensive business. In the 2008 election cycle, the oil and gas industry ranked seventeenth in terms of the size of its political funding.

Rank	Organization	Amount ($)	Democrats	Republicans
1	Koch Industries	942 000	14%	86%
2	Exxon Mobil	552 578	24%	76%
3	Chevron Corporation	498 749	27%	73%
4	Valero Energy	472 372	25%	75%
5	Occidental Petroleum	295 350	26%	74%
6	Marathon Oil	287 625	33%	67%
7	Sunoco Inc.	250 600	59%	41%
8	American Gas Assn.	232 850	42%	58%
9	Weatherford International	200 200	78%	22%
10	Chesapeake Energy	183 045	23%	77%

▲ Oil and gas industry contributions to Republicans and Democrats – 2008 election cycle.

Funding from the oil and gas industry has varied in size since 1990, but what has not changed is the fact that the Republicans have received far greater levels of funding – as the graphs show.

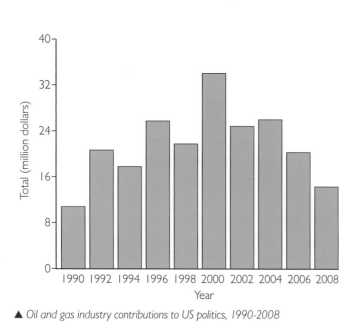

▲ Oil and gas industry contributions to US politics, 1990-2008

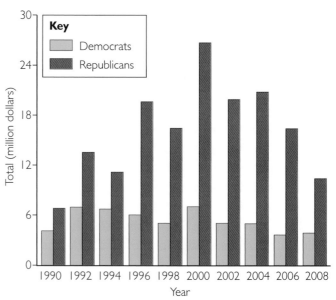

▲ Oil and gas industry contributions to US politics split between Republicans and Democrats, 1990-2008

Oil companies as energy players

Oil and gas companies fund political parties in order to exert their influence. They are major players in the supply of energy and are in business to make a profit.

- Exxon Mobil is one of the world's largest energy companies. It successfully lobbied the US government to gain access to federal lands (such as nature reserves and national forests).
- Chevron Texaco lobbies on all energy issues, including opening up the Arctic National Wildlife Refuge in Alaska for oil drilling (see pages 94-99).

Politicians work in the oil companies' interests:

- In June 2008, the Republicans prevented a bill from being passed by Congress which would have imposed a 25% windfall tax on oil company profits that were considered to be 'unreasonable'. The bill would also have rescinded $17 billion in tax breaks that the oil companies were expecting to receive over the coming decade. This money would have been redirected into tax incentives for renewable energy sources, such as wind and solar power, as well as programmes promoting energy efficiency and conservation.
- Also in June 2008, the Republicans pushed for an increase in oil and gas drilling. As the 2008 Presidential Elections loomed, both Republicans and Democrats blamed each other for rising energy costs and petrol prices topping $4 a gallon. The Republicans blamed Democrat opposition to opening up the Arctic National Wildlife Refuge to oil and gas exploration and drilling for making the fuel problem worse. The Republicans believe that rising energy prices will encourage public support for expanding US oil and gas development into sensitive areas like the ANWR.

▲ Arguments rage over whether the Coastal Plain of the Arctic National Wildlife Refuge in northern Alaska should be opened up for oil drilling. Oil pipelines already scar the pristine Alaskan landscape. Those supporting further expansion of Alaskan oil say that current extraction only affects a few hundred acres of Alaskan land. But those against, use photos like this one to show that a narrow strip of pipeline in the landscape – although not taking up much land – has a huge impact on the scenery.

> Will producing more oil bring the price down?
>
> **What do you think?**

Over to you

1. **a** Use the table on page 33 to rank the OPEC members in terms of their oil production. What does this show about the relative importance of the different producers?
   b Now rank the OPEC members supplying North America with oil. What does the ranking show about the relative importance of North America's oil suppliers? Does the ranking surprise you?
2. In pairs, use the examples of OPEC and political funding in the USA to list the advantages and disadvantages of cartels for consumers and producers.

On your own

3. Complete a table to assess the economic, social and political impacts of oil companies and OPEC on the energy market.
4. Using the BBC News website, or North American news websites (e.g. NBC, CNN), research the 2008 US Presidential Election and the attitudes towards opening the Alaskan oil fields. Produce a table comparing **a** which politicians mentioned Alaskan oil, **b** which party they belonged to, **c** how politicians in other parties responded. Explain any differences you find.

In this unit you'll learn that there is uncertainty over the global energy supply in terms of demand, reserves and peak oil and gas.

Oil – a finite resource

How long will global stocks of oil last? Scientists, led by the Oil Depletion Analysis Centre in London, say that global oil production will peak before entering a steep decline that will have massive consequences for the world's economy and the way we live. Colin Campbell, head of the Depletion Centre, says that the theory is quite simple '… and one that any beer drinker understands. The glass starts full and ends up empty – and the faster you drink it, the quicker it's gone!'

Country	Oil consumption
USA	20.7
China	7.2
Japan	5.2
Russia	2.9
Germany	2.7
India	2.6
Brazil	2.2
UK	1.8

◄ *Oil consumption in 2008 (million barrels a day)*

Increasing demand

In 2008, global oil production and consumption stood at around 80-85 million barrels of oil a day. But who is using it all? The table shows four of the G7 countries, plus the BRICs – Brazil, Russia, India and China. These eight countries alone account for over half of the world's total oil consumption.

However, not only is the world already consuming vast quantities of oil, but demand is still rising:

- By 2010, global oil consumption is predicted to average 89.2 million barrels a day. China's consumption will increase to 10% of the global total, and India's to 3% – but that is still only 13% of total oil consumption, despite the two countries containing 40% of the world's population.
- By 2030, global oil consumption is expected to reach 113 million barrels of oil a day.

Oil is not the only energy source in demand:

- By 2030, total global energy consumption is expected to have grown by 50% – with demand from developing countries surging by 85%, compared to a 19% increase in industrialised countries.
- As the economies of China and India grow (see right) their energy use is predicted to double between 2005 and 2030 – fuelling not only their demand for oil and gas, but for coal as well.
- Electricity generation is expected to nearly double by 2030 – fuelled mainly by coal and natural gas. China accounts for 70% of new coal consumption.
- Nuclear power is also expected to increase by nearly 50% – mostly in China and India.

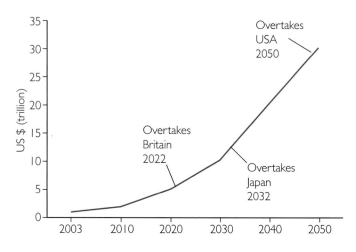

▲ *India's economic growth forecast, 2003-2050*

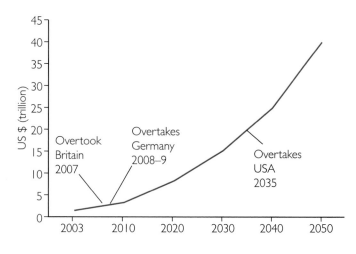

▲ *China's economic growth forecast, 2003-2050*

- The largest increase in demand for oil will come from transport (as people in countries such as India and China purchase more cars).
- The use of renewable energy is expected to increase up to 2030. This includes a mixture of fuels, such as biomass (increasing slowly) and wind, solar and biofuels (increasing rapidly) – see the graphs below.

Although the overall trend for energy consumption is rising, demand is variable. Western countries continue to consume by far the largest share, and, although China and India are increasing their consumption rapidly, the West is expected to dominate consumption patterns for the foreseeable future.

Where are the oil and gas reserves?

The graphs below show where the world's largest reserves of oil and gas are located. Saudi Arabia has the largest reserves of oil, but Russia has by far the largest gas reserves. Proven reserves are those which the oil and gas industry considers can be recovered using existing technologies.

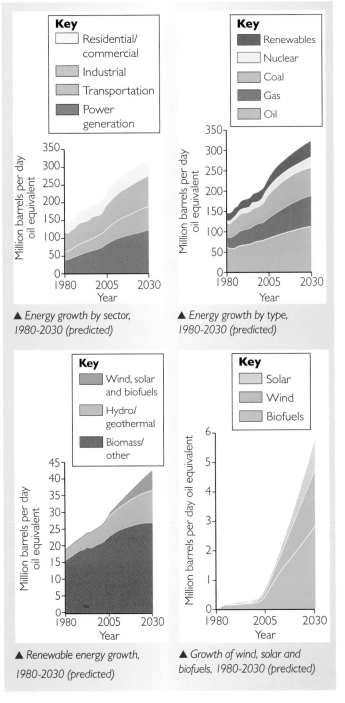

▲ Energy growth by sector, 1980-2030 (predicted)

▲ Energy growth by type, 1980-2030 (predicted)

▲ Renewable energy growth, 1980-2030 (predicted)

▲ Growth of wind, solar and biofuels, 1980-2030 (predicted)

▶ The world's top proven gas reserves, 2007 (trillion cubic feet)

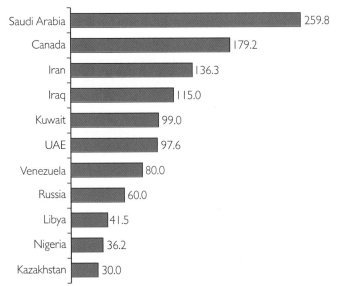

▲ The world's top proven oil reserves, 2007 (billion barrels)

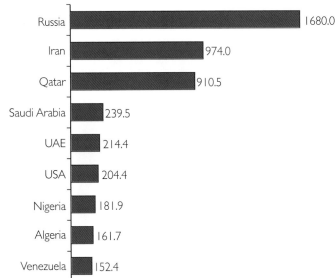

Can oil production increase?

When the price of oil soared in the first half of 2008, basic demand-and-supply economics suggested that consumption would fall and supply would rise. But neither seemed to happen (previous increases in the price of oil also failed to reduce consumption noticeably).

Unlike the OPEC nations, which try to regulate supply to keep oil prices stable, the other oil-producing countries – such as Russia and Mexico – are 'free traders' with an economic incentive to produce oil flat out when prices are high. However, higher drilling costs – and national policies restricting foreign investment – meant that these countries seemed unable to increase their oil output in 2008, and they continued to produce a combined total of around 50 million barrels of oil a day (or about 60% of the world's supply).

Increasing oil production is expensive and takes time to materialise; more wells need to be tapped, and pipeline capacity increased:

● Saudi Arabia is spending US$50 billion to increase its production to 12.5 million barrels a day by 2009. This is still some way short of the 15 million barrels that it needs to produce to meet anticipated future demand.

● OPEC's 13 members plan to invest US$150 billion to expand their production capacity by 5 million barrels a day by 2012 (from OPEC's current 36 million barrels a day).

● To meet anticipated future demand, OPEC will need to pump 60 million barrels a day by 2030.

Even if production capacity does increase, the amount of exploitable new oil discovered since the 1960s has steadily fallen, as the graph shows. However, not everyone is pessimistic. High oil prices have sparked a global dash for new oil fields, with companies searching deep oceans and alternative sources in their hunt for new oilfields (see pages 28-31). However, new oilfields take years to go into full production once they are found. Since 2005, Brazil has discovered huge new oil fields. However, 75% of Brazil's oil reserves are under at least 400 metres of water.

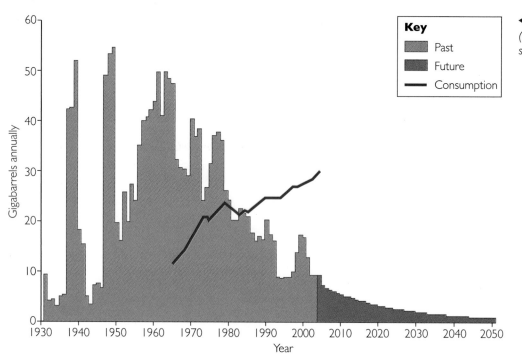

◀ New oil discoveries, 1930-2050 (projected), plus oil consumption since 1965

Peak oil and gas

In 2007, the International Energy Agency (IEA) warned that the world will face a 'supply crunch' by 2012. Forecasters employed by the French and German governments expect global oil production to peak by 2020. While the date of **peak oil** may not be fixed in stone, one thing is certain – it will happen – and (according to statistics produced by BP) oil production in at least 18 countries is already in decline:

- Norway's production has slumped by 25% since its peak in 2001.
- The North Sea is considered to be a dying oil basin, and UK oil production has fallen by 43% in 8 years. The oil field at Prudhoe Bay in Alaska has had a similar decline.
- Mexico's Cantarell field in the Gulf of Mexico was one of the most prolific in the world, but production started to fall in 2006 and then dropped by 9% in 2 years.

The peaking of global gas production is unlikely to occur before 2020. However, problems are already developing in the UK, which used to be self-sufficient in gas but became a net importer in 2004 – and has one of the highest rates of gas depletion of any producing country. It will need to find secure supplies of imported gas to replace falling domestic production.

The impacts of peak oil

Economic growth has always depended on a growing supply of oil. After the peak, many forecasters expect global oil production to fall by about 3% a year. This means that, with growing oil demand, the deficit between the oil we want and the oil we get will expand by more than 4% a year. Within 10 years, we could have just about half the oil supply required to sustain economic growth. This is likely to lead to large spikes in the price of oil, which could cause deep recessions.

> • **Peak oil** (or **gas**) refers to the year in which global production will reach its maximum level and then fall into sustained decline.

▶ *A car storage tower at the Volkswagen plant in Wolfsburg, Germany. Peak oil will impact on our love affair with the car. Not only does it take an average of 20 barrels of oil to produce one car (in the USA), but peak oil will mean a growing deficit of transport fuel.*

Over to you

1　In pairs, produce a Venn diagram to show the social and economic impacts of peak oil and gas.

2　**a** Research a variety of sources to find out how global reserves of oil are assessed, where new reserves are being discovered, and how much oil might still be recovered.

　b On the basis of your research, hold a class debate on the future of our energy supplies. Discuss these topics:
　　- How can we be sure that estimates about the size of known oil reserves are reasonably accurate?
　　- How do we know how long those reserves will last?
　　- How do we know what reserves are still out there to be discovered?
　　- How can we cope with the uncertainty?

On your own

3　Find out how American and British towns and cities are preparing for peak oil. Use this website www.europeanenergyreview.eu and follow the links for peak oil to find the 'Preparing for peak oil' document.

Exam question: Referring to examples, examine the issues when assessing global reserves of energy. **(15)**

> *Which will have the greater impact – climate change or peak oil?*
>
> **What do you think?**

In this unit you'll look at our reliance on fossil fuels – the costs, geopolitical tension, and potential for conflict.

Relying on fossil fuels

The International Energy Agency's (IEA) *World Energy Outlook 2007* may not sound like a gripping read, but it contains some startling projections! As global energy use increases, the IEA predicts that:

- fossil fuels will continue to dominate the energy mix between 2005-2030 (having an 84% share) – with coal use growing most rapidly
- by 2030, global carbon dioxide emissions associated with energy use will rise by 57% – dramatically increasing the threat of catastrophic climate change

The IEA's projections assume that the world will continue with 'business as usual', i.e. that it will continue to rely on fossil fuels as the main source of energy. The IEA also predicts that:

- consumption of natural gas will double by 2030
- oil production will have to increase by about 1.4 million barrels a day to meet growing demand – the equivalent of another Saudi Arabia coming on stream every seven years.
- consuming countries will rely increasingly on imports of oil and gas – much of it from the Middle East and Russia. As demand from China and India grows, the amount of oil they import by 2030 will be more than the combined imports of the USA and Japan today.
- supplies of easily accessible oil and natural gas will no longer keep up with demand after 2015. To close the gap, the world will have no choice but to use energy more efficiently and also increase the use of other sources of energy.

The cost of 'business as usual'

Business as usual comes at a cost – burning fossil fuels is almost certainly driving climate change. The two graphs on the right show the predicted growth in CO_2 emissions in both OECD and non-OECD countries between 1980 and 2030. In non-OECD countries, CO_2 emissions are expected to increase at an average annual rate of 1.9% between 2005 and 2030. As a result, those countries will represent around 95% of the annual growth in energy related emissions in the period up to 2030.

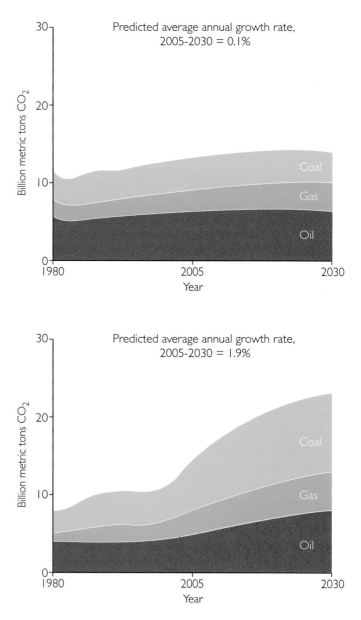

▲ *Energy related CO_2 emissions in OECD (top) and non-OECD (bottom) countries, 1980-2030*

In 2006, the British Government published a review of global warming by Sir Nicholas Stern – former Chief Economist of the World Bank. The Stern Review estimated that, under the business-as-usual model, climate change would cost the world 5-20% of global GDP, whereas efforts to limit greenhouse gas emissions and the impacts of global warming would cost just 1% of global GDP.

Already, global economic losses due to natural disasters (such as extreme weather events, which are likely to increase in magnitude and frequency as a result of global warming) appear to be doubling each decade – with annual losses expected to reach $150 billion before long. In addition:

▲ *The IPCC is increasingly confident that certain weather events, such as hurricanes, will become more frequent, widespread and intense as climate changes - causing serious economic losses*

- a 2-3°C rise in temperature could reduce global economic output by 3% – by reducing crop yields, for example
- if temperatures rise by as much as 5°C, 10% of global output could be lost, with the poorest countries losing more than 10% because they lack the income to pay for coping strategies – such as storing water in large dams.

The importance of the Middle East

Energy insecurity and the global reliance on fossil fuels are leading to tension and the potential for conflict between consumers and suppliers, as you will see on the following pages.

The Middle East is a key supplier of oil, and – as demand rises and other supplies peak – it will become increasingly important.

- The world's current known recoverable oil reserves amount to about 1000 billion barrels. Of those reserves, 71% are in the Middle East.
- By 2025 – without new discoveries – that proportion will have risen to 83% and the Middle East will be the only remaining major reservoir of oil reserves.
- All the major global economies depend hugely on Middle Eastern oil. The Middle East currently supplies 76% of Japan's oil, 26% of Western Europe's and 21% of the USA's.

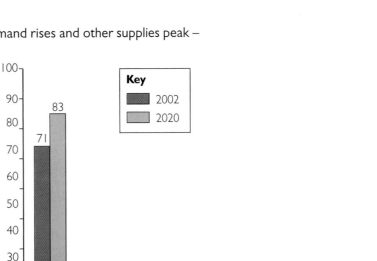

▶ *The world's dependence on Middle Eastern oil reserves*

The Middle East

Security and political stability in the Middle East are vital if the rest of the world wants to rely on its oil supplies. However, these are far from guaranteed, because of terrorism in Iraq and elsewhere – and ongoing tensions with Israel, which often result in conflict or threats of conflict (such as over Iran's nuclear ambitions).

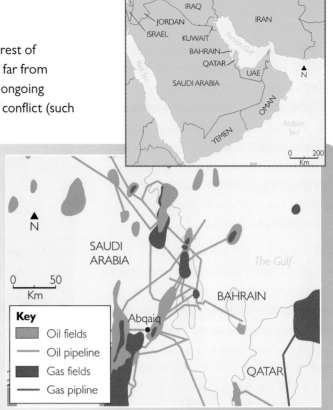

The attack on Abqaiq

On 24 February 2006, terrorists attacked the oil-processing plant at Abqaiq in Saudi Arabia. Two vehicles carrying explosives tried to smash their way into the plant's compound. A two-hour gun battle followed, during which the explosives went off. Two terrorists and two guards were killed but there was no damage to the plant.

Why was this attack important?

- It was the first direct al-Qaeda terrorist attack on a Saudi oil facility.
- Saudi Arabia is the only oil-producing country with excess capacity – nowhere else in the world can pick up the slack.
- Abqaiq is the largest oil-processing plant in the world – producing 6.8 million barrels of oil a day (about 75% of Saudi Arabia's total output).
- A successful terrorist attack could have halved Saudi oil production for up to a year.
- Abqaiq is at the hub of 12 000 miles of oil and gas fields and pipelines – all vulnerable to direct attack (see the map).

Key
- Oil fields
- Oil pipeline
- Gas fields
- Gas pipeline

▲ The oil and gas fields and pipelines around Abqaiq

What are the alternatives?

After the terrorist attack on the World Trade Centre in New York on 11 September 2001 (9/11) (see right), the USA and other Western countries tried to reduce their dependence on Middle Eastern oil. Non-OPEC countries began to gradually increase their oil production – especially Russia and the African producers, such as Nigeria. However, Russia's reserves only represent 5% of the world's total and are already in decline. Africa's reserves amount to 7% of the total and its largest producer, Nigeria, will peak in 2010.

By 2025, many of the largest oil producers outside the Middle East will no longer be significant. Like it or not, the Middle East will have a bigger share of the pie than ever before.

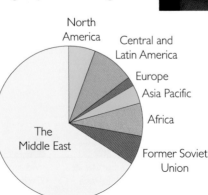

North America
Central and Latin America
Europe
Asia Pacific
Africa
Former Soviet Union
The Middle East

▶ The world's oil producers in 2008

Background

Perceptions and misconceptions

● The term 'Middle East' is problematic. Where exactly do we mean when we talk about 'The Middle East'? The map shows wide variations in perception about this ostensibly simple question.

● Of course, this region is only actually 'East' at all if seen through the eyes of the 'West' – to anyone in Asia, it lies to the west of them.

● The early exploitation of Middle Eastern oil reserves (e.g. in Saudi Arabia) was co-ordinated by Western companies – who at first paid little or no royalties to those who owned the oil. Even the national borders of many of the countries in this region were decided by Western civil servants after the First World War – using a process of maps, rulers and red pens. As a result, the West has often been resented for its seemingly high-handed, arrogant attitude – a resentment which continues today over many policies in Iraq and Afghanistan.

● The West also has a negative image in the eyes of most Middle Eastern countries because of its perceived pro-Israeli policies, which is not always fair.

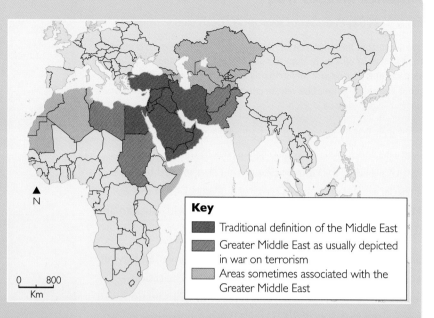

Key
- Traditional definition of the Middle East
- Greater Middle East as usually depicted in war on terrorism
- Areas sometimes associated with the Greater Middle East

0   800
   Km

▲ *Where is the Middle East?*

● In turn, the Middle East has a largely negative image in the eyes of many in the West – who often lump its countries together as universally unstable and undemocratic staging posts for Islamist terrorist attacks against the West and its interests. This is also far from true.

OPEC and the rising price of oil

All of the oil-producing countries of the Middle East are members of **OPEC** (see pages 32-33). This organisation aims to safeguard the interests of its members by ensuring stable oil prices, while at the same time securing them a steady income. Major OPEC producers with small populations, like Saudi Arabia (27 million people), are more concerned with stability, while smaller producers with large populations, like Nigeria (150 million people) are more likely to be interested in increasing their revenue.

Dramatic rises in the price of oil from 2002 onwards (see the graph on page 7) were the result of:

● rapidly increasing demand from developing countries like China and India

● OPEC's reluctance to increase oil supplies and risk a price crash

● declining oil reserves, with many existing reserves becoming more difficult to exploit

● an improvement in the economies of some of the oil-exporting countries, such as Mexico and Indonesia, which led to an increase in their own demand for oil (Indonesia has now ceased to be an oil-exporting country and, in May 2008, announced it was leaving OPEC as a result)

● political instability in the Middle East (especially following the 2003 US-led invasion of Iraq), and also in Nigeria where separatist movements have been threatening oil supplies and kidnapping oil workers

● oil traders in the world's financial institutions speculating on the commodity markets (also see page 33)

These factors have added to the '**security premium**' – the extra cost built into the price of oil to allow for any disruption in supply. When spare production capacity is low and geopolitical tension is high, the security premium rises.

Iraq and Iran - the fight for control

America invades Iraq

Early in 2002, President George W. Bush said:

'It is important for Americans to remember that America imports more than 50% of its oil – more than 10 million barrels a day. And the figure is rising ... this dependence on foreign oil is a matter of national security. To put it bluntly, sometimes we rely on energy sources from countries that do not particularly like us.'

In March 2003, American and allied forces invaded Iraq. Iraq has the fourth largest oil reserves in the world, and its then leader, Saddam Hussein, was considered to pose a threat to the security of Western oil supplies in the Middle East. By 2003, he was making deals with Russian and Chinese oil companies – and the USA acted.

Since the first Gulf War of 1990 (when Iraq invaded Kuwait and had to be forcibly removed by a military coalition – led by America), Iraq had been subject to UN sanctions. Before the invasion of 2003, the USA put pressure on Iraq (through the UN) to admit that it had stockpiled weapons of mass destruction (such as chemical weapons) – or face a military attack. What is now clear to many people is that the main motive behind the invasion was actually access to Iraq's oil reserves.

The USA and Saudi Arabia have a close relationship. The USA is dependent on Saudi Arabian oil, and Saudi Arabia relies on the USA for military and technical support. However, by invading Iraq, the USA may have hoped to reduce its dependence on Saudi oil and increase its energy security by introducing a new supplier. In 2003, it removed its troops from Saudi Arabia, but they remained in large numbers in Qatar, Egypt and Kuwait. Since US troops left Saudi Arabia, there have been several terrorist attacks there – including Abqaiq.

General John Abazaid, who retired in May 2007, said the 'strategic situation' in the Middle East – the rise of extremism and the global dependence on oil – would necessitate a long-term military presence there. However, America might see an inevitable decline in its dominance as China expands and the Russian economy revives.

▲ *'I am saddened that it is politically inconvenient to acknowledge what everyone knows – the Iraq war is largely about oil.' Alan Greenspan, 2007. He was the Chairman of the US Federal Reserve during the Iraq Invasion.*

▶ *American troops in Iraq in 2008 - resigned to a long stay*

Iran and Central Asia

Since 9/11, America's aim – according to the White House – has been to democratise the 'greater' Middle East and Central Asia, and its involvement in Iraq and Afghanistan is part of that process. Iran lies between the two and is a significant energy player and a major **production hotspot**. It has the third largest oil reserves, and the second largest natural gas reserves in the world. Japan and China both import about 13% of their oil from Iran.

Iran has a strongly anti-American government, and America has done its best to isolate it by imposing sanctions and by influencing the energy rich countries of Central Asia. However, to the north of Iran lies Russia, the regional power. An alliance is now developing between Moscow, Tehran and Beijing – built around energy exports from Russia and Iran in exchange for Chinese goods. Oil is becoming a **strategic** resource, i.e. one through which these countries can strongly influence the global economy.

America excluded

America is excluded from the deals between Russia, China and Iran, and is fighting hard to secure oil by means of pipelines/energy pathways running through friendly countries, such as Georgia, Azerbaijan and Turkey (see page 171). Meanwhile China is negotiating for the construction of pipelines from Central Asia to satisfy its own needs and to overcome the problem of its vulnerable sea-lanes (see page 20 on the Strait of Malacca).

India also wants a share of the region's energy. The 'IPI pipeline', or the 'Peace pipeline', is a proposed 2775 km (1724 mile) long pipeline to deliver natural gas from Iran to Pakistan and India.

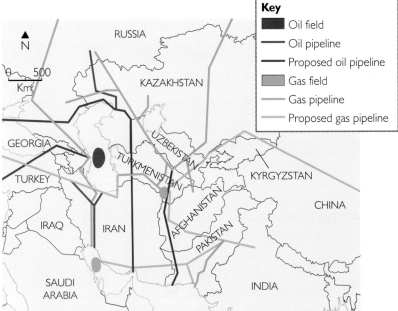

▲ Existing and proposed oil and gas pipelines out of Central Asia

Key
- Oil field
- Oil pipeline
- Proposed oil pipeline
- Gas field
- Gas pipeline
- Proposed gas pipeline

Background

Production hotspots

A 'production hotspot' is an energy producing country or region where there is political instability, such as:

- Iran – which has an anti-Western, anti-Israeli government, and where there is international concern that it might be developing nuclear weapons
- Russia – the government of which has played 'energy politics' with its neighbours and European customers (see pages 170-171)
- Venezuela – where the Government has nationalised the oil industry and diverted oil supplies that previously went to the USA to other Latin American and Asian countries
- Nigeria – which faces internal rebellion in its oil-producing region and problems of corruption

Over to you

1 Draw a table to show the costs and benefits of continuing to rely on fossil fuels as major sources of energy.
2 In pairs, explain why so much time and trouble is spent to maintain the position of oil and gas as key resources.
3 In pairs, research ONE of the production hotspots listed above (you could also research Saudi Arabia or Iraq). Find out about **(a)** its known reserves of energy, **(b)** its main markets, **(c)** its relationships with energy consuming countries. Feed these back in class and discuss the potential conflicts that exist.

4 In what ways has power shifted to the producing rather than consuming countries? What are the possible implications of this in the next decade?

On your own

5 Research the latest IEA World Energy Outlook. How have forecasts changed since this unit was written in 2008.
6 In 400 words, assess how far you think the world can afford to continue with 'business as usual'.

In this synoptic unit you'll investigate some of the approaches that we can adopt to meet our future energy needs and increase our energy security.

Alternatives to 'business as usual'

Not far from the old Silk Road, Chinese Government scientists have begun boring holes deep into the granite bedrock in the first steps towards building what could become the world's largest nuclear waste store. As China hurtles headlong into nuclear power, countries worldwide are looking to secure their energy future and security. Choices range from continued reliance on fossil fuels ('business as usual'), to measures such as reducing consumption by the use of 'green taxes', and carbon credits.

Nuclear power

Resource 1

China embraces nuclear future

China intends to spend $50 billion to build 32 nuclear power stations by 2020. Some analysts say that China will build 300 more by 2050, which will generate about the same power as all the nuclear power stations currently operating in the world today. In comparison, the USA currently has just over 100 operating nuclear power stations.

China's plans are being greeted with both optimism and concern. Nuclear power stations release few greenhouse gases. However, their safety and the issue of what to do with the radioactive waste are problems which the Chinese Government still has to solve.

Adapted from *The Washington Post*, 29 May 2007

Resource 2

Nuclear electricity producers
The top ten producers of nuclear electricity in terms of their percentage of total world production, and the contribution of nuclear power to each country's domestic electricity generation. The data are for 2005.

	Producers	% of total world nuclear electricity generated	% of nuclear in total domestic electricity generation
1	USA	29.2	19
2	France	16.3	79
3	Japan	11.0	28
4	Germany	5.9	26
5	Russia	5.4	16
6	Korea	5.3	38
7	Canada	3.3	15
8	Ukraine	3.2	48
9	UK	3.0	20
10	Sweden	2.6	46

Resource 3

Nuclear power – yes or no?

Why should Asia say NO?	Why should the UK say YES?
Greenpeace China gives the following reasons why Asia should not develop nuclear power:	In 2008, the British Government backed the building of a new generation of nuclear power stations to replace the existing worn out ones. The British Government's view is that:

Why should Asia say NO?

Greenpeace China gives the following reasons why Asia should not develop nuclear power:

- The costs associated with safety, security, insurance, liability in case of accident or attack, waste management, construction and decommissioning are rising rapidly, whereas the cost of solar and wind power is falling.

- Nuclear waste disposal is an unsolved problem (it remains radioactive for up to 10 000 years and no safe containment has been devised yet).

- Nuclear technology is also used in nuclear weapons production, so there is a risk of nuclear weapons proliferation.

- Mining, extracting, processing and transporting nuclear fuel produces CO_2 emissions at every stage.

- Any investment in nuclear power is money denied to developing and promoting renewable energy and energy efficiency.

Why should the UK say YES?

In 2008, the British Government backed the building of a new generation of nuclear power stations to replace the existing worn out ones. The British Government's view is that:

- the UK should maintain the 20% of its electricity generation produced by nuclear power that would otherwise be lost when the old nuclear power stations are decommissioned.

- new nuclear power stations are needed to reduce the UK's dependence on imported energy from the Middle East and Russia, and increase British energy security.

- nuclear power will help the UK to meet its carbon reduction targets and fight climate change.

- the next generation of nuclear reactors will be cleaner, more cost-efficient, and cost less to decommission than the existing ones.

- nuclear power is a tried, tested and broadly safe technology.

Renewable energy

Renewable energy (including large-scale hydro-electric power) accounts for 16.5% of the world's primary energy, and 19% of its electricity generation. If HEP and traditional biomass are excluded, the figure drops to 6% of global electricity generation, but the advantages of a greater reliance on renewables are enormous – and the technical potential is huge.

Resource 4

Renewables – global use, growth rates, targets and potential

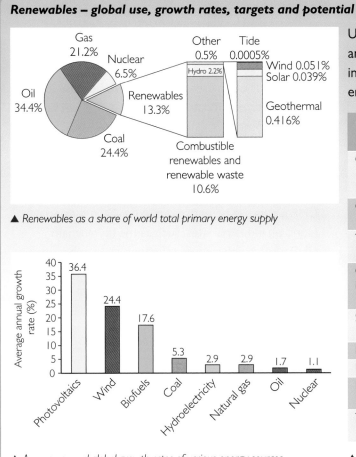

▲ Renewables as a share of world total primary energy supply

▲ Average annual global growth rates of various energy sources, 2001-2006

Use of renewable energy – solar (photovoltaics), wind, and biofuels – grew at a far greater rate than fossil fuels in the first years of the twenty-first century. Renewable energy targets are also growing, as the table shows:

Country/region	Targets for renewable energy ...	Progress so far ...
California, USA	20% of electricity mix by 2010; 33% by 2020	10.9% of in-state mix in 2006
China	15% of total primary energy by 2020	8% in 2006
The EU as a whole	20% of total energy by 2020	7% in 2007
Germany individually	27% of electricity by 2020; at least 45% by 2030	About 14% at the end of 2007
Spain individually	12% of primary energy by 2010	6.9% in 2004
Mali	15% of energy by 2020	N/A
New Zealand	90% of electricity by 2025	70% in 2007
India	10% of added electricity capacity during 2003-2012	N/A
Thailand	21.2% of total energy by 2011	19.8% in 2001

▲ Renewable energy targets and progress in selected countries

Resource 5

Renewable energy – home and abroad

UK

An extra 4000 onshore and 3000 offshore wind turbines could be built across the UK as part of a £100 billion plan to boost the use of renewable energy. Prime Minister Gordon Brown said in June 2008 that the Government's plans represented 'the most dramatic change in energy policy since the advent of nuclear power. The North Sea has passed its peak of oil and gas supply – but it will now embark on a new transformation into the global centre of the offshore wind industry.'

New Zealand

In September 2007, Prime Minister Helen Clark of New Zealand announced that her nation had set a target of becoming 'the first truly sustainable nation on Earth,' with a goal of significantly reducing greenhouse gas emissions – in part through a large increase in the use of renewable energy. New Zealand now aims to get 90% of its electricity from renewable sources by 2025, having achieved 70% by 2007 (much of it from hydro-electric power stations).

Conservation and recycling

Woking – decentralising energy generation

Decentralising energy generation and developing local energy generators, along with energy efficiency measures, has reduced Woking Borough Council's CO_2 emissions by 82% and energy consumption by 52%.

Woking has developed a network of 60 local generators, near to where the electricity is actually used. They are used to power, heat and cool municipal buildings and social housing – as well as town-centre businesses. The sustainable and renewable installations include the use of solar power (PVs) and Combined Heat and Power (CHP).

Woking Borough Council's investment in CHP and renewable energy was £12 million up to the end of 2006. However, between 1990 and 2005, the Council saved £5.4 million in municipal energy bills.

In late 2008, Woking introduced a Low-Carbon Homes Programme to encourage its residents to minimize their CO_2 emissions and water consumption. The first phase of the programme was to convert a detached house as an example to show homeowners what steps they could take themselves to conserve energy and reduce their carbon footprint, as the diagram below shows.

Combined Heat and Power (CHP)

Electricity power stations waste 65% of the heat they generate, but CHP plants can be up to 95% efficient. CHP captures and recycles the waste heat produced in electricity generation.

- CHP still often uses fossil fuels but, because it is so efficient, it cuts emissions and reduces fuel dependency.
- CHP can use different fuels in the same boiler including biomass (e.g. straw, wood pellets) as well as gas, oil and coal.

CHP plants can be enormous, e.g. where they are found on industrial sites, such as the Immingham CHP plant supplying two refineries on Humberside, which is to be expanded to reach the same electricity generating capacity as the Sizewell B nuclear power station. Or they can be smaller and located in town centres, such as Woking and Southampton.

▼ Oak Tree House, Woking Borough Council's low-carbon home uses the following energy conservation methods

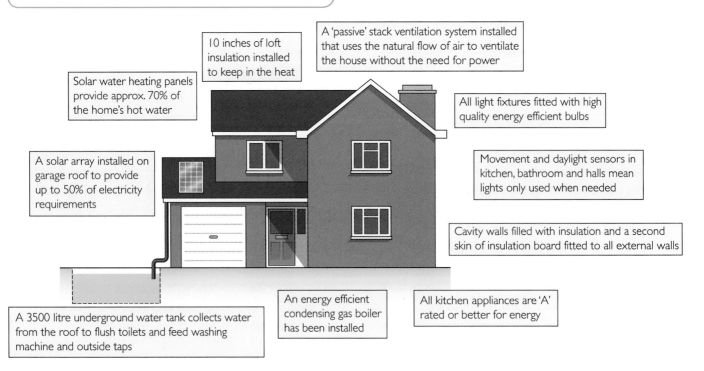

Solar water heating panels provide approx. 70% of the home's hot water

10 inches of loft insulation installed to keep in the heat

A 'passive' stack ventilation system installed that uses the natural flow of air to ventilate the house without the need for power

All light fixtures fitted with high quality energy efficient bulbs

A solar array installed on garage roof to provide up to 50% of electricity requirements

Movement and daylight sensors in kitchen, bathroom and halls mean lights only used when needed

Cavity walls filled with insulation and a second skin of insulation board fitted to all external walls

A 3500 litre underground water tank collects water from the roof to flush toilets and feed washing machine and outside taps

An energy efficient condensing gas boiler has been installed

All kitchen appliances are 'A' rated or better for energy

Green taxation

'Green' taxes are designed to protect the environment, and are aimed at cutting the use of natural resources and encouraging recycling. The British Government's plans to change road tax (Vehicle Excise Duty) are an example of green taxation.

Resource 7

Road tax increase

In 2010-11, an estimated 9.4 million motorists will have to pay more road tax under reforms aimed at punishing 'gas-guzzling' and polluting vehicles. In 2008, the maximum road tax for a vehicle registered between March 2001 and March 2006 was £210. From April 2010, that will increase to £455 for the heaviest polluters.

▲ Vehicles such as Range Rovers and some people carriers emitting more than 255g CO_2 per km will pay up to £455 a year under the new system

It is calculated that the Government will receive more than £1 billion in additional revenue from the scheme by 2011. Friends of the Earth said: 'Increasing VED on old polluting cars will encourage people to choose greener vehicles, cut fuel bills and lower CO_2 emissions'. They also called on the Government to invest the money raised in better public transport.

Resource 8

Green tax revolt

More than 7 in 10 voters in Britain insist that they would not be willing to pay higher taxes in order to fund projects to tackle climate change, according to a new poll. The survey also reveals that most Britons believe 'green' taxes on 4x4s, plastic bags and other consumer goods have been imposed to raise cash, rather than to change our behaviour.

Friends of the Earth blamed the Government, saying: 'The Government is using climate change to identify potential new taxes and revenues, but the public are not seeing anything in return. The Government could put a windfall tax on the big oil companies and use that money to insulate homes, or introduce a feed-in tariff to pay people to produce renewable energy.'

Adapted from an article in The Independent, *2 May 2008*

Synoptic question

a Summarise the broad issues facing countries as they decide on their 'energy futures'. **(12)**

b Analyse the strengths and weaknesses of the pro-nuclear and pro-renewable energy lobbies. **(16)**

c To date, how effectively are international, national and local decision-making each playing their part in deciding 'energy futures'? **(12)**

Can the world follow New Zealand's lead and significantly increase the use of renewable energy?

What do you think ?

Chapter summary

Films, books and music on this theme

Music to listen to

'Temper' by System of a Down

'The Price of Oil' (2002) by Billy Bragg

Books to read

Powerdown: Options and Actions for a Post-Carbon World (2004) by Richard Heinberg

Films to see

'*There Will be Blood*' (2007) by Paul Thomas Anderson

'*Three Kings*' (1999), based on the Desert Storm conflict in 1991 to halt Iraq's invasion of Kuwait (which was an attempt to gain control of Kuwait's oil fields)

'*Jarhead*' (2005), also about the Desert Storm conflict in 1991

Try these exam questions

1 Suggest how the distribution of major oil exporters and importers affects the energy security of some nations. **(10)**

2 The development of alternative energy sources is a possible response to future energy demands. Assess the possible costs and benefits of this approach. **(15)**

What do I have to know?

This chapter is about the conflicts between those who have water, versus those who don't. Water is a human need, but it is not evenly distributed. Physical factors help to determine supplies, while human factors determine how well these supplies are distributed. Growing demand for water does not match supply, and countries often seek supplies across national borders. The potential for conflict is therefore high. High demand can also lead to long-term degradation. The Specification has three parts, shown below, with the examples used in this book.

1 The geography of water supply

What you need to learn	Examples used in this book
• Water supply is controlled by physical factors. • Fresh water supply is finite.	• California's fresh water supply, and its link to climate, river systems and aquifers.
• The growing mismatch between supply and demand, leading to water stress.	• Water demand and stress in California. • The global imbalance – stress, scarcity, water security; The Millennium Development Goals.
• Human activity affects water availability.	• Demand for water in the USA – conflicts within the Colorado Basin. • Quality of water issues in Indonesia.
• Access to water is often controlled by wealth and poverty.	• Links between global water insecurity and poverty / wealth; USA, Indonesia.

2 The risks of water insecurity

What you need to learn	Examples used in this book
• The development, extraction and use of water sources can lead to environmental and supply problems.	• California – pollution of supply, over abstraction, salt-water incursion; environmental issues in the Sacramento-San Joaquin river delta; salinity issues – the Salton Sea.
• There is potential for conflict where demand exceeds supply.	• Water stress in the Middle East (Turkey and Israel) and North Africa (trade in virtual water in Kenya and Egypt)
• Water supply is a geopolitical issue, crossing political boundaries.	• Turkey, Egypt, Kenya – agreements and treaties between areas; water abstraction on the Nile.
• There are risks in developing pathways between areas of surplus and deficit.	• Turkey (agreements with Israel), Egypt (Toshka project) and Kenya (virtual water). • Australia – the Murray-Darling Basin.

3 Water conflicts and the future

What you need to learn	Examples used in this book
• Trends in water demand and supply make the future insecure. Climate change may have a significant impact for some.	• Future water supply and demand in the Murray-Darling Basin, Australia.
• Different players determine water futures; their aims may conflict. • Dealing with water demands e.g. diverting supplies, storage, conservation, restoration.	• Different approaches to water exploitation – privatisation (Bolivia) or public supplies (China's technological fix). • Desalinisation in Israel; water transfer in the Middle East.
• Technology and water supply, e.g. water transfer, desalinisation.	

2.1 California calling

In this unit you'll learn about the problems of supplying water to California.

The end of California dreaming?

From the gold rush in the mid nineteenth century, to the development of the Hollywood movie industry in the 1920s – and the surfing days of the 1960s – Americans have been drawn to the sunshine state of California. Attracted by its climate, millions of Americans have seen California as the place in which to live the 'American dream' of wealth, sunshine and surf – and perhaps fame. California is the world's sixth largest economy, and for many Americans it offers a high-quality lifestyle. But the American dream may be coming to an end for 37.7 million Californians, as they face up to ever-increasing problems with their water supplies. Recent variations in rainfall, the impacts of climate change – and an unquenchable thirst for water – are combining to show just how fragile human relationships with their surroundings can be.

Threats to the Californian dream

Precipitation

- Much of California is arid, with an average annual **precipitation** of between 200-500 mm.
- 65% of that precipitation is lost through evaporation and transpiration, while 13% flows out to sea – leaving only 22% as runoff for human use.
- 50% of California's rain falls between November and March – leading to seasonal shortages.

Population

- California's population has grown from only 2 million people in 1900, to 10 million by 1950, and 37.7 million by 2007. It is likely to reach 45-50 million by 2025.
- It is not just the overall size and rate of growth of California's population that is creating a water problem, but also its **spatial imbalance**. Three-quarters of the demand for water comes from the heavily populated areas to the south of Sacramento, while 75% of the precipitation falls to the north – as the map below shows.
- The increasing demands for water exceed California's natural supplies.

▲ The San Luis Reservoir in July 2007, reduced to 20.8% capacity and down 62 metres from its normal levels. This reservoir supplies water for southern California, the Central Valley farms and Silicon Valley.

Water makes headline news

Californian farmers are operating on the razor's edge. Lack of water, due to dry weather conditions last winter, has cut farmers' margin for error to zero. There may not be enough water in the system right now to begin work for 2008.
California Farm Bureau Federation, August 2007

Water is the lifeblood of southern California, where two-thirds of the state's population lives, and this has been the driest year on record.
PBS News, September 2007

With an antiquated water system, a growing population, and the pressure of global warming, California is in trouble. California needs to act now to avoid a water crisis.
Dr. Reese Halter, AlterNet.org, December 2007

The effects of La Niña, compounded by global warming, will intensify the politics of sharing the Colorado River.
National Geographic, January 2008

▼ A satellite photo of California and northern Mexico, showing the October 2007 bushfires that threatened much of heavily populated southern California

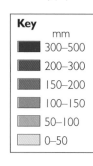
▶ Spatial imbalance – California's driest areas are also its most heavily populated

Key
mm
- 300–500
- 200–300
- 150–200
- 100–150
- 50–100
- 0–50

Sacramento
Nevada
Utah
Pacific Ocean
Arizona
MEXICO
0 250 Km
N

San Francisco
Sierra Nevada Mountains
Coastal Range
Los Angeles
Salton Sea
Colorado River
San Diego

Californian crisis

California's drought of 2000-2007 has forced Californians to wake up to the problem of water supply. California's status as the USA's leading economy, a major supplier of food (and home to three of the country's largest cities), is now at risk. Building and maintaining the Californian dream has been at the expense of the natural environment.

- Wetlands have been drained, natural habitats altered, and fish stocks depleted, to secure water supplies.
- Additional problems are polluted waterways, the over-extraction of groundwater and increasing salinity (especially in the Salton Sea).
- The Bay-Delta region and the Salton Sea have become environmental disaster zones, and the once mighty Colorado River has been reduced to little more than a trickle as it enters the Gulf of California.

Maintaining the dream

Currently, California depends on two major water supply lines, as shown on the map:

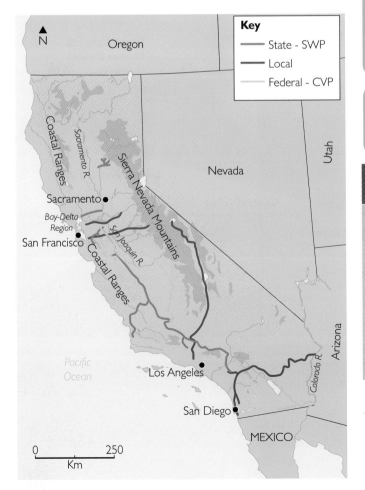

▲ California's water supply system

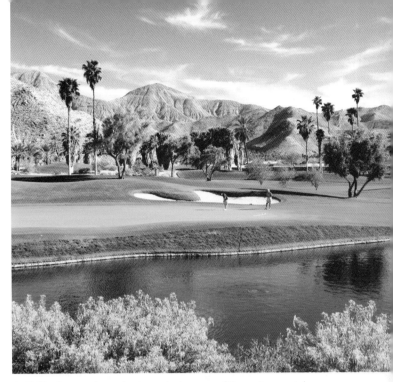

▲ Californians are becoming increasingly aware of the environmental impacts of their growing demands for water. Residents of Palm Springs are beginning to wake up to the fact that they live in the desert and they need to worry about water.

- The **State Water Project (SWP)**, combined with the **Central Valley Project (CVP)**, provides water from the Sacramento-San Joaquin River Delta for central and southern California.

- The **Colorado River** provides 60% of southern California's water via a system of dams and aqueducts.

State Water Project	Central Valley Project	Colorado River
• It contains 20 major dams and reservoirs.	• It contains 22 dams and reservoirs	• It contains 11 major dams/reservoirs
• It irrigates 0.3 million hectares of farmland.	• It irrigates 1.2 million hectares of farmland.	• It irrigates 1.4 million hectares of farmland
• It provides drinking water for nearly 20 million Californians.	• It provides drinking water for 2 million Californians.	• It provides drinking water for 25 million Californians.

The future

There are already conflicts over water between cities, and between farmers and environmentalists (who are trying to cut demand). Already California takes 20% more Colorado River water than envisaged in 1963, and shortages are forecast for the future.

California's natural water supplies

Geographical controls

California is almost twice the size of the UK and has enormous variations in terms of climate and relief. Geographical controls exert a strong influence on the availability of water.

- Mountain chains run parallel to the coast and prevent moist air reaching inland. The **prevailing** moist air flow from the Pacific Ocean is forced upwards by the mountains, cools and condenses – forming **relief rainfall** and, at higher altitude, snowfall.
- Most rain (up to 500 mm) falls in a coastal zone no more than 250 km wide.
- The south and far east of California (which includes Death Valley and the Mojave Desert) receive under 100 mm of rainfall, due to the **rain shadow** effect of the Coastal Ranges and Sierra Nevada Mountains.

- **Surface runoff** occurs over impermeable, saturated or baked surfaces – eventually reaching river channels as **streamflow**.
- Some surface water **infiltrates** through the surface layer and eventually **percolates** through the rocks to become **groundwater**. Nearly a third of California's fresh water comes from groundwater sources, known as **aquifers**.
- Most of the major rivers are fed by snowmelt from the Sierra Nevada Mountains. The Sacramento River flows southwards, meeting the north-flowing San Joaquin River at San Francisco Bay. These two rivers provide huge volumes of water for urban and agricultural use. Further south, river flow is variable and can even dry up.

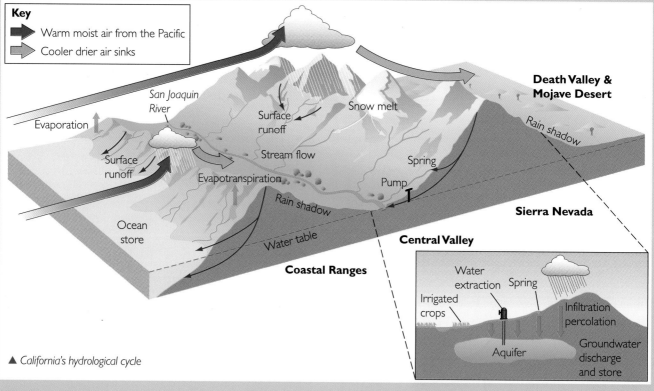

▲ California's hydrological cycle

Weather systems

The weather systems that affect California create three problems:

- A **high-pressure** system over the Pacific Ocean blocks moist air currents from reaching southern California. Occasional shifts in the system allow storms and heavy rain to reach the area.

- **El Niño** events bring above-average runoff and flooding to the south-west, while **La Niña** events bring drought.
- In recent years, extended droughts have meant that groundwater and surface storage levels have decreased.

The Colorado River Basin under pressure

The huge Colorado River Basin drains 7% of the USA and covers an area 1.1 times the size of France. Throughout the twentieth century, numerous treaties and agreements were needed to allocate 'fair shares' of its water to the seven surrounding US states, plus Mexico.

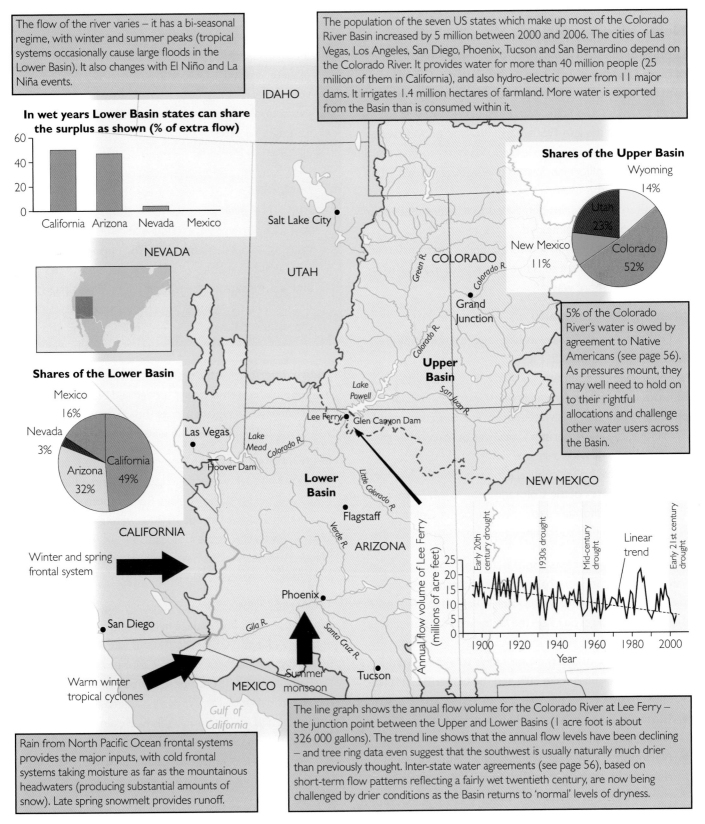

The flow of the river varies – it has a bi-seasonal regime, with winter and summer peaks (tropical systems occasionally cause large floods in the Lower Basin). It also changes with El Niño and La Niña events.

The population of the seven US states which make up most of the Colorado River Basin increased by 5 million between 2000 and 2006. The cities of Las Vegas, Los Angeles, San Diego, Phoenix, Tucson and San Bernardino depend on the Colorado River. It provides water for more than 40 million people (25 million of them in California), and also hydro-electric power from 11 major dams. It irrigates 1.4 million hectares of farmland. More water is exported from the Basin than is consumed within it.

In wet years Lower Basin states can share the surplus as shown (% of extra flow)

Shares of the Upper Basin

Wyoming 14%

Utah 23%

New Mexico 11%

Colorado 52%

Shares of the Lower Basin

Mexico 16%

Nevada 3%

Arizona 32%

California 49%

5% of the Colorado River's water is owed by agreement to Native Americans (see page 56). As pressures mount, they may well need to hold on to their rightful allocations and challenge other water users across the Basin.

Winter and spring frontal system

Warm winter tropical cyclones

Rain from North Pacific Ocean frontal systems provides the major inputs, with cold frontal systems taking moisture as far as the mountainous headwaters (producing substantial amounts of snow). Late spring snowmelt provides runoff.

Summer monsoon

The line graph shows the annual flow volume for the Colorado River at Lee Ferry – the junction point between the Upper and Lower Basins (1 acre foot is about 326 000 gallons). The trend line shows that the annual flow levels have been declining – and tree ring data even suggest that the southwest is usually naturally much drier than previously thought. Inter-state water agreements (see page 56), based on short-term flow patterns reflecting a fairly wet twentieth century, are now being challenged by drier conditions as the Basin returns to 'normal' levels of dryness.

Californian conflicts

Providing sufficient water for the needs of California's growing population – with its affluent, water-heavy lifestyle – and also the needs of agriculture for **irrigation**, has meant the transfer of water both within and between states.

North versus south

Because of the spatial imbalance between the distribution of rainfall and the distribution of population in California – with too many people living too far away from the natural water sources – the State Water Project (SWP) was constructed in the late 1950s/early 1960s to provide water for southern California. However, it has been controversial:

- Northern Californians feared that their water would be 'owned' by the south.
- Southern Californians demanded guarantees that water flows would be maintained.
- The Bay-Delta region is where supplies are transferred, and it has become a sensitive area where multiple users all demand more water.

Sharing the Colorado

In 1901, the Alamo Canal was begun, to bring irrigation water from the Colorado River to the farmers of the Imperial Valley in southern California. However, political tensions with Mexico (through which much of the canal passed), meant that this water supply was not guaranteed. Therefore, the All-American Canal was built (completely on US land) to provide a secure water supply for California's growing agricultural economy. Mexico and the six other US states of the Colorado River Basin began to be alarmed that California's expanding use of the Colorado's water was beginning to threaten their own water supplies.

The resulting agreements about water shares (see the top panel on the right) were based on flow patterns from the early twentieth century, when rainfall was about 10% higher than it is today. Pressures on the Colorado River are now building:

- Mexico takes 10% of the Colorado's total flow. States in the Lower Basin take 50% and the Upper Basin falls short by 10%.
- California takes 20% more than its original 1963 allocation.
- Native Americans are owed 5% (see right), but could claim more because their Reservations extend along the river and two of its tributaries.

Key decisions about sharing the Colorado

- The Colorado Compact (1922) divided the river into two basins – the Upper and Lower.
- The boundary between the Upper and Lower Basins was established at Lee Ferry.
- The US states allocated to each Basin were given the shared right to extract 7.5 million acre-feet of water from the Colorado River each year. The Lower Basin states were also allowed an extra extraction of 1 million acre-feet per year.
- The right to extract 1 million acre-feet of water was allocated to Mexico in 1944.
- However, the specific water shares of individual states were not detailed in the Colorado Compact. Arizona resisted any agreement about this for many years, because of uncertainties about California's growing water demands. The US Supreme Court resolved the Arizona v California dispute in 1963 and water shares were formally agreed (see page 55), with Native Americans also being allocated shares (see below).

Native American water rights

In 1908, the Supreme Court recognised Native American **water rights** across the USA, regardless of whether a tribe had used the water or not.

In 1963, Native American Reservations were granted the use of enough water to irrigate any land where it was practicable to do so. Five Reservations, with a total population of about 10 000, were granted approximately 5% of the Colorado's flow. As a result, Native American tribes have the best water rights along the Lower Colorado River.

The Navajo Tribe's claim is potentially huge. Their reservation is 25 000 square miles and is located entirely within the Colorado River Basin. At its western boundary, the main stem of the Colorado River, plus the San Juan and Little Colorado tributaries, flow through tribal land. The Navajo's claim could be as high as one third of the Colorado's flow. If claimed, it would destroy the 1963 Agreement. While their actual needs are well below this level, they could sell some of the increased allocation as a way of earning much needed income.

Planning ahead

The Colorado Compact was based on 1922 flow levels – but things have changed a lot since then. Lakes Powell and Mead in the Colorado River Basin (see page 55) have not reached capacity since 1999. Rainfall has been declining since the 1980s and the climate is changing. Population levels in the area have also increased dramatically since 1922, especially in California. It now makes no sense to base water allocations on out-dated data.

In December 2007, a new agreement was reached. Instead of sharing the Colorado's flow, the states will aim to divide the shortages. The actual amount of water available will determine the deliveries to each state. California has also been given until 2016 to reduce the amount it extracts by 20%, in recognition of the realities of climate change.

In October 2003, the *Los Angeles Times* reported that 'California has enough water to meet its needs today and tomorrow without new dams, peripheral canals or catastrophic costs.' Instead of increasing overall supplies, attention is now being focused on demand management and developing more-efficient supply techniques:

▲ *Lake Powell in June 2002 (top) and January 2004*

From this ...

California has 883 golf courses, using around 90 billion gallons of water each year. Irrigated pasture for beef herds uses about 5.3 million acre-feet of water – as much consumption as all 37.7 million Californians – including for swimming pools and lawns!

Crop irrigation in California

... to this?

- *Domestic conservation: 30% savings by repairing leaks, metering supplies and efficient appliances. 50% savings by planting California-friendly, drought-tolerant plants and using smart sprinkler systems.*
- *Groundwater banks: Saving storm water for release during dry periods.*
- *Re-using wastewater: Cleaned water from sewage treatment plants currently flows into the sea. It could be re-used for landscape irrigation and industry, or to recharge aquifers for later domestic use.*
- *Saving storm water: Concrete channel storm-drains, which prevent flooding by directing the storm water into the sea, could instead redirect it to urban parks.*
- *Reducing agricultural water usage: Farms currently use 80% of California's clean water. A 10% reduction would double the amount of water available for urban areas.*
- *Smart planning: New housing developments should only be built where supplies of local ground and surface water are adequate for their needs.*

Over to you

1 As a group, discuss how the views of urban and rural communities might differ when it comes to providing water supplies.
2 Devise a public information campaign, based on leaflets, posters and presentations, which is designed to encourage all Californians to use water sustainably.

> *Native Americans hold the key to future water allocations in the Colorado Basin.*

What do you think?

On your own

3 Why does each state now require more water than it was allocated in 1963?
4 Arizona has always been suspicious of California's intentions. Research how far the Central Arizona Project challenges California? These websites will help you:
http://www.capaz.com/index.cfm
http://pubs.usgs.gov/circ/circ1213/introduction.htm
5 How far have historic decisions over water allocations put the Californian dream at risk?

California - environment at risk

In this unit you'll consider the conflict between supplying water to California and safeguarding the environment.

Inherited consequences

During the twentieth century, the main focus of Californians was on acquiring sufficient water for their needs. The environmental consequences of this water acquisition were never really considered. However, some of the harmful consequences of this attitude are now becoming more apparent. The previous disregard of Californians for the environment, its natural processes and its habitats is now putting all three at risk. The Sacramento-San Joaquin and Colorado River Deltas, and the Salton Sea, are all in environmental trouble.

The Sacramento-San Joaquin River Delta

Half a million people live around the 60 low-lying islands of the Sacramento-San Joaquin River Delta (Bay-Delta), but years of neglect have allowed the ageing levees which protect this low-lying area from flooding to decay. Almost half of California's annual runoff flows through this delta, and two-thirds of the state's population depend on it for their water supplies. But there are some serious problems:

- Old, poorly maintained, man-made levees and river banks along the Feather River, Clear and Deer Creeks have allowed salty water to submerge some of the islands and reduce water quality for all users and habitats.
- Several species of fish are on the endangered list – the migratory paths of salmon and smelt are blocked at Stockton on the San Joaquin when fresh water oxygen levels are too low. The big water pumps for the State Water Project (SWP) also suck in a lot of fish. Pumping is stopped 3-4 times a year to allow fish stocks to recover.
- Water treatment works discharge chlorine into these rivers, and the build up of trihalomethanes as a result (which are suspected of causing cancers), mean that safe water supplies cannot be guaranteed.

The CALFED Bay-Delta Program (see right) aims 'to reduce conflicts by developing a sustainable, long-term solution to water management and environmental problems'.

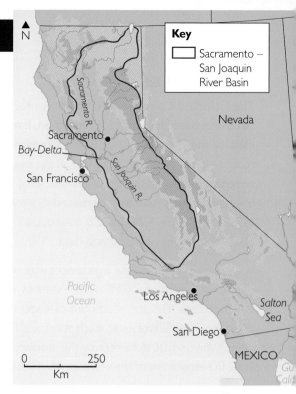

▲ The Sacramento-San Joaquin and Colorado River Deltas, and the Salton Sea

The Colorado River Delta

The Colorado River used to flow into the Gulf of California in northern Mexico, forming a delta. However, since the building of the Hoover and Glen Canyon Dams upstream (see page 55), the flow of the Colorado has been so reduced that the delta has been massively degraded and almost 2 million acres of riverside habitats have dried up.

CALFED's aims for the Bay-Delta

CALFED's aims are:
- a reliable, good-quality water supply
- ecosystem restoration
- improved levees and protected delta islands
- increased water storage
- improved water conveyance
- guaranteed farmers' supplies
- maintained dissolved oxygen levels to preserve fish and the wetland ecosystem

◀ The low-lying islands and agricultural land of the Bay-Delta, much of it protected by man-made levees

The Salton Sea

The Salton Sea is an agricultural sump – it acts as a receiving basin for runoff from irrigated farmland. It formed when canal banks collapsed in 1905, allowing the Colorado River to fill a natural hollow with water. This created a large lake covering 376 square miles, which became a wildlife refuge boasting millions of birds, fish and flora among its fresh and saltwater marshes. Not only that, but it also attracted half a million tourists a year and developed a thriving community. However, the future of the Salton Sea is now threatened:

- 75% of its inflow is from cotton, citrus and sugar beet farms in the Imperial and Coachella valleys – with a toxic mix of pesticides, fertilizers, defoliants and nutrients.
- High evaporation rates mean that the salt content here is 25% higher than in the Pacific Ocean.
- High salinity, algal blooms and eutrophication are thought to have caused the deaths of millions of birds and fish (7.5 million fish died in one day!).
- The New River brings industrial pollutants and sewage from Mexicali in Mexico.
- The lands of the Cahuilla Tribe are at risk from the de-oxygenated and poor-quality water, increased salinity and reduced fish stocks.
- Tourists have stopped visiting because of the high salt content, the stench of decay and the perception that the sea is a toxic waste dump.

Ironically, as farmers develop more water-efficient irrigation techniques, there is less runoff to the Salton Sea, but the inputs of pesticides, fertilizer, etc. are more concentrated. In 2003, the $2 billion Salton Sea Restoration Project faced a choice – maintaining the Salton Sea as an agricultural sump, or restoring its natural habitats and boosting its recreational and economic potential.

Possible choices for the future

The Salton Sea Restoration Project has to choose between:

- reducing the nutrient mix in the water to prevent algal blooms – but if farmers allow lower rates of drainage and runoff to the sea, the water level will fall and the chemical concentration will increase
- creating evaporation ponds to extract the damaging salt before it enters the sea
- diverting freshwater from the Colorado to dilute the salty sea
- limiting fish stocks by using surpluses to make fertilizers and pet food rather than allowing them to die and decay
- allowing the sea to evaporate completely
- cleaning up the New River from Mexico to reduce pollution
- creating a desalination project to produce fresh water for increasing urban demand and then selling the salt as a by-product

▲ Dried salt deposits on the shore of the Salton Sea

Over to you

1 a In pairs, conduct research into the CALFED Bay-Delta Program, using the websites:
http://calwater.ca.gov/index.aspx
http://www.nationalaglawcenter.org/assets/crs/RL31975.pdf
 b Make lists of the economic costs and environmental benefits of the strategies being adopted.
 c Select either the economy or the environment as the focus for a persuasive letter to the editor of a major newspaper explaining why action is needed now. Write 200 words, including a map and data.

On your own

2 a Write about 150 words describing the value of the Salton Sea as seen first by an economist and then by an environmentalist. This weblink might help:
http://www.sci.sdsu.edu/salton/
 b Whose view would you expect the Cahuilla Tribe to agree with? Why?

Exam question: Referring to examples, assess the validity of the statement that 'water conflicts are as much to do with water quality as quantity'. (15)

Is fulfilling water demand more important than environmental protection?

What do you think?

Global imbalance

In this unit you'll learn that there is a global imbalance between water demand and water supply.

Where is all the water?

The amount of water available worldwide is finite. If a 4.5-litre jug represents all the water on Earth, just one tablespoon represents the available freshwater. Freshwater makes up just 2.5% of the water on the Earth's surface (see the diagram on the right) – and only half of that is available for human consumption. NASA estimates that every drop of freshwater has been consumed at least once before – because water flows through a closed hydrological cycle that takes 1000 years to complete (see below). Solar energy causes water to evaporate from the land and sea, which then falls as precipitation. However, most water is locked up as ice or groundwater for anything from 1000-10 000 years, as the graph shows.

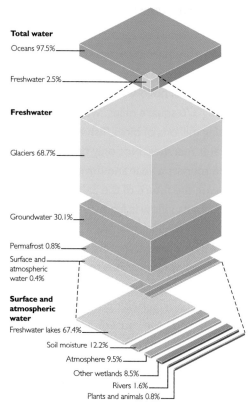

▲ The global distribution of the world's water

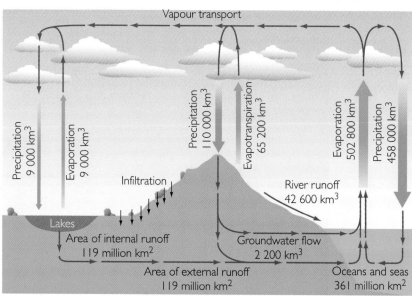

◄ The closed global hydrological system

Is there enough water for everyone?

In theory, there is no global water shortage; in fact, we only use 50% of the available water. However, rapid population growth in areas where water supplies are limited, a deterioration in water quality, and an uneven distribution of global water supplies, means that more and more people are facing severe water shortages. A **world water gap** now exists between the 'haves' and 'have-nots' – as wealthy nations consume greater and greater quantities of water. The water statistics are alarming:

- Out of a global population of around 6.6 billion: 1.4 billion lack clean drinking water, 2.4 billion lack adequate sanitation, 0.5 billion face water shortages every day.
- Every 15 seconds, a child dies from a water-borne disease.
- Half the world's rivers and lakes are badly polluted.
- Half the world's rivers no longer flow all year.
- Food supplies are threatened as water shortages increase.
- Population growth by 2025 will demand a 20% increase in water supplies
- 12% of the world's population consumes 85% of its water.

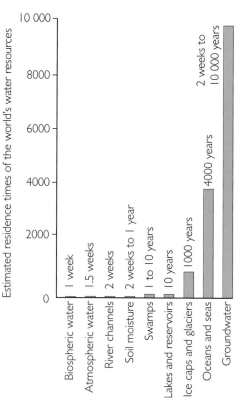

▼ Estimated storage times of the world's water resources

Stress and scarcity

As the world's population continues to grow, the number of people affected by **water stress** and **water scarcity** is expected to rise sharply. When a country's water consumption is more than 10% of its renewable freshwater supply, it is said to be water stressed. This means that it has less than 1700 m³ of water available per person per year, which can cause temporary shortages. When the amount available falls below 1000 m³ per person per year, a country experiences water scarcity, which threatens food supplies, holds back economic development and causes environmental damage.

> 2.3 billion people live in river basins under water stress, with annual per capita water supply below 1700 m³. Of these, 1.7 billion people live in conditions of water scarcity, where annual per capita water supply falls below 1000 m³. By 2025, at least 3.5 billion people – or 48% of the world's projected population – will live in water-stressed river basins. Of these, 2.4 billion will live under conditions of water scarcity.
> **Adapted from the UN report, 'World Water Day, 2007'**

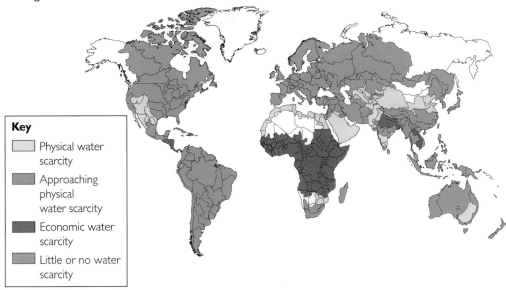

◀ *Global water scarcity by type*

Key

- Physical water scarcity
- Approaching physical water scarcity
- Economic water scarcity
- Little or no water scarcity

- **Water stress** occurs when the demand for water exceeds the amount available during a certain period, or when poor quality restricts its use.
- **Water scarcity** comes in two types:
- **Physical scarcity** exists when shortages occur because demand exceeds supply.
- **Economic scarcity** exists when people simply cannot afford water, even when it is readily available.

Who consumes all the water?

During the twentieth century, global water consumption increased by 600% as a result of population growth and economic development.

- Farming uses 70% of all water and, in some LEDCs, the figure is as high as 90%. Increased global prosperity will demand even higher rates of food production – with increased demands for water to produce it.
- Industrial and domestic consumption increasingly compete with agriculture as societies develop.
- Daily domestic water use averages 47 litres per person in Africa, 95 litres in Asia, 334 litres in the UK, and 578 litres in the USA.

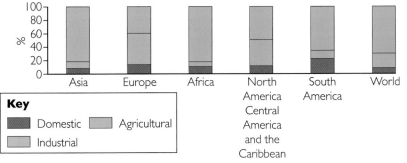

Key

- Domestic
- Agricultural
- Industrial

▲ *Global water use by sector, 1982-1997*

Over to you

1 **a** Work in groups of three. Use atlases to identify the world's major climate zones and areas of high population growth and densities.
 b Use this information to predict areas of water surplus and water shortage.
 c Do current areas of water scarcity match the patterns you have identified?

On your own

2 Why do you think areas of economic water scarcity and physical water scarcity are not always the same?
3 Why does agriculture take so much of the world's water?

In this unit you'll find out that the global water crisis is about more than water shortages.

Global warning

In 2003, the UN's World Water Development Report warned that the balance between human demand and the availability of water was at a precarious point. 'Water may be the resource that defines sustainable development … the lack of freshwater is emerging as the biggest challenge of the twenty-first century.' Access to safe water for people is seen as both a fundamental need and a basic human right, but for too many people it represents part of their daily struggle. In some parts of Africa, women walk for over 8 hours a day to collect water – an average distance of 6 km. In addition, water shortages and a lack of access to safe water cause serious problems for a third of the world's population – and prospects are not good for the future.

Water security crisis

At the second meeting of the World Water Forum (2000), water security was established as a key issue for the twenty-first century. People's health, welfare and livelihoods depend on secure supplies of freshwater, but high demand and the misuse of water resources put these at risk. For example, irrigation for agriculture drains groundwater supplies, and its misuse often leads to contamination from chemicals like pesticides and fertilizers. Regional insecurities vary, but some of the most significant problems are shown in the table on the right.

▼ *The decreasing size of the Aral Sea in Kazakhstan (in 1977, 1989, and 2006), due to the development of large-scale agricultural irrigation projects (using its source rivers) during the time of the Soviet Union – which prevented its natural replenishment. Its salinity and pollution levels are now so high that most fish have died and water supplies in the wider area have been badly affected by wind-blown salt from the dried up seabed.*

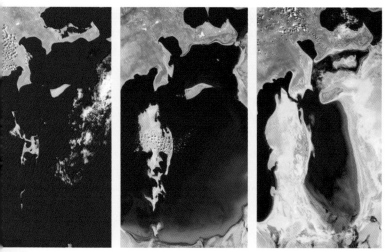

Region	Issues of water insecurity
Asia and the Pacific	Critical health problems: A third of the population lacks access to safe drinking water; there are, for instance, 500 000 diarrhoea-related infant deaths each year in Asia. Water pollution: The level of bacterial waste from human sources is ten times greater than recommended levels. Overuse: • Agriculture accounts for 90% of freshwater withdrawals in South Asia. • Aquifer depletion in Asia has led to a drop in water availability per capita from 10 000 m³ in the 1950s to 4200 m³ in the 1990s. • Withdrawals in West Asia far exceed natural replacement rates.
Africa	Poverty and water scarcity: • 25 African countries will face either water stress or scarcity by 2025. • Over 300 million people lack access to safe water – nearly 51% of people in sub-Saharan Africa are without a safe water supply, and 41% lack adequate sanitation. • There is a lack of groundwater protection from agricultural uses (which make up 88% of total water use).
Europe/ Central Asia	There is: • a lack of access to clean drinking water in Eastern Europe and Central Asia. • increasing water consumption, with half of Europe's cities over-exploiting their groundwater reserves. • a declining water quality in countries with groundwater pollution (from nitrates, pesticides, heavy metals and hydrocarbons), e.g. the Aral Sea (see below), the Mediterranean, Scandinavian lakes.
Latin America and the Caribbean	There is: • groundwater contamination and depletion from the increasing release of heavy metals, nutrients, chemicals and hazardous wastes from mining, agriculture and industry. • poor sanitation because only 2% of the sewage produced in Latin America is treated (so cholera and typhoid are common diseases). • economic scarcity, with a conflict over access to and use of water
North America	Aquifer depletion is steadily increasing, due to population and urban growth, and the expansion of irrigation and industry (e.g. cotton farming in Texas has reduced water supplies). Water pollution from agricultural runoff has contaminated many ground and surface waters.

Health and well-being crisis

The Millennium Development Goals adopted by world leaders in 2000 (see pages 214-217) included a target to halve by 2015 the proportion of people without sustainable access to safe drinking water and sanitation. This target is vital for tackling the global issues of human health and well-being. However, by 2005, only 12% of developing countries had managed to introduce effective strategies to provide access to safe water and sanitation. In some LEDCs, freshwater supplies are often not separated from waste products, so the water becomes contaminated – which can lead to outbreaks of water-borne diseases like cholera and dysentery.

It is possible to transform people's lives by improving levels of sanitation. The diagram on the right shows that increasing freshwater supplies not only improves health but particularly eases the pressures on girls and promotes social inclusion (often girls do not attend school during menstruation because there is nowhere for them to clean themselves).

The World Health Organisation states that 'every dollar spent on improving sanitation generates an average economic benefit of $7', thus reducing poverty and increasing economic development. The UN declared 2008 as the 'International Year of Sanitation', in order to raise awareness of the issue of sanitation, and to accelerate progress towards the MDG target of reducing by half the proportion of people without access to basic sanitation by 2015.

Political crisis

The Central African Republic (CAR) demonstrates how a change in the political situation can threaten water security. There is no shortage of rain in the CAR, but the population of 3.5 million is suffering from water scarcity and poor health. Fighting between government forces and rebels – and attacks by local bandits – forced thousands of people out of their villages to seek shelter in the Bush. Many had to resort to using stagnant pools, and polluted streams and rivers, for their water supply. Those who stayed in their villages often found that their wells were not working because of faulty equipment and few trained technicians.

Many people suffered from diseases like intestinal parasites, diarrhoea and guinea worm after drinking polluted water. Households lost income, food and their children died. In the CAR half the infant deaths were caused by poor sanitation, and the current situation has made the problem worse. As children make up 60% of the CAR's population, the future looks very bleak.

▲ Pasay City (Greater Manila) in the Philippines in 2008. The contamination of freshwater supplies by waste is common in shanty towns.

INTERNATIONAL YEAR OF
SANITATION
2008

The official logo for the International Year of Sanitation, 2008 ▶

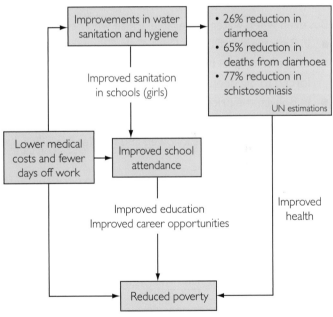

Improvements in water sanitation and hygiene

- 26% reduction in diarrhoea
- 65% reduction in deaths from diarrhoea
- 77% reduction in schistosomiasis

UN estimations

Improved sanitation in schools (girls)

Lower medical costs and fewer days off work

Improved school attendance

Improved education
Improved career opportunities

Improved health

Reduced poverty

▲ The spiral of well-being – sanitation works!

My infant son died last year after suffering acute diarrhoea. He was 4 years old. My 2-year-old daughter has been ill for two months with dysentery. We have no money to treat her. It will be a miracle if she lives. It is what poor people have to live with.
In September 2006, Abida Bibi reported the death of her child as a result of the water crisis in Gujranwala, Pakistan. 250 000 children die each year in Pakistan from water-borne diseases.

Global water crisis – 2

Access crisis

There is a strong link between poverty and lack of water, or, to put it another way, there is a definite correlation between wealth (measured by GDP per capita) and access to safe water and sanitation, as the graphs show. A lack of investment in basic water infrastructure across the developing world means that people like Abida Bibi on page 63 are not alone – children die every day because of a lack of access to basic amenities.

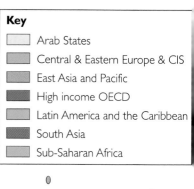

Key
- Arab States
- Central & Eastern Europe & CIS
- East Asia and Pacific
- High income OECD
- Latin America and the Caribbean
- South Asia
- Sub-Saharan Africa

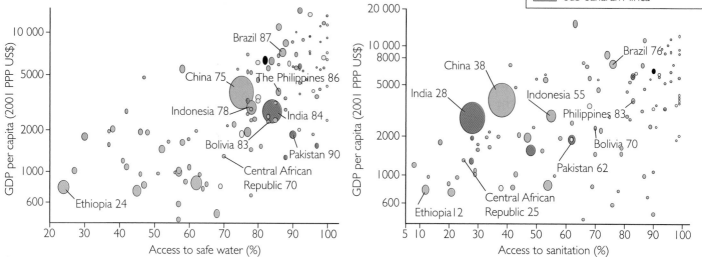

▲ The link between wealth, access to safe water and sanitation

But even when water is available, those who need it may not be able to access it:

- In Rajasthan, India, The Coca-Cola Company was accused of extracting so much water to use in its local bottling plant, that agriculture in the area suffered and farmers were forced to dig deep wells and buy bottled water (see page 281).
- In Bolivia, the prices charged by French-owned water companies were too high for the local people, and 200 000 'chose' not to be connected to the water supply (see page 90). Health problems followed and families took illegal action to siphon off water from the official supply.
- 40 000 people in Detroit, USA, were too poor to pay their water bills and resorted to 'illegal tapping' at night.

Quality crisis

People need access to water, but it has to be safe water. Economic development does not necessarily go hand-in-hand with high water quality.

Indonesia's drive for economic development, and its manufacture of cheap goods to sell to Europe and the USA, has led directly to the pollution of water supplies. The Citarum River in West Java, Indonesia (near the capital, Jakarta), carries the waste from 9 million people and hundreds of factories and farms. Untreated river water is used to irrigate the surrounding paddy fields, to wash clothes, and as drinking water. Poor people risk disease from the filthy water to salvage plastic, wood, glass and anything else of value to earn a few dollars a week (see the photo opposite).

But it is not just manufacturing that pollutes water – farming can also have a detrimental effect on water quality. 40% of the world's food comes from irrigated land, and irrigation often leads to problems of salinity (see the background box) and waterlogging (which stops air getting into the soils and effectively drowns the plants).

There is more to the world water crisis than a shortage of water.

What do you think ?

◀ *The Citarum River, West Java, in 2007 – possibly the world's most polluted river*

Background

Salinity

Salinity can be a problem for irrigated land. Irrigation water contains mineral salts (sulfates and chlorides of sodium, calcium and magnesium), which can become concentrated in the surface layers of the soil as the water table rises. In arid areas, the moisture evaporates from the surface layers of the soil – leaving salt crystals behind. Most plants cannot tolerate the salts and die. Salinisation can destroy the very land that irrigation was meant to improve.

The salt can be flushed out of the soil when it rains – if the layers are not too thick – but clays and nutrients will also be leached away, causing a breakdown of the soil structure. The leached particles contain salts which can then be carried back to rivers – increasing the salinity downstream from irrigated areas.

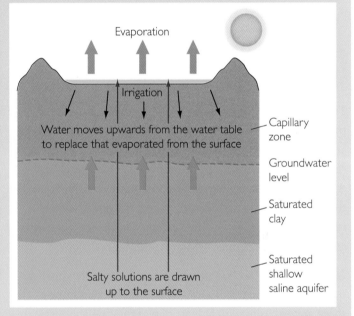

▲ *Salinisation in action*

Over to you

1 Complete this activity as a class.

 a Make a table and, for high-, middle-, and low-income countries, rank the following factors according to their impact on access to safe water and sanitation. So, for example, 1 = worst impact, 9 = least impact.

 Factors: irrigation, urbanisation, population growth, industrialisation, droughts, floods, cost/ownership, climate change, pollution

 b Justify to each other the way you have ranked the factors.

 c What other factors determine access to clean water?

On your own

2 a Use the following websites to make notes on how improvements in sanitation can lead to improved health, personal dignity, social and economic development, and a better environment:

 http://esa.un.org/iys/

 http://hdr.undp.org/en/mediacentre/

 http://www.unwater.org/wwd07/downloads/documents/escarcity.pdf

 http://www.unwater.org/wwd07/downloads/documents/wwd07brochure.pdf

 b In what ways do women stand to gain most from improved sanitation?

2.5 Water insecurity

In this unit you'll learn that there is a real risk of political insecurity as water-stressed regions battle to meet increased demands for water.

Potential for conflict

Agency will mediate in water disputes

A new United Nations body has been announced to help avoid possible future 'water wars'. The Water Co-operation Facility will be based in Paris, and will mediate in disputes between countries sharing a single river basin.

Launching the new Facility, the Director-General said: 'It will provide the water community with the necessary resources, environment, political backing and legal mechanisms to anticipate, prevent and resolve water conflicts.'

Disputes over water are nothing new – the word rival comes from the Latin word rivalis, meaning one using the same stream as another.

Adapted from a BBC news article by Tim Hirsch, 21 March 2003

▲ The basin of the Limpopo River includes the four countries of Botswana, Zimbabwe, South Africa and Mozambique. Due to water extractions in the upstream countries, the Limpopo – which used to flow continuously – is now dry for up to eight months of the year when it reaches Mozambique (pictured). This basin is a potential source of water conflict in the future.

The risks of real water shortages are growing, and, as they do, so does the potential for conflict. Governments around the world are realising that – although water is renewable – it is not an infinite resource. Demand is rising due to population growth and economic development, and the risk of demand outstripping supply has already been realised in many regions. International disputes, agreements and treaties reflect the urgency of the situation.

263 rivers, along with many aquifers, cross or form political boundaries around the world – and 90% of all countries share water basins with at least one of their neighbours.

As the map below shows, the flow of many major river basins has been affected by dams, reservoirs and diversion schemes in order to meet increasing human demands.

However, conflicts can arise when patterns of economic development on either side of a political boundary are uneven – with borders becoming zones of tension as scarcity, stress and quality issues build up. Several 'hotspots' have emerged where water is contested, as the panel opposite shows. The biggest risks arise when one country threatens its neighbours' water supply. Small wonder that talk of **Water Wars** has been increasing.

▶ The effect of water management, such as dams, on the world's major river basins

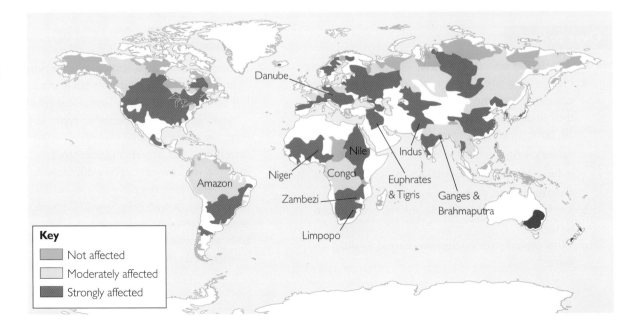

Danube
Nile
Indus
Niger
Congo
Amazon
Euphrates & Tigris
Zambezi
Ganges & Brahmaputra
Limpopo

Key
- Not affected
- Moderately affected
- Strongly affected

Contested resources – potential Water Wars

- Turkey v Syria and Iraq – The Euphrates and Tigris Rivers provide massive potential for economic development in Turkey (see pages 70-71), but their exploitation could leave the Middle East dry.
- China v India – The Brahmaputra River could be diverted to ease scarcity problems in southern China, but this would reduce supplies to India.
- India v Pakistan – Indian dams on the Indus River threaten to cut off irrigation waters to Pakistan.
- Egypt v Ethiopia v Sudan v Uganda – The Blue and White Nile Basins supply Egypt with vital water, but 85% of this water comes from the other countries upstream, where population growth and increased demand could threaten Egypt's supplies (see page 77).
- Angola v Namibia v Botswana – The Okavango Basin is fed by rivers originating in Angola and Namibia. Increasing exploitation of those rivers by these countries could threaten the Okavango Delta in Botswana (an important wildlife area for ecotourism and a major source of revenue for the country).

Water – the key players

World Trade Organisation and TNCs

Countries need secure water supplies to further their economic development. Many developing countries have benefited from international aid to improve their water provision. Most aid projects try to work with the needs of the local people, and also follow international guidelines to gain the agreement of all riverside land users (known as **riparians**). However, three of the most controversial schemes of the twenty-first century – China's Three Gorges Dam, Turkey's South-eastern Anatolia Project (see pages 70-71), and India's Narmada Project – have proceeded without the restrictions imposed by outside pressure or guidelines.

The WTO is now encouraging countries to open up their economies to private investment – in return for debt relief. Countries wishing to develop major water schemes have been turning to private companies for finance. Therefore, as countries follow the WTO guidelines, control of their water infrastructures is being transferred to multinational companies, such as Veolia/Vivendi, Ondeo and Bechtel, who see water as a business – just like any other. Water supplies may be improved, but the local consumers have to pay for

it. Water Riots, rather than Water Wars, could become a common sight as conflicts arise over the price of water (see page 90).

The United Nations

Water is a vital resource, and countries can bargain with their neighbours for control of and access to it. The UN's World Water Assessment Programme (WWAP) was established in 2000 to monitor changes in demand for water and the likelihood of international tensions. It is the role of international institutions such as the UN to try to find peaceful solutions when conflicts arise. In 2000, the UN Secretary General, Kofi Annan, warned that national rivalries over water could lead to violence.

The graph shows that, of all the disputes over water between 1948 and 1998, most ended peacefully – just 43 involved military acts, and 18 of these were between Israel and its neighbours. If a water-poor country like Israel has the military capacity to attack a water-rich neighbour, tensions may rise (see pages 72-75).

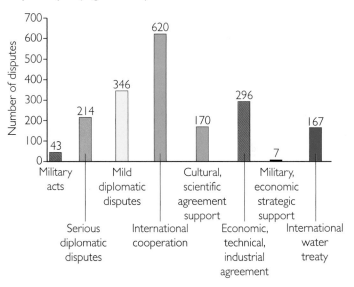

▲ Water Wars or peaceful solutions? River basin disputes, 1948-1998

Water – a global commodity

The annual profits of the global water industry are now about 40% of those of the oil industry. The major water companies have become powerful TNCs:

- Suez – 125 million customers in 40 countries (subsidiaries Ondeo and United Water)
- Veolia/Vivendi Universal – 110 million customers in 50 countries
- Bechtel-United Utilities – 140 million customers in 50 countries

Water stress – the Middle East and North Africa

The Middle East and North Africa are probably the driest and most water-scarce regions on Earth – with 5% of the world's population and just 1% of its freshwater. The amount of water available varies from 1200 m³ per person per year in parts of Iran, to just 200 m³ per person per year in Jordan. Forecasts suggest that there will be an average of 500 m³ per person per year across the whole Middle East and North Africa region by 2025 (see the graph).

Population growth, increasing affluence (with demands for swimming pools and golf courses) – and the development of irrigated farmlands – are increasing the pressure on water supply. Groundwater is being extracted faster than it can be replenished, and fossil reserves are being tapped to meet farmers' needs – they use 89% of all of the water extracted in order to irrigate their crops (see the pie chart).

Mounting pressure

In 1991, Boutros Boutros-Ghali, then the Egyptian Foreign Minister – and the future UN Secretary General (1992-96) – warned that 'The next war in the Middle East will not be about politics, but over water'. The Middle East faces a number of issues that may lead to further instability in the region:

● An overall scarcity of water (see the map opposite), and also poor access for many to what there is.

● Declining oil reserves, with a future drop in oil revenues – which finance much of the economic development in the region.

● A rising youthful population and increasing demands on scarce water and food resources and infrastructure (see the table).

The countries of the Middle East can currently use the revenue from their oil exports to pay for expensive **desalination** plants to provide extra water, and also to pay for food and water imports, but there are real fears over future water shortages.

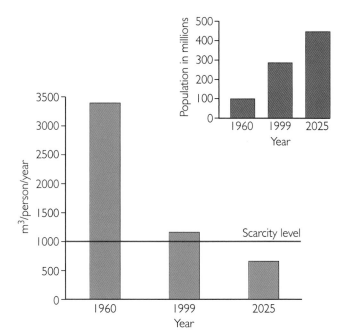

▲ The correlation between increasing population in the Middle East and North Africa and reduced water availability per person per year, 1960-2025. On average, a million people require a billion m³ of water a year, which means that, currently, the Middle East can only meet two-thirds of its needs - depending on desalination plants and imported water for the remainder.

▼ The share of available water used by each sector in the Middle East and North Africa

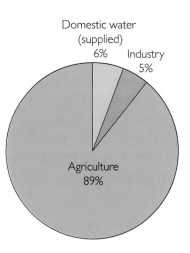

	Population mid-2007 (millions)	Annual growth rate (%)	Expected growth by 2050 (%)
Iraq	29.0	2.5	114
Saudi Arabia	27.6	2.7	80
Syria	19.9	2.5	75
Jordan	5.7	2.4	71
Egypt	73.0	2.1	61
Israel	7.3	1.5	50
Iran	71.2	1.2	41
Lebanon	3.9	1.5	27
Turkey	74.0	1.2	20

▲ Projected population growth in the Middle East by 2050

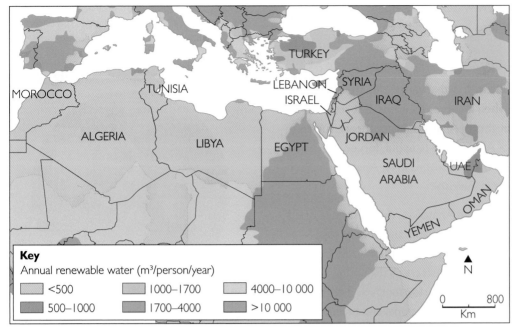

◄ *Levels of water stress and scarcity in the Middle East and North Africa in 1995. (Remember that, using accepted definitions, levels under 1700 m³ per person per year mean water stress and levels under 1000 m³ mean water scarcity.)*

Key

Annual renewable water (m³/person/year)

<500	4000–10 000
1000–1700	
500–1000	>10 000
1700–4000	

0   800
Km

Those Middle Eastern countries without oil or gas reserves depend on modern high-yielding crops – which consume vast quantities of water – to generate wealth and feed their populations. So, for example, Turkey and Israel use their natural water supplies in order to improve productivity and hence economic development. However, because their economic development strategies involve high levels of water consumption, they require access to rivers which flow into or out of neighbouring countries. As a result, their water policies may have serious impacts on Syria, Iraq, Jordan, and Lebanon – destabilising the region further.

No single country in the Middle East can resolve its water problems without having an impact on another country – and climate change will only make the problems worse. There is a need for international co-operation on water-management schemes – but in order for these to work, there needs to be stability in the region.

Conflict

Demand for water already exceeds natural supply in many Middle Eastern countries, with most suffering from official water stress or scarcity (see the map above). The amount of renewable water available per person has halved since 1960 – and is projected to halve again by 2050. Turkey and Israel have become major focal points as far as water management and international relations are concerned.

- The Euphrates and Tigris Rivers originate in Turkey, but also supply Syria and Iraq with water. Turkey has been accused of wanting to cut off Syria and Iraq's water supply as a result of its GAP scheme (see pages 70-71).
- Threats to Israel's water probably culminated in the Six-Day War in 1967. Syria and other Arab states objected to Israel's National Water Carrier Project, and tried to destroy it. Israel retaliated by bombing their attempts to divert the River Jordan's tributaries away from Israel.
- Successive droughts across the whole region from 1990-2005, triggered fears of increased tensions.
- The bombing of Lebanese water pipelines by Israel in 2006 highlighted regional sensitivities over water. Again, Israel feared that its water supplies were under threat.

Over to you

1 Refer to the list of potential Water Wars on page 67, and use atlases to construct a series of wall charts displaying the human and environmental characteristics of the potential Water War zones. (You should include economic fact files, images and data on water resources to complete this.)

On your own

2 Use the data in this unit, and any additional research, to construct a series of maps to show which countries in the Middle East and North Africa face the most serious challenges as global climate changes.

Managing water insecurity

In this unit you'll find out how water-management strategies have raised tensions in the Middle East.

Turkey's GAP

Turkey is a large country (780 000 km²), with a varied climate and abundant rainfall – so it should be well off when it comes to water. But regional variations in rainfall, summer droughts in Anatolia (in southeast Turkey), and shortages in the main cities of Ankara and Istanbul, forced the Turkish government to embark on the $32 billion Southeastern Anatolia Project (also known as GAP from its name in Turkish).

GAP is an attempt to improve incomes in Anatolia – the least developed part of Turkey – by developing an integrated water and energy supply system. Turkey aims to become the 'breadbasket' of the Middle East by 2015, through increasing the production of cash crops (like nuts, fruit and vegetables), alongside the traditional wheat, barley, lentils and cotton. Socio-economic development, with better educational and healthcare facilities – plus additional jobs – is also intended to stop the migration of young people from the region.

While its intention has been to improve the well-being of its own people, Turkey's attempts to increase its water supply and power generation have created a **geopolitical** issue. The GAP project has upset neighbouring countries and also global institutions. Syria and Iraq are unhappy about the project, because it involves damming the Euphrates and Tigris Rivers, which provide both countries with much of their water. The new dams will restrict river flow while their huge reservoirs are filling up, but water flow is supposed to return to normal when they are full. However, they fear that in future Turkey could choose to withhold water for political reasons.

Fact file

Turkey's water budget:

- The total input is an average of 643 mm of rainfall per year.
- 55% of this is lost as evaporation and transpiration, 14% goes underground as groundwater, and 15% is unusable – leaving only 16% available for use.

The aims of the GAP project are to:

- construct 22 dams, 19 hydroelectric power plants and two water-transfer tunnels, to provide 22% of Turkey's electricity by 2010
- provide irrigation for 1.7 million hectares (representing 20% of Turkey's cultivable land and supporting 9% of its population)
- diversify agriculture into cash crops
- stimulate agro-industrial urbanization (intensive, business-based farms which free up workers to move to and work in the cities)
- stop the migration of young people from the region
- help the southeast Anatolian economy to grow by 400%
- help the Turkish economy as a whole to grow by 12%

> We don't want this dam. This is where I belong. I was born here. My identity is here in this town. Our grandfathers and great-grandfathers were born here.
> **A Kurdish resident of Hasankeyf**

- **Geopolitics** is the study of the ways in which political decisions and processes affect the way space and resources are used. It is the relationship between geography, economics and politics.

> Neither Syria nor Iraq can lay claim to Turkey's rivers, any more than Ankara could claim their oil. This is a matter of sovereignty. We have a right to do anything we like. The water resources are Turkey's; the oil resources are theirs. We don't say that we share their oil resources, and they cannot say that they share our water resources.
> **The Turkish President, speaking in 1997, and taking a hard line over the issue of water rights**

▼ The Southeastern Anatolia Project (GAP)

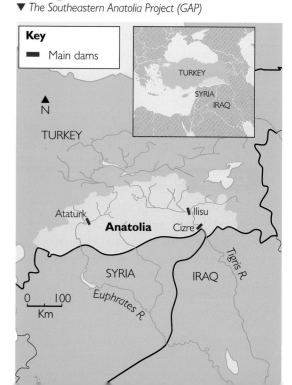

Key
— Main dams

Is GAP too big?

By 2008, most of the GAP project had been completed, but the final elements (particularly the Ilisu Dam) have been attacked by the international community, the World Bank, and environmentalists. The building of the Ilisu Dam, on the River Tigris (65 km upstream from the Iraqi/Syrian border), has caused controversy. It is the biggest part of GAP and highlights the difficulties involved in securing water for the Middle East. Funding for the giant dam became difficult when several foreign governments and engineering companies pulled out after the potential negative impacts became clear (see below).

▲ *The historic town of Hasankeyf will be submerged*

- The ancient town of Hasankeyf will be flooded, along with its historic buildings.
- The remains of the twelfth-century bridge and important cave churches and dwellings in the cliffs will be lost.
- 34 000 residents, mostly Kurds, will be forced to move.
- There will be an increased risk of malaria and water-borne diseases around the lake.
- There is also the risk of it becoming a major waste dump, because effluent will spill into it.

Blair urged to drop support for Turkish dam

The British Government has been urged to scrap plans to provide £200 million in credits towards the £1.47 billion Ilisu dam, which is set to drown one of the world's oldest towns and flood nearly 80 000 people out of their homes.

The Swedish construction giant Skansa has pulled out of the deal, saying that it 'will not participate in construction projects when they result in serious risks to the environment or society'.

The World Bank has shunned the scheme because it violates its ethical and environmental codes. An international report about dams says that local people should participate in decisions about dams in their area – rather than having them imposed on them – and that ethnic minorities should be protected in particular. The Ilisu dam would displace tens of thousands of Kurds, who say that they have never been consulted about it.

The report also undermines one of the major justifications for dams – that they provide green energy which does not contribute to global warming. Research shows that rotting vegetation covered by the waters of the dam's huge reservoir will release carbon dioxide and methane, which both help to cause climate change.

Adapted from *The Independent on Sunday*, 12 November 2000

A geopolitical time bomb?

By 2004, Turkey had bowed to the international pressure and reviewed and amended the Ilisu part of the scheme to reduce the impacts on local settlements and take into account major social and environmental considerations, such as the risks of waste dumping and the relocation of communities. However, some human rights issues, such as the forced relocation of some Kurdish villages, remained unresolved. Also, Iraq and Syria still needed guarantees about river flow levels – but it seemed that Ilisu would now go ahead.

Austrian, Swiss and German companies formed a consortium to build the Ilisu Dam and, in March 2007, their respective governments approved funding for the project.

Turkey has agreed to release water from Ilisu for Syria (but not Iraq) at a rate agreed by international law. However, just downstream from Ilisu and north of the border, Turkey intends to build the Cizre Dam to collect additional water for irrigation before the Tigris crosses the border – leaving Syria and Iraq to argue over what is left. Therefore, it is not difficult to understand why Turkey's neighbours are upset.

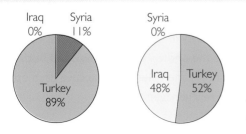

◀ *The percentage that each country contributes to the annual flow of the Euphrates (left) and Tigris Rivers*

Israel's water insecurity - 1

Israel's issues

Israel has a problem and the facts are simple:

- It consumes 500 billion litres more water than it receives naturally.
- The population is growing – with an annual growth rate of 1.5%.
- Droughts are increasingly common and are lasting longer.
- Internal competition for the limited water supplies is increasing.
- It is often in dispute with its neighbours.

Israel is haunted by the fear of water shortages. Disputes over water have been common in the Middle East, and any action taken by Israel to secure water supplies could affect the political stability of the region. In order to maintain peace, there need to be secure **water pathways** to help guarantee supplies. But Israel's relations with its neighbours mean that those pathways are frequently under threat.

Background

Geopolitical timeline for the Middle East region

- 1948-49: The modern state of Israel was formed.
- 1953-55: A US Special Envoy dictated regional water shares between Israel, Jordan, Lebanon and Syria – giving Lebanon and Syria the smallest shares, even though they provided most water.
- 1967: The Six-Day War – now seen as an Israeli reaction to Syrian attempts to divert the Jordan River. It ended with Israel gaining greater control over the river and securing enhanced water supplies by capturing the West Bank from Jordan, the Golan Heights from Syria, and the Gaza Strip from Egypt (see the map opposite).
- 1974: Syria built the al-Thawra Dam and upset Iraq because it reduced the flow of the Euphrates by 25%.
- 1979: Egyptian peace treaty with Israel. President Sadat of Egypt said: 'The only matter that could take Egypt to war again is water'.
- 1980-88: The Iran-Iraq War, largely over territorial disputes in the Shatt al Arab waterway.
- 1993: The Oslo Accords between Israel and the Palestinians led to an Israeli withdrawal from part of the Gaza Strip and West Bank, and Palestinian self-government in those areas. This was an interim agreement on the way to a final settlement.
- 1994: The Jordan peace treaty with Israel initiated regional cooperation over the sustainable use of the Yarmouk River.
- 1996 and 2000: Syria tried to gain access to Turkish water to use in negotiations with Israel. Both attempts failed.
- 2000: A summit at Camp David in the USA, where Israel offered to give up land to the Palestinians but not water.
- 2004/5: Turkey agreed to ship water to Israel in return for high-tech military support (the Manavgat Project).
- 2005/6: The Litani River disputes between Lebanon and Israel.

▲ Contrasting lifestyles – water to spare in the Israeli settlement of Maale Adumim, in the Israeli-occupied part of the West Bank, and black rooftop storage tanks which help to combat unreliable water supplies in the town of Talfit in the Palestinian-controlled part of the West Bank.

● **Water pathways** are the routes taken between sources of water (e.g. rivers, lakes, groundwater) and where it will be consumed. The usual routes are by pipelines or artificial canals.

Whose water is it?

Shifting territorial borders (Gaza and the West Bank) has not made Israel's job of providing its population with water any easier. By 2050, it will have to provide an estimated 11 million Israelis living in a desert environment with their daily water needs. Many Palestinians live in areas which Israel once occupied, but Israel still claims most of the water in those areas as its own – denying the Palestinians access to it and causing further conflict. There is a big disparity in the use of, and access to, water (as the photos above show), and both Israelis and Palestinians accuse each other of mismanagement.

Israel's water sources

Israelis consume more water than any other country in the Middle East – 2200 billion litres each year. Israel's natural supply is only 1700 billion litres, and the over-consumption has become unsustainable. The shortfall is currently made up from importing water from Turkey, from recycling sewage and from desalination plants (see page x). Israel's three main sources of water are all beginning to show signs of degradation, with irreversible contamination by seawater, chemical pollution and over-pumping.

Three areas each provide 25% of Israel's natural water supply, the remainder coming from smaller aquifers:

- The Sea of Galilee (also known as Lake Kinneret), which is fed by the River Jordan and tributaries in the Golan Heights.
- The Mountain Aquifers, which are mostly located in the West Bank. They are shared between Israel (80%) and the Palestinians (18%). The remaining 2% is unavailable.
- The Coastal Aquifer, which supplies 80% of Israelis with water in the most economically developed area of Israel. Israel controls more than 90% of this aquifer.

Water disputes

Syria wants its borders to be reinstated to where they were before 1967, as part of the price of peace. This would mean that the Golan Heights would be returned to Syria and 25% of Israel's water supply would be compromised. Israel fears this for a number of reasons:

- It feels that Syria's water-management systems are inferior and could contaminate Lake Kinneret.
- The threat from Turkey's GAP project could force Syria to divert the River Jordan away from Israel to ensure its own water supplies.
- Israel would be at the mercy of a long-time enemy for a quarter of its water supplies.

Israel is also facing other water dilemmas. Overuse and misuse are stretching water resources to their limits and putting supplies at risk:

- The Mountain Aquifers are mostly located in the disputed West Bank, where urban growth has meant increased pumping and declining quality.
- The Gaza Strip is showing signs of salt seepage as water levels in the Coastal Aquifer fall. Israel thinks that the Palestinians there have over-pumped supplies, causing seawater to seep through into the Aquifer – which then threatens Israel's supplies.

Israel fears that the Palestinians in Gaza and the West Bank will use too much water and create a natural disaster of salt seepage and shortages. The Palestinians think that Israel's own water demands leave them with too little water. If Israel gives in over increased Palestinian access to water supplies, it will have to rely on inferior Palestinian technology and unlikely political goodwill to protect its water. Suspicion and fear cloud the geographical issues, and conflicts over access to and control of supplies have to be resolved before water supplies are safe for anyone.

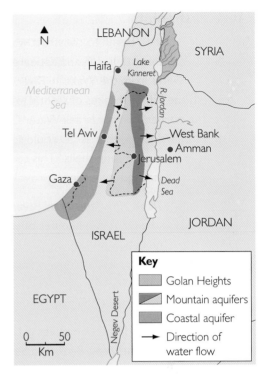

▲ Israel's water sources. Since 1967, Israel has met much of its requirements from the occupied territories – clinging to the Golan Heights as a crucial source of water.

Palestinians get less than 60 units of water a day [the international minimum is 150], and Israeli domestic use alone is 300-800 units. It is worse in Gaza, where a lot of the water is not fit to drink, and that is why we have a lot of health problems. **A resident of Gaza**

Israel's water insecurity – 2

More flashpoints

A In 2002, Lebanon began to construct water pipelines along its border with Israel, and appeared to be ready to divert water from the Wazzani River, which flows towards the fish-farm lakes of several Jewish kibbutzim over the border. Former Israeli Water Commissioner, Dan Zazlaavsky, said that 'Israel could solve this water diversion with a few tank shells.' This comment helps to illustrate the strength of feeling it provoked.

B Between 1967 and 2006, the Litani River in southern Lebanon was the focus of a number of armed disputes with Israel. The Lebanese suspect that the 2006 occupation by Israeli troops of the disputed border zone between the two countries was mostly to do with controlling this river (see the map on the right).

C Suspicions surround Israel's motives in building the 'Dividing Wall' in the West Bank (see below). It has led to the destruction of wells, separated Palestinian villages from their farmland and water sources (see right), and freed up water supplies for Israelis. The line of the barrier appears to follow the edges of the Western Mountain Aquifer, which flows towards Israel.

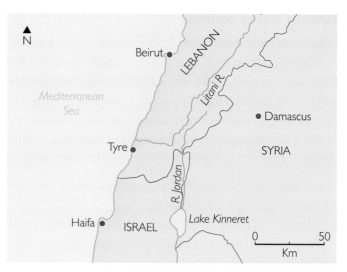

▲ Israel has long coveted the Litani River in southern Lebanon as a source of water, because it could meet 40% of Israel's water needs

▲ The Dividing Wall separates Palestinian communities from their farmland

Israel as a blueprint for the future?

Global demand for water will continue to grow and, without international agreements to share resources – or increased supplies – alternative ways of satisfying each country's needs have to be found.

Israel's technological expertise is among the most advanced in the world, thus ensuring a high standard of living for its citizens. Israel and southern California have similar climates, but Israelis use far less water than Californians. Political uncertainty, together with the natural geography of the country, have forced Israel to become more water efficient.

The Israeli government took control of water in 1959, when it developed plans for the National Water Carrier system. Revolutionary, drip-feed irrigation systems were developed to take water to the Negev Desert settlements – and also reduce water consumption, evaporation and salinisation.

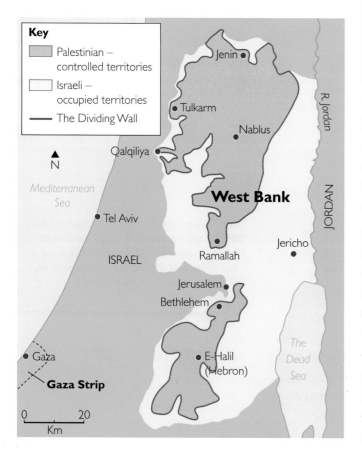

Key
- Palestinian – controlled territories
- Israeli – occupied territories
- The Dividing Wall

There are several elements to Israel's current water-management strategies:

Managing limited supplies – reviewing current uses and developing more-efficient techniques.

● Recycling sewage water for agricultural uses could lower overall consumption. 65% of all crops (and 90% of the cotton crop) are already produced in this way.
● Reducing agricultural consumption and shifting the economy to high technology, which would mean relying on greater food imports instead (see below).
● Better water-treatment plants and stringent conservation techniques.
● Demand management by charging 'real value' prices for water to reflect the costs of supply and also of ecosystem management.

Acquiring new supplies

● Importing 50 million m³ of water per year by ship from Turkey (the Manavgat Project agreed in 2004).
● Piping seawater from the Red Sea and Mediterranean to new inland desalination plants. The Desalination Master Plan (see below) envisages providing 25% of Israel's supplies by 2020.

Expanding virtual water supplies

● Importing water-rich foods saves vast amounts of water.

> ● **Virtual water** is water which is transferred by trading in crops and services which require large amounts of water for their production. By importing a tonne of wheat from a water-rich area, a water-stressed area saves 1000 m³ of water.

▶ The desalination plant at Ashkelon, which provides a reliable and predictable supply of water – it already produces 5-6% of Israel's needs (100 million m³ of water a year). This could reach 25% by 2020, with the opening of the additional plants of Hadera, Ashdod, Caeserea and Palmatin. However, each plant takes 3 years to build, is very expensive, and uses so much energy to operate it that it needs its own dedicated power station! They are also prime terrorist targets, which is a big concern for Israel.

Over to you

1 As a group, make a list of the criteria needed to judge the strengths and weaknesses of the ways in which Turkey and Israel have been managing their water supplies.
2 Consider how far different nations should be able to control shared natural watercourses – do all users have rights?
3 How far do you think international relations influence decisions about water management?

> *Dams – once the solution to the water supply issue – are fast becoming a problem.*

What do you think ?

On your own

4 Carry out some extra research into GAP using the website: http://www.adiyamanli.org/ataturk_dam.htm and select one other major water supply scheme (eg: Bujagali/Uganda; Kisha-Ganga/India-Pakistan; Nam Theun/Laos; Pangue/Chile)
Construct a series of flow charts to illustrate the positive and negative impacts of these schemes on the following: farming, settlements, heritage, local rivers, health, wealth, soils and the environment.

Exam question: Referring to examples, explain why future water supplies for many regions are increasingly insecure. **(15)**

In this synoptic unit you'll begin to investigate the link between water, poverty and wealth.

An African water crisis

Debt and despair became synonymous with many African countries in the 1980s and 1990s. The population grew at an alarming rate, poverty drove people to use marginal land, soils were eroded, and Africa went into debt crisis. When droughts then struck and incomes fell further, many countries looked to their rivers as a means of progress and escape.

From the 1980s, major irrigation schemes were introduced to encourage people to grow cash crops to help boost exports (such as green beans and cut flowers from Kenya), so that the countries could get out of debt. The result was a huge additional demand for water.

Before the 1980s, African rivers had been dammed as part of large-scale development projects, such as the Aswan Dam in Egypt. The economies of Egypt and Ghana (which built the Akosombo Dam) grew significantly as a direct result of the dams. Now other countries want to do the same – sometimes by harnessing rivers shared with their neighbours – like Togo and Benin's plans for the Mono River, and Uganda and Ethiopia's ideas for the White and Blue Niles.

However, water in sub-Saharan Africa is already in short supply, even without these additional demands. When the growing impacts of climate change are considered, the future looks bleak. The map in Resource 1 opposite shows that a number of African countries will almost certainly be officially water scarce by 2025, which will threaten:

- their economies (agriculture accounts for a high percentage of GDP in many African countries – see Resource 2)
- the millions of people employed in agriculture.

In the following resources, you will assess the impacts of this water crisis on Africa.

Africa's water future

The challenges facing Africa in the twenty-first century include:

- increasing the number of people with access to safe water and sanitation
- raising people out of poverty
- managing and maintaining water resources.

This unit looks at three approaches:

- The trade in virtual water (e.g. roses from the Lake Naivasha region in Kenya, Resources 3 and 4)
- Major irrigation schemes (e.g. Toshka, Resources 5 and 6)
- Small-scale projects (e.g. WaterAid, Resources 7–10)

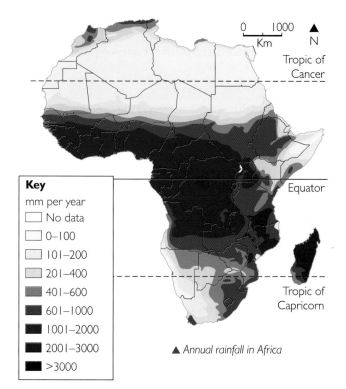

Tropic of Cancer

Equator

Tropic of Capricorn

Key

mm per year

- No data
- 0–100
- 101–200
- 201–400
- 401–600
- 601–1000
- 1001–2000
- 2001–3000
- >3000

▲ *Annual rainfall in Africa*

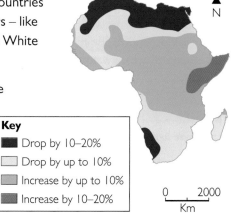

Key

- Drop by 10–20%
- Drop by up to 10%
- Increase by up to 10%
- Increase by 10–20%

0   2000
Km

▲ *How rainfall is predicted to change by 2030*

Resource 1

Freshwater stress and scarcity

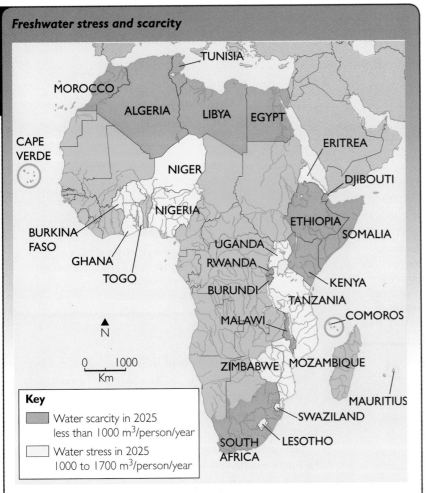

Key

- ▨ Water scarcity in 2025
  less than 1000 m³/person/year
- ▢ Water stress in 2025
  1000 to 1700 m³/person/year

▲ Those African countries which the UN believes will be suffering from water stress or scarcity by 2025

Whose water is it?

One of Africa's great rivers is the Nile. Ten countries lie within the Nile Basin – and 40% of Africa's population. Egypt has traditionally benefited most from the Nile, but 85% of the river's flow originates from the Blue Nile in Ethiopia. While Ethiopia's economy lagged behind, Egypt was able to support 73 million people and develop its irrigated farms – but now the pressures are mounting.

After many years of civil war and famine, Ethiopia has emerged as a growing economy – and it is increasingly tapping into the resources of the Nile. With a similar-sized population to Egypt – but a faster growth rate – it also wants to develop irrigated farming, and it is likely that it will need some of Egypt's supply of water. Sudan also has plans to exploit the Nile and wants to increase irrigation for its own economic development.

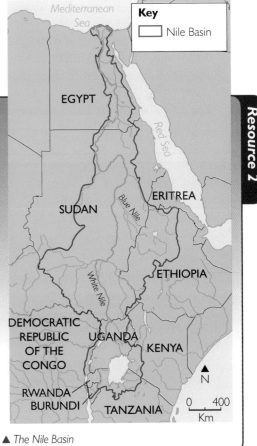

Key

- ▢ Nile Basin

▲ The Nile Basin

Nile Basin countries

	Population (millions) 2006	Population growth rate per year (%) 2006	Population with access to safe water (%) 2006	Agricultural production (% of GDP) 2005
Burundi	7.9	2.5	79.0	34.8
D.R. Congo	3.6	2.8	58.0	5.6
Egypt	72.8	2.1	98.0	14.9
Eritrea	4.5	2.5	60.0	22.6
Ethiopia	79.0	2.8	22.0	47.7
Kenya	35.6	3.2	61.0	27.0
Rwanda	9.2	2.5	74.0	42.3
Sudan	36.9	2.6	70.0	33.7
Tanzania	38.5	2.9	62.0	44.5
Uganda	28.9	3.3	60.0	32.7

▲ Selected data from the UN Human Development Report for countries sharing the Nile Basin

What is the water used for?

The trade in virtual water

Exporting water

What do rural Africans think as they pass fields of cash crops of sunflowers, roses or coffee, as they trek 5 km a day to collect water? African countries find it difficult to feed their own people or provide safe drinking water or sanitation, yet – from Egypt to Kenya – precious water is used to produce export crops for European markets. Cut roses provide an income for farmers unable to compete with the prices of crops such as sugar, which European farmers receive subsidies to grow.

In a sense, African countries are exporting their water in the very crops they grow. Kenya needs water, but also needs to trade in virtual water (see page 75) through the crops it produces. Environmental pressure groups argue that, each time European customers buy salad crops or flowers grown in Africa, they are making water shortages worse. Africa exports water and Europe exports drought back. However, as the quotes below show, it is a complicated situation.

People want to buy ethically and do their bit for climate change, but often do not realise that they can support developing countries and also reduce carbon emissions. Recent studies show that flowers flown from Africa can use less energy overall than those produced in Europe, because they are not grown in heated greenhouses.
Hilary Benn, Britain's International Development Secretary, 2007

A product grown outdoors in a warm country and flown to the UK may have no higher carbon footprint than a product grown out-of-season in Europe in a heated greenhouse.
Sir Terry Leahy, Tesco's Chief Executive, 2007

To you it is a bag of salad. But to farmers in Kenya starved of water, it may spell destitution. In Kenya, food items grown for export include lettuce, rocket, baby leaf salad, mangetout, peas and broccoli. Even a small 50-gram salad bag wastes almost 50 litres of water. A mixed salad containing tomatoes, celery and cucumber, as well as lettuce, would require more than 300 litres of water to produce it.

Jeremy Laurance, Health Editor, *The Independent*, 2006

Half of all cut flowers sold in British supermarkets come from Kenya, where the volume of exports to Britain grew 85% between 2001 and 2005. The demand for water from nearby Lake Naivasha is unsustainable.

Naivasha is being sacrificed. Almost everybody in Europe who has eaten Kenyan beans or Kenyan strawberries or gazed at Kenyan roses has bought Naivasha water. It is sucking the lake dry. It will become a turgid, smelly pond with impoverished communities eking out a living along bare shores.

David Harper, a biologist at The University of Leicester, quoted in *The Independent*, 2006

Lake Naivasha at risk

Lake Naivasha is the centre of Kenya's flower-growing industry – but its success has created problems too. The following extracts and quotations come from: *Naivasha: Withering Under the Assault of International Flower Vendors*, published online in January 2008 by Food & Water Watch and The Council of Canadians.

Kenya is a begging country. We are among the top countries on the World Food Programme list for food donations, even though we have a freshwater lake that would allow us to grow food to feed ourselves. Yet we grow flowers and then ship them 5000 miles to Europe, so that people there can say 'I love you, darling' and then throw them away three days later. To me that is an immoral act.
Isaac Ouma Oloo, activist and ecological safari guide

The flower farms that ship more than 88 million tons of cut flowers a year – worth $264 million in the Lake Naivasha region – pose serious ecological problems, including the overuse of pesticides and fertilizers.
Food & Water Watch

Its unsustainable use could drain Kenya's rivers and Lake Naivasha – harming both the ecosystem and the economy. Labourers and their families arrive to work in greenhouses and processing plants. The population rose from 7000 in 1969 to 300 000 in 2007.
The Council of Canadians

Major irrigation schemes

Toshka – the 'Project of the Millennium'

Egypt has embarked on a controversial scheme to irrigate areas away from the densely populated Nile Valley (see the map). The $70 billion Toshka Project is part of the long-term plan to boost food production for Egypt's growing urban population, and for export. It will be funded by international investment – with most funding to date coming from loans by Saudi Arabia and the United Arab Emirates. Transnational companies like Cadiz from California will oversee production, processing, harvests, storage and marketing.

Planners claim that this mega-project will deliver 'A Second Egypt' and 'A New Valley', and will support 16 million people. 95% of Egypt's population lives within 16 km of the Nile, and Cairo is densely populated – with up to 40 000 people per km². The Toshka Project aims to relieve pressure on the Nile Valley and Cairo.

Toshka's aims are to:
- provide food, electricity and jobs for 16 million Egyptians in new towns in the desert, and also relocate people to farms in irrigated areas
- use pumps and canals to transfer water from Lake Nasser into the Western Desert
- increase Egypt's irrigated area by 30% (2000 km²)

- recharge underground aquifers
- grow high-value crops – olives, oil seeds, citrus fruits, vegetables
- maintain a low use of pesticides and fertilizers
- create sustainable communities with self-sufficient smallholdings for families
- improve roads, railways and telecommunications
- promote tourism

Toshka – impacts and doubts

- Lake Nasser is silting up, so long-term water supplies may not be reliable.
- The open transfer canals (pictured) may fill up with desert-blown sand, and water in them will also evaporate in the desert heat.
- The irrigation water could lead to salinisation.
- The water pumps will use huge amounts of electricity, the supply of which is unreliable.
- Waterlogging could occur in the Toshka basin.
- There would be a need for fertilizers.
- There is a risk of the spread of water-borne diseases in the stagnant canal water.
- The Toshka Project would fail if upstream countries like Ethiopia and Sudan take more water from the Nile, causing water levels to fall.
- The finance is insecure and will lead to debt.

WaterAid makes a difference

Small-scale projects

WaterAid

WaterAid is an international charity with a mission to overcome poverty by enabling the world's poorest people to gain access to safe water, sanitation and hygiene education. For example, WaterAid has set out to improve the livelihoods of people in Ethiopia, using techniques that are appropriate to local conditions, affordable and easy to maintain.

The aim in rural areas is to achieve 100% access to clean water and sanitation. The provision of clean water is also supported by hygiene education, which leads to improved health and, in particular, reduces the number of children who die from diarrhoeal diseases.

WaterAid works closely with local communities and involves them in the planning, building, managing and evaluation of projects. Attention is now focusing on engaging women in projects – and those who are marginalised, such as the elderly and those suffering from HIV/AIDS.

WaterAid's projects enable communities to achieve a better quality of life and escape from poverty, as Resources 8–10 show.

Two successful WaterAid projects in Ethiopia ...

Mobile toilets

There is a big difference between the life I was living before and the life I have now. Before I was homeless and I didn't have an income to support myself. Now, with the toilet, I have an income and am saving money every month. Between 70 and 80 people use this toilet every day. Sometimes, if people have no money, I will let them in for free – so that we can keep the area clean. My life has been improved very much from before. I would like to see these toilets in other areas of the city too, as they keep the areas clean and also offer opportunities to more people.

Eskender Tadesse (pictured) runs a mobile toilet and kiosk, set up by WaterAid. These public toilets – on the streets of Addis Ababa (Ethiopia's capital) – help to keep the environment clean and also provide formerly homeless people with a salary through the kiosks.

Resource 9

The difference clean water brings

Before we only had unprotected sources of water. My family suffered badly. My three-year-old daughter died from this water. There were parasites, which gave us illnesses and stomach problems. Many children used to die, but now this has changed and children don't die from these diseases. Before we used to have to go to the health clinics all the time – often every day. I used to spend time walking there and hours just queuing to be seen, but now I can save my time and money. I have bought 20 chickens and one goat from the money I have saved. With the time I can work on my maize and pepper crop.

Mehari Abraha, Tigray province, Ethiopia. WaterAid has provided protected spring-fed wells to ensure clean water supplies for this family and others in the area.

... and one in Uganda

Resource 10

It is about teaching too

I use tools such as picture cards to explain sanitation messages to the community. I tell people they have to maintain their water sources regularly and prevent animals from using the source. I explain to people how they can construct latrines from local materials. Without latrines, drinking water becomes contaminated with faeces. I got involved in teaching these messages because I love cleanliness and want the best for others.

Mr. Massabidi (pictured) now promotes good hygiene practices to members of his community in the village of Nabinaka in the Wakiso District of Uganda, following discussions with WaterAid's local partner, YIFODA.

Even without climate change, pressures on limited water resources are increasing and North African countries will run short of water by 2025. Up to 350 million people will be at risk of water stress by the 2020s.

As an advisor to the UK Overseas Development Agency, you are responsible for drawing up a report advising on the most appropriate way of helping African nations to avoid conflicts over water resources. Use the following exam question to help formulate your report and present it to the rest of the class.

Synoptic question

a Explain why African countries see mega-projects as a key part of their economic development programmes. **(10)**

b Assess the likely environmental and socio-economic impacts of the different proposals for coping with water shortages in African countries. **(16)**

c How serious is the issue of water stress in **i** Africa as a whole, **ii** the Nile Basin specifically? **(14)**

In this unit you'll find out that even wealthy countries have water problems.

Water – different hemispheres, different approaches

The UK's national water grid – rejected!

In 1972, the UK Government considered establishing a national water grid to connect areas of the country with surplus water to areas with shortages. It made geographical sense to link the country's major drainage basins together to ensure that the heavily populated – but dry – southern and eastern areas did not suffer water shortages. Drought in the mid-1970s made this a popular idea, but subsequent wet years – and the privatisation of the regional water authorities in 1988-9 – meant that the water grid scheme was abandoned.

However, by 2006, a national water grid was back on the agenda, as extended dry periods and the implications of climate change hit home. But two big obstacles now stand in the way of a national water grid:

- The privatised water industry, now mainly owned by different multinational companies, would be unlikely to work together on such a plan, or agree to the expense necessary.
- The shareholders of the private water companies would also be unlikely to back such a scheme, because company profits (and shareholders' incomes) would have to fall to help pay for it.

Australia's National Plan – accepted!

In contrast to the UK, the Australian Government announced a National Plan for Water Security (The Plan) in January 2007, which focussed on the Murray-Darling Basin (MDB). Australia recognised the need for national management of its water for several reasons:

- Australia is an arid country – rainfall is seasonal, unreliable, unevenly distributed, and amounts are declining.
- Australians are the greatest per capita consumers of water in the world, and population growth in the south and east is putting pressure on supplies.

Prior to The Plan, each Australian state managed its own water resources, but the Government felt that the 'old way' of managing water and the Murray-Darling Basin had 'reached its use-by date'. The rest of this unit looks at why a national management plan was needed.

The MDB – an asset under threat

The Murray-Darling Basin:

- is the size of France and Spain combined
- covers 14% of the Australian land mass
- provides 75% of Australia's water (85% of the country's irrigation water)
- provides 40% of the nation's farm produce, worth A$13.6 billion
- is home to 2 million people, with millions more outside the Basin – in the cities of Sydney, Melbourne, Brisbane and Adelaide – who all depend on it for food and water.

Like many of the world's great river basins, the MDB is dying. There has been a five-fold increase in water extractions since the 1920s – matching Australia's five-fold increase in population. The reduced water flow is now leading to damaged ecosystems.

The MDB is under threat from increasing and competing demands, which have not always been well managed or coordinated. Difficulties arise because the Basin is so large and comprises several different natural environments:

- Queensland is sub-tropical with rainforests.
- The New South Wales eastern ranges are cool and humid.
- Victoria is more temperate.
- The Eastern edges of South Australia are hot semi-arid areas.

The Murray-Darling Basin

The natural environment determines water availability across the Murray-Darling Basin. Seasonal and local variability mean that some areas experience surpluses and others deficits. Transfers between, and regulation of, the basin's water sources are vital.

The MDB has a number of water sources:

- The Great Artesian Basin and Murray Groundwater Basin provide groundwater supplies. However, extraction is increasing by 4% each year.
- Average rainfall across the MDB is 480 mm, but it varies between areas – frontal systems bring rain to the northern areas and the Great Dividing Range brings relief rain.
- The graph shows rainfall variability from 1900-2000, which reflects the impact of El Niño on Australia's weather. El Niño brings dry westerly winds across Australia, leading to droughts in the east. Variability in rainfall creates variable runoff patterns.
- Diversions from the Snowy Mountains and the Glenelg River provide additional inputs to the Basin.

The western areas have a water deficit because potential annual evaporation rates are higher than average rainfall, whereas surpluses exist to the east.

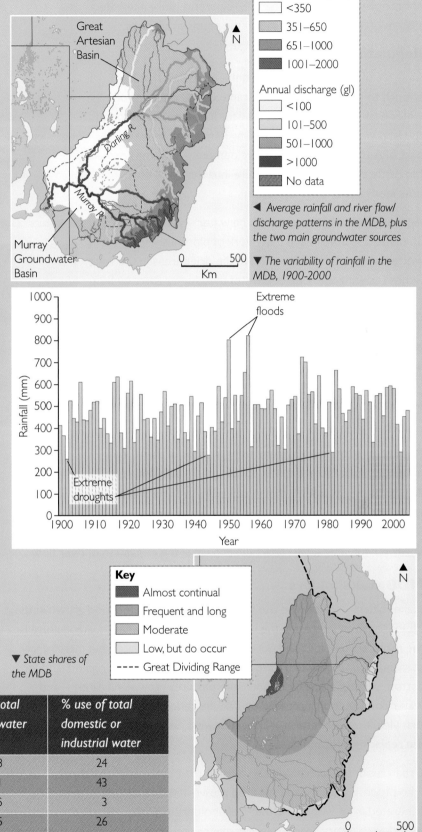

Key

Annual rainfall (mm)
- <350
- 351–650
- 651–1000
- 1001–2000

Annual discharge (gl)
- <100
- 101–500
- 501–1000
- >1000
- No data

◀ *Average rainfall and river flow/discharge patterns in the MDB, plus the two main groundwater sources*

▼ *The variability of rainfall in the MDB, 1900-2000*

▼ *State shares of the MDB*

Key
- Almost continual
- Frequent and long
- Moderate
- Low, but do occur
- ---- Great Dividing Range

▲ *The likelihood of drought in the MDB*

State	% of MDB area	% use of total irrigation water	% use of total domestic or industrial water
New South Wales	56.6	58	24
Victoria	12.3	31	43
Queensland	24.5	5	3
South Australia	6.4	5	26
Australian Capital Territories	0.2	1	4

MDB – the need for change

Competition for water and a growing concern for the environment meant that the MDB required a more integrated system of management to satisfy the needs of all users.

Regulating river flow

Only 6% of Australia's rain falls in the MDB, but that is enough to yield between 14 000 and 53 000 gigalitres of water each year – much of it used for irrigation. Irrigation of the land west of the Great Dividing Range was an essential part of Australian economic development – and transformed the cattle and sheep ranches there into the 'food bowl' that now feeds millions of Australians.

Agriculture reallocates most of the water – via a system of 30 dams, 3500 weirs and a network of pipelines – from the wetter headwaters of the MDB to the drier areas in the west – thus reducing the total outflow of the Murray River by 80% at its mouth.

As a result:
- natural floodplains no longer flood
- red gum trees are dying through lack of water along thousands of kilometres of riverbanks
- and between 50-80% of native wetland bird and fish species are now extinct in this area

Regulation of the river system, using a series of dams, was needed to guarantee water supplies and smooth out irregular flow patterns, caused by the variable climate. This means storing water in times and areas of surplus – for release when and where shortages occur. River levels need to be high enough to sustain the economy and the natural environment. The table shows the impact of the regulation of river flow.

MDB stakeholders

Stakeholders – those who use MDB water and consume food grown in the MDB – have an interest in the way the resource is managed. Agriculture is the major stakeholder and not only takes the most water, but increasingly demands more. Between 1996 and 2002, the amount of irrigated land for cotton, cereals, grapes, rice and dairy pastures grew significantly – and water consumption increased by 29%. Irrigation boosts profits, and crops that yield the best returns – like cut flowers, vegetables, grapes and nuts – are replacing beef and sheep.

Urban residents both inside and outside the MDB – including those who process, package and sell food grown in the MDB – receive income generated from within it. Other stakeholders include:

- industrial users; manufacturing, mining
- aquaculture, freshwater fishing
- those offering recreational activities
- local and regional governments
- environmental groups
- international heritage and conservation agencies

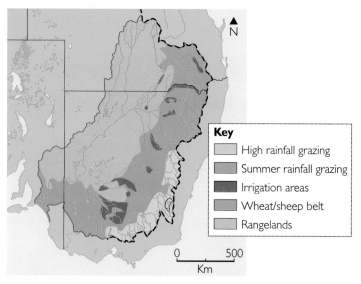

▲ Agricultural land use in the MDB

Key
- High rainfall grazing
- Summer rainfall grazing
- Irrigation areas
- Wheat/sheep belt
- Rangelands

0   500
Km

	Average flow pre-regulation (gigalitres)	Average flow post-regulation (gigalitres)
Runoff	23 850	23 850
Inter-basin transfers	0	1200
Diverted flow for irrigation	0	11 580
Evaporation	0	1430
Wetlands/floodplain	10 960	6970
Outflow to sea	12 890	5070

▲ The effect of regulating the MDB

MDB farmland fact file

Farmers dominate the cry for more sustainable water management in the MDB, because they are the largest stakeholders. The MDB contains:

- 89 million hectares of farmland
- 42% of all Australian farms
- 25% of all Australian cattle
- 50% of all Australian sheep
- Just 1.4% of the MDB uses 75% of Australia's irrigation water
- 95% of the MDB is non-irrigated farming, dependent on rainfall
- Farmers' yields in the MDB were worth A$13.6 billion in 2007.

Environmental degradation

The MDB is experiencing environmental degradation as a result of changes in agriculture and a piecemeal approach to regulating the river system. The use of technological fixes, such as pesticides, fertilizers and irrigation, has enabled Australia's agricultural sector to grow, but they have been at the expense of the environment. If things continue unchecked, the system will become unsustainable.

Environmental degradation

Salinity (the process is explained on page 65)

- Salt concentrations are highest when the river flow is low.
- Salinity levels along the Murray River are increasing towards the west as a result of irrigation and drought.
- Farmers are losing produce as a result of the increased salinity, and natural habitats are being destroyed.
- The salt-rich water table is corroding roads, bridges, building foundations and pipes in the cities of Wagga Wagga and Shepperton.

Eutrophication

- Nitrates and phosphates in the fertilizers used by farmers are washing off the land and encouraging the growth of algal blooms and eutrophication.
- This process occurs in stagnant or slow-moving water – reducing the oxygen in the water to 'lifeless' levels.

Groundwater

- Groundwater supplies are becoming depleted as farmers extract 4% extra each year.

- Irrigation is raising the water table and causing waterlogging, which leads to reduced oxygen levels and the drowning of plants.

Soil degradation

- The structure and fertility of the soil is declining because of overcropping.
- There is acidification because of overfarming, leading to infertile soils.
- There is wind and water erosion as a result of vegetation clearance.

Ecosystem loss

- Native vegetation and wildlife habitats, such as the Barmah-Millewa Forest, are being cleared to create more farmland.
- Natural heritage sites, such as the Macquarie Marshes and Narran Lakes (see page 86), are being lost because of over extraction of water for farmers.

Cultural loss

- The removal of the indigenous Aboriginal population as farmers take more land.
- The loss of heritage sites/tourist sites as land is converted to cereal production.

The mouth of the Murray River – near Adelaide in South Australia – is silting and salting up, and was closed in 2002-3. It costs A$7 million a year to dredge the mouth to keep it open, so that species can move between the river and the sea.

The marina owner at Hindmarsh Island argues that it is already too late to save the river. 'We need water. We need it today.' he said. 'The low water level has taken a toll on tourism and recreation, with most boats at the marina more than a metre below their boarding jetties.'

Across the Lower Lakes on Mundoo Island, fourth-generation cattle farmer, Colin Grundy, has been importing water by truck to keep his stock alive.

One resident of Narrung said that their water supply was so bad that they could not drink it. 'We are the unlucky ones that ran out of rainwater and have to use Council water – and it is the colour of the lake. It is brown, disgusting water and it smells really bad.'

MDB - conflict and management

Emerging conflicts

Different stakeholders have different views about how water in the MDB is managed and used, and, as a result, a number of conflicts are arising in the MDB. Two of these are discussed below – and each one raises some serious questions.

1 The Macquarie Marshes area

A Ramsar protected area (see page 129), upstream in New South Wales

The situation:
- Irrigated cotton farms are replacing non-irrigated grazing land.
- The value of grazing land has fallen because of droughts.
- Farmers are capturing the limited rainfall and storing it in their own reservoirs.
- The runoff to streams has been reduced.
- The NSW Government has licensed farmers to increase groundwater extraction.
- Trees and marsh vegetation are dying.
- Bird, fish and plant species are all suffering.

The conflict:
- Support the Ramsar status or the needs of farmers?

Big questions raised:
- Should more water be directed into the marshes to support them?
- Who should manage the water supply?
- How can nature pay for its share?

2 Mildura-Hattah

The site for a proposed waste dump, downstream in Victoria

The situation:
- This area has been identified as a potential site for the dumping of toxic waste.
- There would be a danger of possible seepage into the surrounding soil and groundwater.
- There would be a negative impact on local property values.
- There would be a negative impact on recreation and the local tourism economy.

The conflict:
- Toxic dump or nature and tourism?

Big question raised:
- Where should toxic waste go – it has to go somewhere?

The need for management

Up to 1987, individual states were mainly concerned with securing flows within their own part of the MDB – overseen by the River Murray Commission. Between 1987 and 1992, all of the states and the National Government signed up to the Murray-Darling Agreement, which made basin-wide strategies and more integrated approaches possible. Quotas for users, trading of allocations, and guidelines on irrigation techniques, were introduced. However, some states, like Queensland, still tended to go their own way – expanding irrigation areas and disregarding downstream impacts.

One of the biggest changes was a controversial policy to allocate water to the river basin. Called 'The Cap', it represented a basin-wide approach, agreed by state governments and the national government, to create 'environmental flows'. It limits basin extractions to those recorded in 1994, in order to help the environment. But angry farmers have referred to it as the 'Water for fish and ducks programme', because it allocates water back to riverside ecosystems.

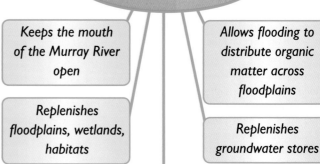

Keeps river banks moist to encourage plant growth to stabilise them

Allows a downstream flow of nutrients to sustain habitats

The benefits of environmental flows

Keeps the mouth of the Murray River open

Allows flooding to distribute organic matter across floodplains

Replenishes floodplains, wetlands, habitats

Replenishes groundwater stores

Flushes stagnant pools

All change in 2007

In 2007, Australia was in the grip of its worst drought on record – and the MDB was the worst affected area. In 2006, just 1317 billion litres of water flowed into the MDB – almost 25% less than the previous minimum in 1902. The future of the entire MDB was at risk, and local conflicts between different users paled into insignificance. Politicians realised that an integrated management system was needed to restore the Murray-Darling Basin and ensure reliable water supplies in the future.

2007 saw a major change with the National Plan for Water Security, and – after battles between Victoria, South Australia and Queensland over allocations – the Murray-Darling Plan began to be implemented in March 2008, with the new Murray-Darling Basin Authority (MDBA) in charge. The main points of the Plan are outlined on the right.

The future of the MDB

The sheer size of the MDB will make environmental recovery difficult, and it may take 50-100 years for the rivers to benefit fully from environmental flows, reduced salinity and secure access rights. However, the new policies mark a major shift in emphasis from 'taming and damming rivers', to balancing economic, social, environmental and political aspects, as shown below. From March 2008, the MDBA managers began attempting to calculate the true costs of water usage in the MDB – and also lay the foundations for a more sustainable future.

Secure access rights for all users
Meeting economic needs
Environmental flows
Managing biodiversity
Controlling erosion
Climate change?

Maintaining the quality of drinking water
Protecting the beauty of the landscape
Planting trees instead of crops – but which crops?
Importing food?

The future of the MDB – in the balance

The Murray-Darling Plan

> The system is completely empty, so we need to put a huge amount of water aside just to get back to normal.
> **Mike Young, Adelaide University**

The deal includes:

- giving the MDBA overall management control of the Basin
- a new lower cap on the amount of water extracted from the Basin
- provision for critical human needs – allowing South Australia to store water in upstream reservoirs to ensure that there is enough drinking water for Adelaide and other towns
- the continued setting of annual water allocations by states
- the national government to invest up to $1 billion in phase 2 of the Food Bowl Modernisation Project in Victoria to reduce water loss caused by inefficient irrigation techniques and leaks
- the first ever direct purchase of water for the environment by the national government

▲ *The main aspects of the March 2008 agreement*

Management decisions made today reflect the priorities of twenty-first century society

What do you think?

Over to you

1 Discuss the following points in class:
   a How does the size of a drainage basin determine the way that it is managed?
   b How might approaches at local, regional and national scales vary when considering the future management of the MDB?
2 a Complete a conflict matrix to evaluate the attitudes and values of the main stakeholders involved in the management of the MDB (farmers, domestic consumers, industrial consumers, recreation industry, fishing industry, environmental groups, state governments, national government).
   b Assume the role of one of the stakeholders and explain your views, what the nature of the conflict is and why it is difficult to resolve.
   c Who has the greatest claim on MDB water?

On your own

3 a What makes a river basin healthy?
   b Why do people have different views about this?
   c Will the idea of 'environmental flows' force Australians to rethink the way they use the MDB?
4 Can the MDB meet the needs of the present without undermining the prospects of future generations?

The price of water

In this unit you'll learn about how China's economic growth is putting its water supply at risk, and also about the privatisation of water supplies elsewhere.

China – economic growth at any cost?

As the almost-dry Yangtze River 'flowed' through Chongqing in 2007, five million people faced water shortages. Chongqing was not the only Chinese city to suffer in this way. Wang Yongli, a water conservation engineer, summed up the situation: 'We have a water shortage, but we have to develop – and development is going to be put first. Water in Shijiazhuang is as scarce as it is in Israel, but in Israel, people regard water as more important than life itself – In Shijiazhuang, it is not that way. People put the economy first.'

China is changing. The Communist Government there expected China's farmers to produce enough food to feed the entire population – 1.3 billion people (and growing). However, since the 1980s, economic change and rapid industrialisation has also meant rapid **urbanisation**. China's farmers can no longer keep pace with China's growing urban population and demand for food, so China has become a net importer of food – and world food prices have risen as a result.

As part of China's economic growth, some of the country's most productive farmland in the southern regions has been converted into massive industrial zones – which, in turn, has forced the northern regions to grow more food to compensate for it. Northern China, which receives far less precipitation than the south, now faces water deficits, its aquifers are dangerously low, and farms are short of irrigation water. On top of that, farmers around Beijing have been denied access to reservoirs, because all of the stored water is destined for industrial and residential purposes. The World Watch Institute estimates that 300 Chinese cities faced water shortages in 2008, and that groundwater in northern China is drying up.

Economic growth and the demand for water

The UN Food and Agriculture Organisation (FAO) reported in 2006 that China as a whole was running out of water, and that the current systems of rapid industrialisation and water supply were unsustainable. Alarmingly high rates of economic growth suggested problems ahead with the water supply. The bar chart shows how China's economic growth is drastically increasing or multiplying its need for water.

The same amount of water needed to generate $200 from farming, could generate $14 000 if used for industry. Jobs in industry are better paid than farming and, by producing more goods and having more money, people are encouraged to consume more. An increasing population drives economic growth still further and water consumption grows ever faster.

▲ The mighty Yangtze (no longer) flowing through Chongqing in 2007 (the Yangtze is the longest river in Asia and the third longest river in the world, after the Nile and the Amazon). China was suffering drought in 2007, but previous over extraction of water from the Yangtze to support China's economic development, meant that the drought's effects were hugely increased.

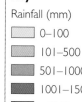

Key
Rainfall (mm)
- 0–100
- 101–500
- 501–1000
- 1001–1500
- >1500

Beijing
Shijiazhuang
Yellow R.
Chongqing
Yangtze R.
Shanghai
Three Gorges Dam

▲ A precipitation map for China

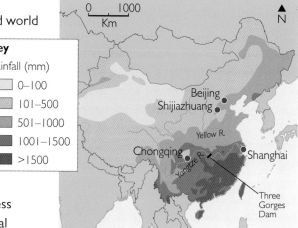

▲ Predicted growth in demand for water by sector by 2030, at 5% p.a. growth in GDP

The impact of economic growth

Not only are many of China's rivers drying up, but many have also become so polluted that they are ecologically dying. Major pollution incidents occur every day and, in some cases, the rivers are so polluted that they are 'no longer fit for human contact'. Wastewater and industrial effluent are regularly dumped into open rivers, and also contaminate groundwater stores. China's rapid industrialisation is outpacing environmental control measures, and its 20 000 riverside chemical plants, which frequently pollute the rivers with effluent, are a symbol of the way in which economic change is threatening environmental well-being:

- 80% of China's rivers no longer support fish.
- The Yellow River contains so much toxic waste and heavy metal that these poisons are entering the human food chain through the contaminated water.
- Arsenic, lead, cadmium and chromium have all been found in rice and vegetables grown using contaminated river and groundwater. They have been linked to developmental diseases, mental retardation and stunted growth in children.

There are other problems too:

- Farmers in the North China Plain are having to drill ever deeper for water, with the result that the Yongding, Yishui, Xia and Hutuo Rivers are all now dry.
- The Yellow River (China's second longest river after the Yangtze) now only flows for an average of 165 days each year.
- Northern cities, like Shijiazhuang, are also now beginning to industrialise, like those in the south, but it is at the expense of rivers and water sources (see top right).
- The Yangtze River (see bottom right) could be ecologically dead by 2010, and 186 riverside cities, including Shanghai, will not be able to depend on safe supplies of water.

Rainfall, and therefore water supply, is not evenly distributed across China – as the precipitation map shows. Increasing industrialisation and economic growth is creating an imbalance of water, which is creating more problems.

The technological fix

Beijing – in the drier north of China – was running out of water, and something had to be done. The massive Three Gorges Dam (the world's biggest) will now control flow on the Yangtze, store water and enable a surplus to build up to be transferred to the drier north – once the South-to-North Water Transfer Project has been constructed. Projects such as the Three Gorges Dam (see page 19) and the South-to-North Water Transfer Project are both controversial and expensive, but China's economic growth can fund them and they are seen as a way of solving the country's water supply problems and securing China's future economic prosperity – despite the environmental consequences of both projects.

Drying north
The example of the city of Shijiazhuang

- The water table is falling by 1.5 metres a year, and is now 200 metres deep.
- The city's population has grown from 300 000 in 1950 to 2.3 million in 2007.
- GDP in the city is growing at 10% per year.
- The amount of water available per person is 3% of the world average.
- Farmers are losing out to industrial and urban demands, which are increasing 8 times faster than agricultural demand.

Poisoned south
The example of the Yangtze River

- 23.4 billion tonnes of industrial waste from oil refineries, tanneries, paper mills, chemical plants and raw sewage are dumped in the river each year.
- Agricultural fertilizer/pesticide runoff delivers 1500 tonnes of nitrogen and 4.6 tonnes of arsenic to the coast every day – feeding the growth of blue-green algae and starving coastal waters of life-supporting oxygen.
- Shipping discharges from 25% of the world's container traffic enter the river.
- Industrial accidents go unrecorded and leak mercury among other poisons.
- Low environmental regulations permitted 90% of all discharges to be untreated in 2006.

Water, privatisation and the future

Privatising water supplies

Water has now gone global, because decisions about the availability of, and access to, water are increasingly made by TNCs rather than national governments. To the governments of MEDCs, passing the responsibility for modernising their ageing water infrastructures over to private water companies seemed like a good idea in the late twentieth century. The governments would save money – and charging people a commercial rate for the water they actually used, might reduce consumption.

Around 7% of the world's population currently receives its water from private companies. By 2015, this figure is expected to reach 17%. *Fortune* magazine declared that 'water could be to the twenty-first century what oil was to the twentieth'. The World Bank now actively encourages LEDCs to sell off their water systems to private companies in return for debt relief. For example, the IMF – along with the Inter-American Development Bank – claims that only private investment of $50 billion each year can guarantee supplies of fresh water for Latin American countries like Bolivia.

Water riots in Bolivia

However, water **privatisation** has not always worked out. The Bolivian city of Cochabamba became an important case study in the arguments about publicly and privately owned water supplies. Agua del Tunari, a subsidiary of the US-owned TNC Bechtel, took over Cochabamba's water system in 1999. The company then immediately raised water prices to pay for a major dam project at Misicuni – and to ensure a 16% profit margin. Cochabamba's poor faced a stark choice between paying 20% of their wages for water supplies or feeding their children. They took to the streets and rioted for four days; after 170 injuries and the

▲ *Rioters vent their fury against water privatisation during street riots in Cochabamba in 2000*

death of a 17 year old, the Bolivian government cancelled Agua del Tunari's contract. Anti-privatisation riots have since occurred in Peru, Panama, Brazil and Colombia as people saw the cost of their water increase by 20-100%.

Water – private or public supplies?

The Pacific Institute (an American think tank) claims that 'water is far too important to human health and the health of our environment to be placed entirely in private hands'. Social organisations, such as the FAO, Oxfam, and WaterAid, define water as a 'public good that should be managed by national governments', and claim that there is evidence that poorer communities are by-passed, access gaps widen and quality suffers if water is managed by private companies.

The table below shows the differences, since 2000, between privately run and publicly run water systems in Bolivia. Throughout Bolivia people continue to install their own pipes, dig 8-metre-deep wells, and develop a chaotic system to provide themselves with a basic human right – water.

La Paz and El Alto *Privately run by SUEZ of France as part of World Bank/ IMF conditions for debt relief*	Cochabamba *Publicly run since 2000 by SEMAPA, after the riots against privatisation*
• There is corporate efficiency.	• There is Government inefficiency.
• Coverage is 100%, so every street has a water pipe installed, although many houses cannot afford to get connected.	• The coverage of pipes to individual houses is 45%.
• The service area only extends to the city boundary.	• Connections have been increased by 16% (9000).
• Connections have been increased by 50% (78 000).	• Water supplies are only available for 2 hours a day, 3 days a week, so residents have to try to store water for use at other times.
• It costs the equivalent of $450 for a poor person to get connected (when wages average $17 a month), so less than 20% are connected.	• SEMAPA are unable to supply 55% of the population, who have to dig their own wells and beg for water.
• 200 000 people have been excluded since 2005 due to cost.	

▲ *Comparing private and public water supplies*

Water – the future

Global demand for water is rising, and – in the future – water stress and scarcity are likely to get worse across the world for three reasons:

- Continued population growth
- Increasing economic development
- Climate change

Demand for water already outstrips supply in many parts of the world and, as the table below shows, there are a number of factors which will influence demand and supply in the future. How nations respond to the future challenges of demand and supply may reflect one of the 3 scenarios:

- Business-as-usual – A firm belief that there will not be a water crisis and that no changes are needed.
- Technology, economics and privatisation – Accepting that problems exist and having faith in the market economy to solve them using technological fixes.
- Values and lifestyles – A radical shift in attitudes, requiring education, international cooperation and behavioural changes. The diagram below illustrates one way of providing additional water in both MEDCs and LEDCS – rainwater harvesting.

Factor	Effect
Public opposition to large dams	Fewer dams will be built if environmental and social opinions dominate policy decisions, which will lower the potential for water storage and increased supply.
Virtual water or local water	National food security issues may reduce the dependency on imported food/virtual water in favour of home grown produce.
Efficiency and waste reduction	For example, precise irrigation and leakage reduction, together with rainwater harvesting, would mean less need for major infrastructure projects, because supply would be increased and demand would be lowered without them.
Technological advances	For example, the development of cheaper desalination techniques, and the acceptance by the public of GM foods that need less water to grow them, would increase the supply of water and also lower demand.
Irrigation	Agriculture already uses 70% of all water worldwide for irrigation and this demand will increase as the population grows or food prices will rise as shortages occur.

▲ Water – factors affecting future demand and supply

Rainwater harvesting in MEDCs (top) and LEDCs ▶

Over to you

1 Discuss as a group whether it is possible to achieve economic development without sacrificing the quality of water supplies.
2 In pairs, draw up arguments for and against the privatisation of water in countries at different levels of development.

> Is water to the twenty-first century what oil was to the twentieth century?

What do you think?

On your own

3 Investigate and research China's South-to-North Water Transfer Project and assess its strengths and weaknesses.
4 Watch the video 'Bolivia for Sale' available at http://www.communitychannel.org/content/view/776/118/ and write a 300-word essay on why the poor may not benefit from privatization.
5 Consider the nature of the challenges posed by the three future scenarios outlined above, and comment on how each one might be achieved.

Chapter summary

What key words do I have to know?

There is no set list of words in the Specification that you must know. However, examiners will use some or all of the following words in the examinations, and would expect you to know them and use them in your answers. These words and phrases are explained either in the Glossary on pages 330-333, or within this chapter.

aquifer	precipitation	urbanisation
desalination	prevailing	virtual water
El Niño	privatisation	water pathways
geopolitical	rain shadow	water rights
groundwater	relief rainfall	water scarcity
high pressure	riparian	water stress
infiltration	salinity	water wars
irrigation	spatial imbalance	world water gap
La Niña	streamflow	
percolation	surface runoff	

Films, books and music on this theme

Music to listen to

'California dreamin' (1965) by The Mamas and the Papas

'Water' (2005) by Blue King Brown

Books to read

The Grapes of Wrath (1939) by John Steinbeck

The Great Thirst – Californians and Water (1992) by Norris Hundley

Films to see

'The Water Front' (2007) 'What if you lived by the largest body of fresh water in the world but could no longer afford to use it?' available at waterfrontmovie.com

Try these exam questions

1 Using named examples, assess the role of different players and decision makers in trying to secure a sustainable 'water future'. **(15)**

2 Referring to examples, assess the potential for water conflict in areas where demand exceeds supply. **(15)**

What do I have to know?

This chapter is about the pressures placed on biodiversity by economic development. Biodiversity is a key resource; it provides goods and services for people. However, people value these differently. When valued and exploited for its economic potential, biodiversity is threatened. Reconciling economic development and the need to maintain biodiversity is a challenge, together with the threats posed by climate change and alien species. The Specification has three parts, shown below, with the examples used in this book.

1 Defining biodiversity

What you need to learn	Examples used in this book
• Defining biodiversity.	• Ways of defining biodiversity – genetic, species and ecosystem diversity.
• Processes and factors influencing biodiversity, e.g. endemism, climate, and human activity.	• The processes that determine biodiversity, and their relative importance, e.g. in the tundra.
• The global distribution of biodiversity, with pivotal areas.	• Global patterns of biodiversity. • Threatened 'hotspots', e.g. the Daintree in Australia.
• The value of ecosystems in terms of biodiversity and ecological resources. • How ecosystems are not equally valued.	• The perceived value of biodiversity to different interest groups, e.g. the Daintree, Alaska's ANWR.

2 Biodiversity threats

What you need to learn	Examples used in this book
• Global distribution of areas under threat, e.g. hotspots, threatened species. • Biodiversity is threatened globally (e.g. climate change) and locally (e.g. by economic development). • Disrupting energy flow and nutrient cycles, e.g. by introducing alien species. • Links between economic development and ecosystem destruction.	• The Daintree (major study) • Mangroves in South-East Asia (major study) • Threats to the Caucasus, southwest Australia, Atlantic Forest (South America). • Deforestation in West Africa, Central America, Brazil and Queensland. • Nile perch (Lake Victoria), foxes and cats (southwest Australia) • Eco-regions, e.g. the Daintree, Sunda Shelf

3 Managing biodiversity

What you need to learn	Examples used in this book
• Sustainable yield and the 'safe' use of ecosystems. • The balance between conservation, management and development.	• Sustainable yields and mangroves – shrimp farming in Thailand versus sustainable uses, e.g. Yadfon project. • Balancing conservation and development in the Daintree.
• The role of different players in managing biodiversity and resolving conflict.	• Players in Alaska's ANWR, the Daintree, and mangroves in South-East Asia; why they often conflict.
• Ways of managing biodiversity, with their advantages and disadvantages.	• Costs and benefits of strategies, e.g. Millennium Development Goals, the Millennium Ecosystem Assessment biosphere reserves. • Sustainable management in mangroves and the Daintree.
• The future of biodiversity is uncertain and may involve difficult choices.	• MEA scenarios. • Options for the Daintree, ANWR, and mangrove development.

The great Alaskan wilderness

In this unit you'll learn about the Arctic National Wildlife Refuge in Alaska, and the threat facing its biodiversity and people's well-being.

The Arctic National Wildlife Refuge

The Arctic National Wildlife Refuge (ANWR) is a vast, wild, unspoilt and pristine place that has been called 'America's Serengeti', because it is the most biologically diverse Arctic region in the world.

The ANWR is in northern Alaska, part of the USA. It runs from the mountains of the Brooks Range north to the Arctic Ocean. Most of the refuge is designated as 'Wilderness', but part of it – the 1.5 million acres of Coastal Plain – is not protected in this way. It could be opened up for oil exploration and drilling, as Prudhoe Bay was in the 1960s and 1970s.

▲ 'America's Serengeti' has a diverse and thriving wildlife population

▲ The Arctic National Wildlife Refuge

A biological heartland

Scientists consider the Coastal Plain to be the 'biological heartland' of the ANWR. Nearly 200 species of wildlife are found there, including:

- caribou – a herd of over 120 000 animals travels hundreds of miles from Canada's Porcupine River region to the Coastal Plain (see above), where the females give birth in the spring and nurse their calves
- polar bears – in the autumn pregnant females look for den sites on the Coastal Plain in which to give birth and nurse their cubs
- marine mammals, such as the bowhead whale, which depend on the rich coastal waters off the ANWR's coast
- birds – during the brief summer over 135 species of birds gather on the Coastal Plain for breeding, nesting and migratory stop-overs (including snow geese, tundra swans, red-throated loons, snowy owls, eider ducks and scoters)
- grizzly bears, wolves, musk oxen, and many other species.

The North Slope stretches from the Chuck-chi Sea in the west to the Canadian border in the east – over 500 miles. Most of Alaska's oil development takes place on the North Slope, and the ANWR is at its eastern end.

Wilderness at stake

The Coastal Plain is the last 5% of Alaska's vast North Slope in which oil exploration and development have not yet been allowed. It is a fragile and vulnerable ecosystem. If disturbed by human activity, it could take many years to become re-established.

The tundra ecosystem

The ANWR lies in the Arctic tundra and has a polar climate.

Climate

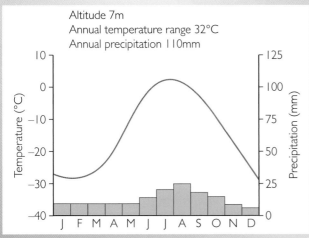

Altitude 7m
Annual temperature range 32°C
Annual precipitation 110mm

▲ A climate graph for Barrow, Alaska, in the Arctic tundra

- Winters are long, dark and cold. North of the Arctic Circle, darkness can last all day.
- Summers are short. The angle of the Sun is so low that temperatures struggle to get above freezing, and rarely reach 6 °C (the minimum temperature needed for continuous plant growth).
- Precipitation is low (the area is classified as cold desert).
- Winds are strong and cold, and cause severe wind chill.

Vegetation

Plants in the Arctic tundra have adapted to the extreme cold and lack of moisture.

- They are low-lying to avoid the strong winds.
- Most of them have small leaves to limit transpiration and short roots to avoid permafrost (see below).
- They have a short life cycle (from germination to producing seeds), which is adapted to the short available growing season.
- The dominant species are lichens, mosses, grasses, cushion plants, and low shrubs.

The tundra ecosystem has very low **organic productivity**.

> - **Organic productivity** or primary productivity is a measure of how quickly vegetation grows, i.e. the rate at which organic matter is produced.

Soils

Soils develop on **permafrost** (permanently frozen subsurface which is impermeable). Soils are frozen in winter and thaw slightly in summer. They are often waterlogged due to the low temperatures and low evaporation rates. In addition, they are shallow, infertile, acidic, peaty and have an accumulation of organic matter.

Background

What is biodiversity?

Biodiversity means biological diversity. It is the variety of all forms of life on Earth – plants, animals and micro-organisms. Biodiversity refers to:

- species (**species diversity**)
- variations within species (**genetic diversity**)
- interdependence within species (**ecosystem diversity**)

Geographers usually concentrate on species and ecosystem diversity.

Ecologists estimate that there are up to 30 million species worldwide, ranging from bacteria to animals and higher plants. However, as the pie chart shows, only 1.4 million have been identified so far. For example, in South America it is estimated that nearly half of all freshwater fish have still not been identified.

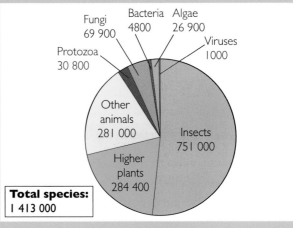

Fungi 69 900
Bacteria 4800
Algae 26 900
Viruses 1000
Protozoa 30 800
Other animals 281 000
Insects 751 000
Higher plants 284 400

Total species: 1 413 000

▲ Living species currently identified

Wilderness vs. development

Oil

Prudhoe Bay is located on the northern coast of Alaska, near the ANWR. Although oil was discovered there in 1968, production did not begin until 1977. Production peaked at about 2 million barrels a day in 1988, but has since declined. It is estimated that all the oil will be extracted by 2015, so a new source is needed.

The ANWR is about 19 million acres or 77 000 km² in size (nearly as big as Scotland). At its widest point, the Coastal Plain is about 100 miles across and about 30 miles deep. It consists of 1.5 million acres – and a lot of oil is believed to be located there.

Serengeti or wasteland? It all depends on your point of view. ▼

This is no Serengeti. The Coastal Plain is a frozen, barren land for nine months of the year. At its best, it is a flat, boggy, treeless place where temperatures can drop as low as -40 °C.
Arctic Power

Why does the USA want oil from the ANWR?

The USA imports vast amounts of oil – a million barrels a day come from Iraq, the rest from elsewhere. The USA wants to become less reliant on imported oil and increase its energy self-sufficiency. Views vary about the environmental value of the Coastal Plain.

Who are the players?

A wide range of people and organisations are interested in whether the Coastal Plain is developed or not. They include:

Politicians

There are two main political parties in the USA – Republicans and Democrats.

Republicans – generally support the development of the Coastal Plain for oil. However, Republican President George W. Bush (whose family was funded by oil) was not able to get enough votes in Congress to push through the legislation to allow this development.

Democrats – are to the 'left' of the Republicans in terms of economic and social matters. They oppose the development of the Coastal Plain.

Arctic Power

Arctic Power is a non-profit organisation with over 10 000 members. It was founded in 1992 to expedite congressional and presidential approval of oil exploration and production within the Coastal Plain of the ANWR. Its members come from a wide range of backgrounds and include miners, fishermen, tourism operators, banks, teachers, service industries, etc.

Alaskan residents

75% of the population supports developing the Coastal Plain. 75% also say they live in Alaska because of the state's pristine environment. They believe that a healthy environment and modern oil development are compatible, and that oil development has been carried out responsibly on the North Slope for nearly three decades. The oil has benefited both Alaska and the rest of the USA.

▲ The Inupiat people live on the North Slope. One village, Kaktovik (pictured), is the only community within the ANWR. The Inupiat strongly support onshore oil and gas exploration in the Coastal Plain.

The Inupiat

The Inupiat people of north Alaska believe that the oil industry around Prudhoe Bay has brought benefits, including jobs. They say: 'We live as our elders taught us, and rely on the land and resources for our physical, cultural and economic well-being. We have seen how development can exist alongside our natural resources and way of life. We believe we can only afford to keep most of our land as wilderness if we can develop a smaller area such as the Coastal Plain of the ANWR for oil'.

The Gwich'in

For thousands of years the Gwich'in people have relied on the Porcupine River Caribou Herd for food, clothing and tools. The Coastal Plain of the ANWR is the birthing place and nursing ground for the caribou – they migrate there every year. The Gwich'in call the Coastal Plain 'Iizhik Gwats'an Gwandaii Goodlit' meaning 'The Sacred Place Where Life Begins'.

The Gwich'in believe that the proposed oil and gas development in the Coastal Plain is a threat to the very heart of their people. Their village chiefs held a traditional gathering for the first time in over a century about the threat to the caribou birthing and nursing grounds. The Gwich'in believe that drilling for oil would violate their people's human rights, because of the impacts it would have on their culture and way of life.

▲ Gwich'in children, wearing traditional caribou hide clothes, dance during the 2001 Gwich'in gathering. The Gwich'in live 150 miles south of the Coastal Plain, but oppose its development.

Environmental groups

These include the Center for Biological Diversity. This is a non-profit conservation organisation with over 40 000 members. They believe in the value of diversity and are dedicated to the protection of endangered species and wild places. They work to protect biodiversity across the Alaskan Arctic, including the ANWR and off-shore marine waters.

Development – for and against

Arguments in favour of developing the ANWR

Many people support the development of the Coastal Plain. Some of their reasons are as follows:

The ANWR would only produce about 3% of the USA's daily consumption of oil. Is it worth going ahead with development?

What do you think ?

1 **Only 8% of the total ANWR is being considered for exploration** Only the 1.5 million acres of the Coastal Plain is being considered for development. If oil is discovered there, only 2000 acres would be directly affected by drilling.

2 **The economic impact** Alaska, and the USA as a whole, would benefit from oilfield development as a result of, for example, royalties, taxes, and lease rentals (rent to allow drilling).

3 **Jobs** Between 250 000 and 735 000 jobs would be created. The jobs would be all over the USA and not just in Alaska.

4 **It is the USA's best chance for a major new discovery of oil and gas** It is estimated that there are between 9 and 16 billion barrels of oil under the Coastal Plain. 16 billion barrels equates to 30 years' worth of the USA's imports from the Middle East.

5 **Imported oil is too expensive** In 2007, the USA imported an average of 60% of its oil at enormous expense.

6 **Oil production in the North Slope oilfields is in decline** The North Slope oilfields provide the USA with nearly 16% of its domestic oil production, but this has been declining since 1988 and is now only around 700 000 barrels a day.

7 **There would be no negative impacts on wildlife** Oil and gas development and wildlife have successfully co-existed in Alaska's Arctic. For example, the Central Arctic Caribou Herd, which migrates through Prudhoe Bay, has grown from 3000 to 32 000 animals since the 1970s.

8 **Advances in Arctic technology** These have reduced the 'footprint' of Arctic oil development. (The footprint is the surface area occupied by development.) If the Prudhoe Bay oilfield was built today, its footprint would be 64% smaller.

9 **Alaskan support** More than 75% of Alaskans favour exploration and production in the ANWR.

Arguments against developing the ANWR

Many people are also against developing the Coastal Plain. Two of their arguments are as follows:

The impact on biodiversity
- Caribou – The US Fish and Wildlife Service concluded that oil development in the Coastal Plain would cause a major decline or displacement of the Porcupine River Caribou Herd. The herd would be displaced to areas where there are more predators, poorer forage and more mosquitoes (which can drain over 2 pints of blood a week from caribou).
- Polar bears – 'Denning' polar bears are very sensitive to industrial activity. Females may abandon their dens if disturbed, which can be fatal for cubs.
- Birds – Some of the birds which gather on the Coastal Plain are highly sensitive to human disturbance. Snow geese, for example, are disturbed by helicopters and aircraft up to 4 miles away.

▲ *Snow geese. The US Interior Department says that human disturbance, along with the destruction of feeding areas, could prevent the birds from building up the fat reserves needed to migrate, thus threatening their survival.*

The impact on the environment

- At Prudhoe Bay, there are mountains of sewage sludge, scrap metal, rubbish, and over 60 waste sites that leak acids, pesticides, solvents and diesel. It is a disaster for the environment. It could be the same on the Coastal Plain, which could be turned into a vast landscape of unsightly derricks, roads and pipelines.
- There is a risk of oil spills from pipelines and oil tankers. The *Exxon Valdez* oil spill in 1989 was an environmental disaster caused solely by human error which affected the entire ecosystem, and there are real fears of further major oil spills.

▲ The Exxon Valdez *supertanker spilled 11 million gallons of crude oil into the sea off south Alaska, killing a lot of important wildlife like this grey whale. The effects of the spill can still be seen today. It can take 50-60 years for Arctic vegetation to recover from oil contamination.*

What is being done?

Protecting biodiversity

The Center for Biological Diversity has:

- drawn attention to the plight of polar bears affected by plans to open the Coastal Plain for oil drilling. Listing polar bears as an endangered species is expected to be used by environmental groups to delay, or block, oil and gas development on the Coastal Plain.
- worked to designate critical habitat for the endangered bowhead whale
- secured the designation of more than 26 million acres of protected ocean and shoreline for the threatened spectacled and Steller's eider ducks.

'Locking-up' the Coastal Plain

This means designating the Coastal Plain as a wilderness, which would permanently protect the area from oil and gas development, while continuing to allow vital subsistence uses. Legislation is attempted every year to designate the Coastal Plain as wilderness, but fails due to lack of votes.

The future

The November 2008 federal elections in the USA were crucial. President Barack Obama and his new Democrat administration are now likely to begin the process of permanently 'locking up' the Coastal Plain.

However, the USA is also likely to increase its oil imports dramatically in the next few years. If there is any future world oil shortage, it may have to reconsider locking up the Coastal Plain – because it simply cannot do without oil.

Over to you

1. **a** Use the information in this unit to draw and complete a conflict matrix to show different groups who are likely to agree/disagree over the development of the ANWR.
   b Choose two conflicting groups and explain why they are in conflict over the development.
2. Complete the table below to assess the short, medium- and long-term impacts of developing the ANWR.

Impact	Short-term	Medium-term	Long-term
Economic			
Social			
Environmental			

3. Debate the statement: 'America should be developing energy efficient technology and alternative forms of energy, instead of relying on oil.'

On your own

4. Research to find out more about the issue of oil in the ANWR. Find out:
   - when the ANWR was created
   - why the Coastal Plain was not classified as 'wilderness'
   - any other arguments for and against drilling for oil (watch out for bias).
5. Use the information in this unit, and your research in activity 4, to answer the question: 'How far do the risks to biodiversity and human well-being outweigh the benefits to be gained from developing oil exploration in the ANWR?'

In this unit you'll find out about the variations in, and distribution of, biodiversity and what influences it.

Giant rat found in 'lost world'

December 2007

A giant rodent five times the size of a common rat has been discovered in the mountainous jungles of New Guinea. The 1.4 kg Mallomeys giant rat was found on a visit to the Foja Mountains, part of the Mamberamo Basin, the largest pristine tropical forest in the Asia Pacific region.

'With no fear of humans, the giant rat apparently came into the camp several times during the trip' said Kristofer Helgen, a scientist with the Smithsonian Institution in Washington D.C.

In 2005, the area visited was described as a 'lost world' after scientists discovered dozens of new plants and animals in the dense jungle.

Global variations in biodiversity

It was perhaps no surprise that the giant rat was found in the tropical rainforests of New Guinea. Biodiversity is greatest in the tropics and declines towards the poles.

- Tropical rainforests contain over 50% of the world's species in just over 7% of the world's land area.
- They also account for 80% of all insects and 90% of primates.
- Brazil, Indonesia and Madagascar contain over 55% of the world's mammals.
- Tropical America has about 85 000 species of flowering plants, compared with about 11 300 in Europe.

The diagram below shows the global variation in species 'richness' by latitude for terrestrial mammals, amphibians and threatened birds (there is no data available for global bird species). Some of the pattern shown can be explained by variations in land mass (where there is a large land area there are more species), but, even so, species richness is much higher in the tropics than you would expect if it was based just on land area.

▼ *Global variations in species richness by latitude*

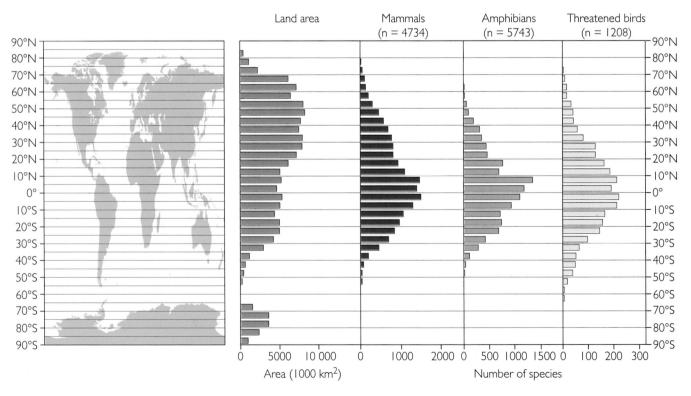

Biodiversity and biomes

Biomes are large global ecosystems. Although they get their names from the dominant type of vegetation found within them, such as tropical rainforest, each biome contains communities of plants and animals and can be linked to soil types.

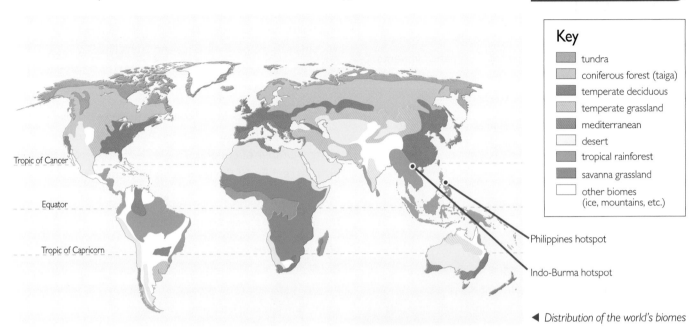

Key

- tundra
- coniferous forest (taiga)
- temperate deciduous
- temperate grassland
- mediterranean
- desert
- tropical rainforest
- savanna grassland
- other biomes (ice, mountains, etc.)

Tropic of Cancer

Equator

Tropic of Capricorn

Philippines hotspot

Indo-Burma hotspot

◀ *Distribution of the world's biomes*

Although the map above shows distinct areas of, for example, tropical rainforest, biomes are not uniform in terms of their biodiversity. There are variations both within and between the same type of biome, so, for example, tropical rainforests are not the same everywhere – they vary according to their geographical location. Even within one area, e.g. the rainforests of North Queensland in Australia, biodiversity varies with some areas much richer than others.

Areas with high concentrations of biodiversity are known as **biodiversity hotspots** (also see pages 104-105). **Pivotal areas** are areas with concentrations of hotspots – for example in South-East Asia (see the map on page 105). They are regarded as very important because of their sheer variety of species.

How productive are biomes?

Not only do tropical rainforests have the highest biodiversity, they are also the most productive biome. **Net primary productivity (NPP)** is a measure of how quickly vegetation grows, i.e. the rate at which organic matter is produced. Tropical rainforests produce the greatest amount of organic matter, due to their large **biomass** – resulting from constant high temperatures, heavy rainfall and a year-round growing season.

Biodiversity hotspot – Indo-Burma

The Indo-Burma hotspot covers more than 2 million km² of tropical Asia. Six large mammal species have been discovered there in the last 12 years alone. The area has high numbers of endemic freshwater turtle species, and almost 1300 different bird species.

Biodiversity hotspot – the Philippines

More than 7100 islands make up the Philippines hotspot. It has been identified as one of the world's most biologically rich countries. Many endemic species are confined to forest fragments that cover only 7% of the original extent of the hotspot. This includes over 6000 plant species and many birds. Amphibian endemism is unusually high and includes species such as the panther flying frog.

● **Net primary productivity energy** is calculated as the energy fixed by photosynthesis minus that lost through respiration. It is measured in kilograms per square metre per year.

● **Biomass** is the total amount of organic matter in a given area.

What influences biodiversity?

Background

A range of factors and processes influence biodiversity. They include climate, endemism and human activity.

Climate

Climate is the main factor controlling the distribution of biomes and, therefore, biodiversity. Climatic elements include precipitation, temperature, light intensity, and winds.

Precipitation The influence of precipitation depends on a number of factors. For example:

- whether it is seasonal
- whether rainfall is reliable
- the type of rainfall (e.g. steady rain or short, heavy bursts of rain)
- whether rain falls in the growing season

Temperature has a major influence on vegetation. Plants begin their growth at about 6°C, photosynthesise more effectively above 10°C, and only begin to suffer stress above 35°C.

Light intensity affects photosynthesis. Tropical ecosystems receive most incoming radiation and have higher energy inputs than ecosystems nearer the poles.

Winds increase the rate of evapotranspiration and the wind-chill factor.

Endemism

In terms of biodiversity, if something is **endemic** it is unique to one place or region and not found naturally anywhere else. Oceanic islands are particularly rich in endemics, because their flora and fauna have evolved in isolation from neighbouring land masses. For example, over 90% of Hawaii's plants are not found anywhere else.

Levels of endemism in Australia are particularly high. Compared with other countries, Australia not only has a large number of endemic species but these also form a high percentage of the total number of species.

The diagram below compares a number of biomes in terms of their species richness and endemic species.

Human activity

Biodiversity is under attack from human activity. Human activities are changing ecosystem structures, fragmenting them, and altering land use. For example:

- habitat change (including loss and degradation)
- over-exploitation
- the introduction of invasive species
- pollution
- climate change

Unit 3.3 looks at these in more detail.

▼ Comparing biomes

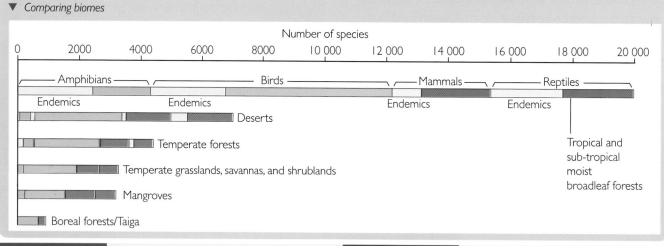

Over to you

1 Work in pairs to produce a PowerPoint presentation on the global variations in biodiversity. You need to refer to climate, endemism and human activity.
2 Discuss in class: What makes a biodiversity hotspot? Why are they important?

On your own

3 Select one of the biomes above. Using atlas data, research **a** its climate characteristics, **b** factors which affect Net Primary Productivity, **c** specific human impacts which are taking place on that biome.

3.3 Biodiversity under threat – 1

In this unit you'll investigate some of the threats to biodiversity.

Dead as a dodo

No photos exist of the dodo, because it was hunted to extinction in the late seventeenth century. The dodo's diet included the seeds of the Calvaria major tree, which is only found on Mauritius. As the seed passed through the dodo, its tough seed coat was abraded and roughened – necessary for its germination – before being excreted. There are now no animals or birds left on Mauritius which do this job, so no new seeds germinate and only a few very old trees survive. The Calvaria major now also faces extinction.

▲ *The dodo weighed 25 kg*

What are the threats to biodiversity?

Extinction is a natural event. However, as a result of human activity, species and ecosystems are threatened with destruction on a scale rarely seen before. Up to 60 000 plant species may become extinct by 2050 if present trends continue.

An increasing global population, rising consumption and economic activity, and globalisation are all contributing to biodiversity loss. The main threats, to biodiversity are: habitat change, climate change, the introduction of **alien species**, over-exploitation, and pollution.

> ● **Alien species** are those which are not native to an area but have been introduced, almost always by human activity.

The diagram below shows the impacts of these threats or drivers on biodiversity in seven ecosystems over the last century, plus their current trends.

Habitat change

The conversion of land for agricultural use has been the main cause of habitat change. Cultivation (including land used for crops and livestock, shifting cultivation, and freshwater aquaculture, i.e. fish farming) now covers 25% of the Earth's surface, and a further 10-20% of grassland and forest is projected to be converted to agriculture by 2050. This conversion is leading to a major loss of habitat for huge numbers of species, resulting in a loss of biodiversity.

The area of land covered by tropical rainforests has already halved, and the current rate of loss is alarming. In the minute it took you to read about habitat change, approximately 100 acres of rainforest were destroyed. Further loss will have a major impact on biodiversity.

Over-exploitation

In terms of marine biodiversity, over-fishing twinned with more-efficient fishing technology (such as the use of drag nets), is the main threat. Demand for fish as food for people and as a feed for aquaculture is increasing, resulting in an increased risk of major collapses of fish stocks.

Over-hunting of animals has been a significant cause of the extinction of hundreds of species, and has caused others to become endangered (such as the tiger).

▼ *Global trends of the threats to biodiversity*

	Habitat change	Climate change	Invasive species	Over-exploitation	Pollution (nitrogen, phosphorous)
Temperate forest	↘	↑	↑	→	↑
Tropical rainforest	↑	↑	↑	↗	↑
Temperate grassland	↗	↑	→	→	↑
Desert	→	↑	→	→	↑
Inland water	↑	↑	↑	→	↑
Marine	↑	↑	→	↗	↑
Polar	↗	↑	→	↗	↑

Key

Driver's impact on biodiversity over the last century
- ☐ Low
- ▨ Moderate
- ▧ High
- ▩ Very high

Driver's current trends
- ↘ Decreasing impact
- → Continuing impact
- ↗ Increasing impact
- ↑ Very rapid increasing impact

Biodiversity under threat - 2

Invasive species

The invasion of non-native species has been responsible for at least half the extinctions since 1600, and islands have suffered the most. The loss or addition of one species can affect the entire ecosystem.

Lake Victoria in Africa has large numbers of endemic species. The Nile Perch was introduced into Lake Victoria to increase the viability of commercial fishing. But in terms of biodiversity, it has been a disaster. By consuming many native species, it has wrought havoc on the whole ecosystem, and many species are now extinct.

Pollution

Although pollution may not cause mass extinction, locally its impacts can be devastating. For example, the use of nitrogen fertilisers in farming has increased dramatically over the last 50 years. Runoff into rivers and wetlands leads to increased algal blooms and **eutrophication** in inland waters and coastal areas. If the nitrogen ends up in temperate grasslands, shrublands and forests it leads directly to lower plant diversity.

> ● **Eutrophication** is the process by which fertiliser causes rapid algal and plant growth and the depletion of oxygen available for fish and other aquatic species.

Climate change

Recent changes in climate have already had significant impacts on biodiversity and ecosystems, including changes to species distributions, population sizes, the timing of reproduction or migration, and an increase in the frequency of pest and disease outbreaks. Many coral reefs have undergone major **coral bleaching** episodes, due to rises in sea temperatures.

The risk of floods, droughts and rises in sea level are other aspects of climate change that will impact on biodiversity. By the end of this century, climate change and its impacts may be the dominant driver of biodiversity loss.

> *Why worry about a few species becoming extinct when there are millions more waiting to be discovered?*
>
> **What do you think ?**

Hotspots under threat

Conservation International is based in Washington D.C. This organisation recognises a total of 34 biodiversity hotspots around the world, as shown on the map. They are areas of mind-boggling species richness, which are under constant assault from human activity. About half of all plant and animal species on Earth are found in these hotspots, which originally covered 15.7% of the Earth's surface. Only about 10% of the original habitat remains.

The Caucasus

The deserts, savannas, arid woodlands and forests that comprise the Caucasus hotspot contain a large number of endemic plant species. This hotspot is home to the two species of threatened Caucasian turs, or mountain goats.

▲ *The Caucasian tur is endemic to the high mountains of the Caucasian hotspot*

Threats

● Recent economic and political crises in the region are intensifying forest clearances for fuelwood.
● Illegal hunting and plant collecting further threaten biodiversity in this region.
● The lower plains are experiencing the greatest destruction.

Endemic plant species	1600
Endemic threatened birds	0
Endemic threatened mammals	2
Endemic threatened amphibians	2
Extinct species	0

▲ *Biodiversity in the Caucasus*

Southwest Australia

The forest, woodlands, shrublands and heath of southwest Australia are characterised by high numbers of endemic plant and reptile species. The western swamp turtle, which hibernates for nearly 8 months of the year, is one of the most threatened freshwater turtle species in the world.

Threats

- Agricultural expansion and high levels of fertiliser use have been the main cause of habitat loss.
- The introduction of invasive alien species, such as foxes and cats, are a major threat to native fauna.

Endemic plant species	2948
Endemic threatened birds	3
Endemic threatened mammals	6
Endemic threatened amphibians	3
Extinct species	2

▲ Biodiversity in Southwest Australia

Atlantic Forest

The Atlantic Forest has over 20 000 plant species and 950 kinds of birds, yet less than 10% of the original forest remains. Over 24 critically endangered vertebrate species are clinging to survival in this region.

Threats

- Habitat loss has been caused by the development and expansion of sugarcane, and later coffee plantations. The region has been losing habitat for hundreds of years.
- Urbanisation due to the rapid expansion of Rio de Janeiro and Sao Paulo.

Endemic plant species	8000
Endemic threatened birds	55
Endemic threatened mammals	21
Endemic threatened amphibians	14
Extinct species	1

▲ Biodiversity in the Atlantic Forest

◀ Biodiversity hotspots (the orange areas). The most remarkable places on Earth are also the most threatened.

Over to you

1 Look at the diagram on the first page of the unit. Which are the main threats to biodiversity? Why are they increasing so rapidly?

2 How far do you agree with the statement that 'By the end of this century, climate change and its impacts may be the dominant driver of biodiversity loss'?

On your own

3 Research a threat to a named ecosystem e.g. pollution. Find out if the threat is increasing and what is being done about it.

4 Write a report of 350 words on hotspots under threat. Describe the distribution of threatened hotspots and the threats that they face.

The Daintree rainforest

In this unit you'll learn about the Daintree rainforest, the tropical rainforest ecosystem and ecosystem services.

Cassowary crisis

The cassowary is a large bird – 1.75 metres tall on average. It is related to the emu and cannot fly. You are unlikely to meet a southern cassowary unless you go to the tropical rainforests of north-east Queensland, Australia. In 1993 there were as few as 54 birds left in the Daintree rainforest but, since they became protected in 1999, their numbers have increased and there are now estimated to be 500.

The cassowary is vital to the rainforest ecosystem. It acts as a 'seed disperser' for over 100 species of rainforest plants with large fruits. Without the cassowary, new plants would concentrate around the parent plant and would not spread throughout the rainforest.

The Wet Tropics

The Daintree rainforest is in northern Queensland, Australia. It is part of a huge stretch of rainforest known as the Wet Tropics, which runs parallel to the Queensland coast. This coast runs alongside the Great Barrier Reef.

The Wet Tropics region has the highest levels of biodiversity and regional endemism in Australia. The Wet Tropics and Great Barrier Reef are together a major biodiversity hotspot.

The Wet Tropics became a World Heritage Site in 1988. It is one of a handful of places in the world that meet all four criteria for World Heritage listing:

- as an outstanding example of the major stages in the Earth's evolutionary history
- as an outstanding example of significant ongoing ecological and biological processes
- as an example of superlative natural phenomena
- containing significant habitats for the conservation of natural biodiversity.

The Wet Tropics covers almost 900 000 hectares (nearly half the size of Wales) and is adjacent to another World Heritage Site – the Great Barrier Reef – which was listed in 1981. The reef contains 400 types of coral, 1500 species of fish, and 4000 types of mollusc. The combination of reef, a spectacular scenic coast, and rainforest is unique to this area.

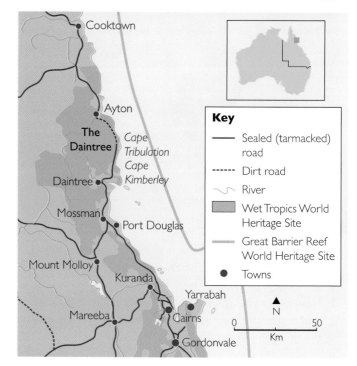

▲ *Cassowaries are able to kill people with their razor-like claws. These birds have become a symbol of the Daintree rainforest. Already endangered because of rainforest clearance throughout Queensland, tourist traffic now increases the risks they face.*

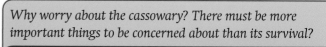

Why worry about the cassowary? There must be more important things to be concerned about than its survival?

What do you think ?

▲ *Two World Heritage Sites*

Why is the Daintree rainforest so special?

The Daintree rainforest, at 135 million years old, is the oldest rainforest in the world. There are plant and animal species living there which are older than human life itself. Australia has less than one thousandth of the world's tropical rainforests, but they are some of the most fragile and important ecosystems in the world.

The Daintree rainforest:

- is home to the greatest number of plant and animal species which are rare, or threatened with extinction, anywhere in the world.
- is home to one of the highest populations of primitive flowering plants in the world (of 19 primitive plant families on Earth, 12 are found here).

- contains over 3500 vascular plants.
- has Australia's largest range of ferns.
- has the highest number of endemic mammals of any region in Australia.
- has almost half of Australia's bird species, and 13 species found nowhere else in the world.
- has nearly a quarter of Australia's frog species, more than 20 of which are endemic.
- has a greater diversity of freshwater fish than any other region in Australia.
- has over 65% of Australia's butterfly and bat species.
- has 28 of Australia's 36 mangrove species.

Not bad for an area that takes up less than 0.2% of the landmass of Australia!

▲ The idiot fruit – one of the most primitive flowering plants. It is found only in the Daintree. In 1970, four cattle belonging to a local farmer were unexpectedly found dead. Autopsies showed the remains of large seeds in their stomachs. The seeds – from the idiot fruit – killed the cattle by producing a poison similar to strychnine.

▲ The Ulysses butterfly is an icon of tropical northern Australia. Its spectacular iridescent metallic blue wings can be seen from a great distance.

The tropical rainforest ecosystem

Climate

Tropical rainforests are found in places with an equatorial climate, with:

- a low daily range of temperature. Temperatures rarely drop below 22 °C at night, or go above 32 °C during the day.
- a low annual temperature range.
- high annual rainfall (over 2000 mm) in intense convectional storms. The Daintree has about 120 days of rain per year.
- high humidity.
- a year-round growing season.

Vegetation

- The trees are deciduous, but the rainforest looks evergreen because the year-round growing season means that they can shed their leaves at any time.
- The vegetation grows in distinct layers. The tallest 'emergents' can grow up to 50 metres in height.
- Only about 1% of sunlight reaches the forest floor. Shrubs and other plants are adapted to the lack of light.
- There are as many as 200 species of tree per hectare (the size of a rugby pitch).
- Rainforests are the most productive terrestrial ecosystem (see page 102).
- Mangroves fringe the coast of the Daintree rainforest.

Soils

- The soils have a thick litter layer (due to the continuous leaf fall), but thin humus layer as a result of rapid decomposition.
- The soils are red in colour, due to the accumulation of iron and aluminium.
- There is rapid **leaching** of nutrients.
- The bedrock is intensely weathered due to the hot, wet, conditions.

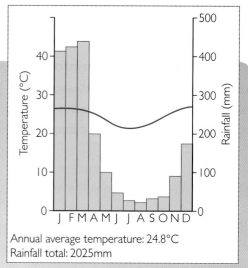

Annual average temperature: 24.8°C
Rainfall total: 2025mm

▲ *A climate graph for northern Queensland*

▼ *Vegetation layers in a tropical rainforest*

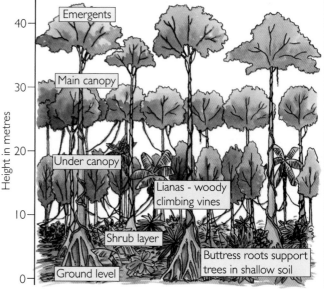

- **Leaching** is the downward movement, and often loss, of nutrients (minerals) in solution in the soil.

Background

Ecosystem services

Ecosystem services are the benefits that people get from ecosystems. They can be classified into four groups:

- **Provisioning services** These are the *products* obtained from ecosystems, including things like: food, fibre (e.g. wood, cotton, etc.), fuel, genetic resources, natural medicines and pharmaceuticals, fresh water.
- **Regulating services** These are the *benefits* obtained from the regulation of ecosystem processes, e.g. air quality, climate, water, erosion and natural hazard regulation – along with water purification and waste treatment.
- **Cultural services** These are *non-material benefits* that people obtain from ecosystems, e.g. spiritual and religious values, knowledge systems, cultural heritage values, recreation and ecotourism.
- **Supporting services** are those that are necessary for the production of all other ecosystem services. Their impacts on people are indirect, or occur over a very long time. They include soil formation, photosynthesis, nutrient cycling and water cycling.

The Daintree and ecosystem services

The Daintree, like all rainforest ecosystems, provides a range of benefits, as the diagram shows. Ecosystem services are a key factor in determining human well-being.

Medicine
25% of drugs include products that come from rainforests. Tropical rainforests have provided chemicals used to treat diabetes, malaria, heart conditions, rheumatism and arthritis. In the Daintree, many tropical plants have been identified as having anti-cancer properties.

Carbon sequestration
Trees absorb carbon dioxide and give out oxygen, thus removing greenhouse gases from the air. The carbon dioxide is stored in the leaves, branches, roots and stems of plants – helping to reduce pollution and regulate climate.

Tourism
The Daintree attracts nearly half a million visitors a year, both from home and abroad. Visitors come for the scenery – the unique combination of coast, rainforest and mountains – the biodiversity in terms of the huge range of plant and animal species in the Daintree, and to visit the Great Barrier Reef.

People of the rainforest
Rainforest aboriginal people are the original owners of the Wet Tropic rainforests. To them, the rainforest is a series of complex 'living' cultural landscapes. This means that the natural features of the rainforest are interwoven with the peoples' religion, spirituality, economic use (food, medicines, tools) and social and moral organisation.

Logging
The commercial timber industry began in the Daintree in the 1930s. During the 1980s, the Daintree rainforest was at the centre of arguments between conservationists and the timber industry. The conservationists argued that continued logging of the rainforest was unsustainable.

Over to you

1 Endemism and climate have both influenced biodiversity in the Daintree rainforest. Which has been the more important? Justify your answer.
2 a Work in three groups. Each group needs to draw a spider diagram on one of the following:
   • How a tourist perceives the Daintree.
   • How an economist perceives the Daintree.
   • How an environmentalist perceives the Daintree.
   b Feed back your results to the rest of the class.
   c Complete a conflict matrix to show those issues that people would agree or disagree about.

On your own

3 a Classify the following ways in which ecosystems can benefit people under 'provisioning', 'regulating', 'cultural' and 'supporting': timber logging, tourism, medical treatments from rainforest plants, absorption of CO_2 from the atmosphere, traditional aboriginal lifestyles.
   b Identify ways in which **each** of these types of benefit is under threat from human activity.
4 Choose one regulating service. Find out how changes to a tropical rainforest affect the way the regulating service works. Write a 300-word report on your findings.

3.5 Threats to the Daintree

In this unit you'll find out about the threats to the Daintree rainforest ecosystem, and nutrient cycles.

Daintree and tourism

The Daintree rainforest is so old that it has survived global climate changes, rising and falling sea levels and cyclones. However, the most dangerous threat facing it now is tourism – and tourism is big business in the Daintree. The Daintree region is situated within the Far North Queensland tourism region in Douglas Shire. It is approximately 110 km north of Cairns by road, and 40 km from Mossman and Port Douglas.

A survey carried out for the Australian Tropical Research Foundation in 2002, found that tourism and recreation in the Daintree region (the part of the Daintree National Park north of the Daintree River to Cape Tribulation) was worth A$141.7 million a year. Tourism and recreation has created about 3500 jobs.

▲ A twenty first century portal to a 135 million year old world

Why do people come?

The table shows that most people come to the Daintree for the scenery, the views and the rainforest itself.

Reason for visit	% of visitors
Scenery/views	89%
Rainforest	87%
Cape Tribulation	80%
The fact it is a World Heritage Area	49%
Wildlife	30%
Getting away from people	27%
Remoteness/isolation	20%
Four-wheel drive (4WD) experience	8%
On the way to Cooktown/Cape York	8%
Other	14%

▲ Reasons for visiting the Daintree

Once they are in the Daintree, most people's activities centre on the rainforest and the coast, because the Great Barrier Reef sits alongside this coast. Activities include:

- bush walking
- four-wheel drive tours
- river cruises
- horse riding
- fishing
- reef diving
- snorkelling

An underwater view of the Great Barrier Reef

How many tourists are there?

Tourism is the fastest growing industry in the world, and the number of visitors to Australia and the Daintree is growing rapidly.

	Australia	Queensland	Far North Queensland tourism region
International tourists	4.9 million	1.95 million	0.8 million
Domestic tourists	72.9 million	16.3 million	1.4 million

▲ Tourist numbers in 2004

In 1983, about 17 000 tourists visited the Daintree. By 2002, this had grown to 436 000 tourists a year. 306 000 of these were day visitors, and 130 000 were overnight visitors staying 3.8 days on average. The number of overnight visitors has been increasing at a faster rate than the number of day visitors. This is reflected in the construction of accommodation, particularly in Cape Tribulation, since 1991. By 2004, visitor numbers were up to 500 000. The table above shows the number of international and domestic tourists for the whole of Australia, Queensland and the Far North Queensland tourism region.

Tourism infrastructure

70% of tourists travel to the Daintree independently. The other 30% come on organised tours, travelling by coach. 99% of visitors cross the Daintree River by ferry to enter the Daintree.

- The ferry carries 21 vehicles and operates from 6.00 am to midnight every day.
- In 2008, it cost A$20 return.
- Traffic peaks from April to September.

The road linking the ferry to Cape Tribulation has been tarmacked, which has increased the number of visitors (approximately 40% of the total visitors) accessing the area with either a hire car or four-wheel drive (4WD). North of Cape Tribulation, the road is not tarmacked, is only accessible using 4WD, and is called the Bloomfield Track (see below).

▲ The Daintree ferry carries, on average, about 700 vehicles a day

Accommodation is available for tourists at specific places within the Daintree.

	Daintree	Douglas Shire
Number of ccomodation establishments (including B&B)	20	58
Number of beds	1278	8822
Number of camping pitches	176	770

▲ Accommodation in the Daintree, 2000

- In 1999, 13% of visitors thought that there was too much accommodation in the region. This increased to 21% in 2001.
- The percentage of visitors staying in four- and five-star accommodation in Cape Tribulation has decreased since 1999, while the number of visitors staying in three- to four-star and budget accommodation has increased.
- 34% of visitors were under 34 years old in 2001; Cape Tribulation is well known as a destination for backpackers.
- In 1999, 37% of visitors thought there was a need for more walking tracks in the Daintree. This had increased to 40% of visitors by 2001.

Daintree and development

Growth and change in Port Douglas

Port Douglas is close to the Daintree and has been affected by the increasing numbers of tourists to the region. Its population is only about 4000 but, as visitor numbers have grown, the village has changed and this has affected its character. The type of accommodation in the village has changed and grown rapidly, as the table shows.

Increasing numbers of visitors have caused other changes:
- A large supermarket was built in the centre of the village in 1999. Small local shops said that people no longer shopped with them.
- A property boom led to rising house prices. Some people benefited from selling their land to developers. Others find it difficult to buy a house or an apartment.
- There are increasing numbers of proposals to build new resort complexes. In 2008, five new resorts were either being built, or were planned.

There is a real fear that development will spread, and some people worry that it is only a matter of time before hotels, restaurants and shops spread further into the Daintree.

▲ A new hotel development in Port Douglas. One of the planning requirements is that developments should be no higher than coconut palms.

	December 1995	December 1999	December 2007
Apartment	568	994	1973
Hotel	782	1121	639
Motel	83	83	59

▲ Accommodation in Port Douglas. The figures show the number of bedrooms available.

The Daintree

The Daintree is incredibly sparsely populated. There are approximately 350 dwellings there, and it was estimated that, by 2000, there were 550 residents. Tarmacking the road from the ferry to Cape Tribulation opened up the region, and small areas of forest were divided into over 1000 plots for sale. Some have been occupied and the owners have been sensitive to environmental concerns. But others have been bulldozed and turned into cattle ranches. One species – a member of the red cedar family of trees – became extinct in 2000 when a landowner cleared a plot of land. If land clearing is not reduced, 85 rare plant species now on private land will also become extinct.

▲ The planned site for a new McDonalds restaurant in Port Douglas. Having given permission in 2000 for the McDonalds to be built, the local council changed its mind because it was concerned about the level of change and development in the town.

However, there is some good news. There are three main limits to development:
- The ferry crossing the Daintree River limits traffic levels (and therefore visitor and population levels).
- There is no mains electricity north of the river. People living there have to use a Remote Area Power System (or RAPS), which means that they have to provide their own generator or solar-powered system.
- Local services can only support a small local population. There is no mains water, or sewerage disposal system, and few shops and services.

Deforestation

Habitat change, especially the conversion of forest to cultivated land for agriculture, began in China about 4000 years ago. But it was largely during the twentieth century that the tropical rainforests came under attack. Tropical rainforest originally covered 15 million km². About half of that has now gone, and the chances are that by the end of this century they will be reduced to 10-25% of their original extent. The map below shows changes in global forest cover between 1980 and 2000.

Central America

Central America once had about 500 000 km² of rainforest cover. By the late 1980s, this had fallen to an estimated 90 000 km². Over 30% of the rainforest in Honduras has been lost since 1960 – with more than 800 km² being lost every year for ranches, banana plantations, small farms and fuelwood. The expansion of fruit plantations is partly due to the country's need to earn foreign currency to repay debt. Exports of hardwood also add to the money earned to repay debt. Honduras spends 20% of its export earnings on debt repayments every year.

West Africa

In West Africa, nearly 90% of the original tropical rainforest has gone – and what remains is heavily fragmented and degraded. Today, unspoiled forests in West Africa are restricted to one patch in Côte d'Ivoire and another along the border between Nigeria and Cameroon.

Most deforestation results from subsistence agriculture, although logging, commercial hunting and the development of infrastructure have also contributed to the loss of forests.

Deforestation hotspots (note that this map shows all forests, not just tropical rainforests) ▶

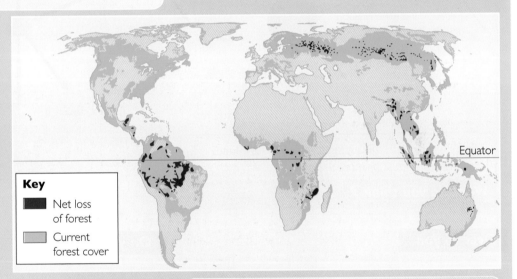

Key
- ■ Net loss of forest
- ■ Current forest cover

Equator

Brazil

It is claimed that in the Amazon one hectare of forest is cleared every second. One of the reasons for this is peasant farming. Since the 1970s, the Brazilian government – concerned about overcrowded cities in the south of the country – has encouraged landless citizens to move to the rainforest and given them land to clear for subsistence farming. Other causes of deforestation in the Amazon include mining, logging and cattle ranching.

Queensland, Australia

Deforestation associated with the cultivation of sugar cane began in the 1860s in eastern Australia, and continued right through until the 1980s and early 90s.

In Queensland, most land clearance (including deforestation) is now undertaken to clear land for pasture (86%), with the remainder cleared for crops (10%), mining, infrastructure and settlement (4%). The area of most intensive clearance activity has shifted from central to southern Queensland.

Nutrient cycles

Background

Ecosystems depend on two basic processes: the **recycling of nutrients** and the **flow of energy** (see page 126).

Nutrients are continually circulated within ecosystems. Plants take up nutrients from the soil, which are then used by the plants themselves or by the animals that consume them. When the plants or animals die, they decompose and the nutrients return to the soil.

The lush appearance of a tropical rainforest is deceptive. It is a fragile ecosystem dependant on the constant recycling of nutrients. Deforestation breaks the nutrient cycle. Humus is not replaced and the soil rapidly loses its fertility

and is easily eroded. The rainforest cannot become re-established and the soil becomes too poor to use.

Nutrient cycles can be shown by simplified diagrams, called **Gersmehl's nutrient cycles**, which show the stores of nutrients (biomass, litter and soil) as well as the transfers between them.

The size of the nutrient stores in the diagram is proportional to the quantity of nutrients stored. The thickness of the transfer arrows is proportional to the amount of nutrients transferred.

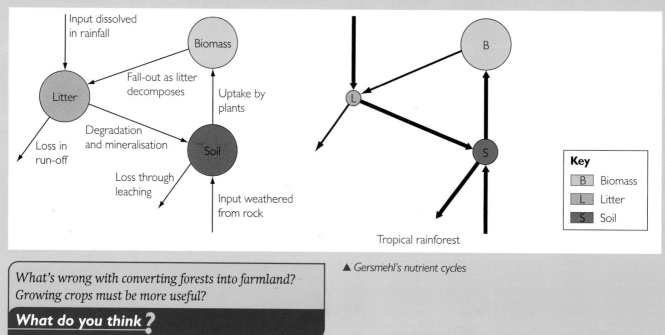

▲ Gersmehl's nutrient cycles

> What's wrong with converting forests into farmland? Growing crops must be more useful?

What do you think ?

Over to you

1 a In groups of 2-3, draw a spider diagram to show how tourism and development pose threats to the Daintree. Classify the threats and present your findings to the class.
   b Write up your findings on a side of A4, listing and classifying the different threats.
   c Explain what you believe are the biggest threats, and why.
2 Deforestation is a major threat to biodiversity.
   a Classify the causes of deforestation as social, environmental and economic.
   b Write 100 words on the consequences of deforestation.

On your own

3 Look at the diagram of Gersmehl's nutrient cycles.
   a Work out how deforestation affects the nutrient cycle in the tropical rainforest.
   b Draw a version of the nutrient cycle to show the changes identified in a.
4 Draw an outline map of the Daintree area from Cape Tribulation to Cape Kimberley. Research and mark on:
   • The main residential sites within the Daintree
   • The main tourist sites and facilities
   • The Daintree River, ferry crossing, road from Cape Kimberley to Cape Tribulation, Bloomfield Track beyond Cape Tribulation, the Great Barrier Reef.

In this unit you'll investigate different ways of managing the threats to the Daintree rainforest.

Managing the Daintree

The Daintree rainforest has survived for over 135 million years, but it is now under threat from human activity. There is a range of strategies and players involved in managing biodiversity in the Daintree, but sometimes they come into conflict.

The Wet Tropics Management Authority

The Wet Tropics Management Authority was formed in 1990, and is responsible for managing the Wet Tropics as a World Heritage Site. The Wet Tropics World Heritage Area covers nearly 900 km². Within this area there are National Parks (including the Daintree), State Forests, and both privately and publicly owned land.

The Authority is based in Cairns and its main functions are:
- developing and implementing plans and policies
- researching and monitoring – enhancing understanding of the importance of the World Heritage Area and monitoring the state of the Wet Tropics
- developing management agreements with landholders and Aborigines
- education through visitor centres and education programmes
- funding for particular outcomes
- promoting the Wet Tropics.

Douglas Shire Council

Until 2008, Douglas Shire Council was the local council for the Daintree – determining issues like planning permission. In the late 1990s, the local council, land developers and farmers wanted more development in the Daintree, which supported a population of 2400 then and more later.

However, local people wanted to see the population reduced to 1200. In 2000, the council voted to gradually reduce the population in the Daintree, as a way of balancing economic growth and biodiversity. To achieve this:
- in 2003, it increased the price of the ferry crossing by $4 to finance land buy-back (see page 117). Tour operators objected, saying that tourists would be unwilling to pay the extra fee. However, tourist numbers have continued to increase.
- it rejected proposals – supported by tour operators – to build a bridge across the Daintree River or introduce an extra ferry, on the grounds that additional tourists would threaten the rainforest.

However, a Queensland review of local government abolished Douglas Shire Council in 2008 and the Daintree became part of Cairns Regional Council. Local people fear that the kind of commercial development which has occurred in Cairns will now take place in the Daintree.

▲ *The Wet Tropics World Heritage Area*

▲ *Cairns – could what has happened here happen in the Daintree?*

> *Should more people be allowed to live in the Daintree?*
>
> **What do you think ?**

Can the threats to biodiversity be successfully managed? - 2

The Rainforest Co-operative Research Council

The Rainforest Co-operative Research Council produced a report on the future of the Daintree in 2000, which is still current. Their findings were that unless action was taken, the area would see an increase in residential development, a loss of biodiversity, and a reduction in its attractiveness to tourists. Below are the strategies that they proposed to protect the environment and build a sustainable community.

Community development

- Between 1200-1400 people to live in the Daintree rainforest and be involved in land stewardship and conservation.
- Base employment on tourism, organic farming, tropical horticulture, and small business opportunities.
- Settle around 600 blocks of land.

Biodiversity conservation

- Adopt settlement and land-management practices on private land to protect the outstanding biodiversity (rare and/or threatened species, habitats and ecosystems).
- Identify biodiversity hotspots for conservation where no development should occur.
- Identify threats from wild animals (including pigs) and weeds.
- Around 540 blocks to remain unsettled.

Douglas Shire Council

- Introduce planning controls for biodiversity conservation.
- Ensure that settlement densities are sustainable.

Electricity supply

- When settlement densities have been reduced to a sustainable level, use underground cables to extend the mains power supply as far north as Cooper Creek.
- People north of Cooper Creek to continue to use their Remote Area Power System.

Indigenous people

- Recognise the rights of aboriginal peoples to own land and promote their culture within the forest.

Water supply and waste management

- Keep water extraction from streams and underground supplies within sustainable limits.
- Use the best available domestic technology for waste disposal.

Roads and ferry

- The ferry should provide essential access and remain the gateway to the area.
- Improve tourist visitor facilities south of the river and recreation facilities north of the river.
- Reduce forest cut-backs – the road to Cape Tribulation should be a green 'tunnel' with 'windows' through the rainforest to mountain and coastal scenery.

Tourism

- Increase tourist numbers to 550 000 visitors a year to support the local economy.
- Increase the number of tourists staying for several nights and revisiting the area.
- Monitor the impacts of tourism to ensure that it is sustainable.

Financial issues

- Use ferry income to assist community services infrastructure and conservation initiatives.
- Require a financial commitment from federal, state and local government to:
  - establish the Daintree Land Trust to support land acquisition and compensate those who lose land
  - meet the cost of priority purchase and financial incentives for conservation
  - subsidise the electricity supply.

Other players

Many organisations support the preservation of the Daintree, including those on this page.

Australian Rainforest Foundation

The Australian Rainforest Foundation (ARF) is a not-for-profit organisation dedicated to education, research and habitat rehabilitation for Australia's rainforests. It is involved in a variety of projects.

- With **Operation Big Bird**, the ARF plans to create a 250 km wildlife corridor to help protect the cassowary. This corridor will link Cairns with the southern coastal town of Cardwell. Wildlife corridors help to conserve biodiversity by enabling species to move, feed, breed, disperse and colonise. Cassowaries are good 'indicator' species – if they can be protected, so can a range of other species sharing their habitat.
- The ARF has received funding from the Australian government for a range of conservation initiatives in the Daintree aimed at reducing the impact of development. Funding has been used to **buy-back** land from those who originally bought it, in order to reduce the available land for development. The funding is also used to support conservation, and the ARF works with landowners to achieve this.

> - **Buy-back** is a Queensland (state) government initiative which the ARF is implementing. The Australian national government has also provided funding for this.

▲ *The 250 km wildlife corridor*

Wildlife Preservation Society of Queensland

This is a community-based, not-for-profit conservation group. It is committed to an ecologically sustainable future for people and wildlife. They supported a ban on development in the Daintree in 2004.

Australian Tropical Research Foundation

This was created in 1993 to oversee the operation of the Cape Tribulation Tropical Research Station and the Wet Tropics Visitor Centres. These facilities encourage research and conservation by increasing understanding of the Australian tropical rainforest ecosystem, and the global importance of rainforest and reef systems.

Over to you

1 a Can the threats to biodiversity in the Daintree be successfully managed? Draw a Venn diagram with four overlapping circles, labelled: Biodiversity protection, Limiting development, Education, Economic incentives. Use the Venn diagram to classify all the proposals or suggestions in this unit.
  b Choose one category, e.g. limiting development. How successful do you think the proposals will be? Explain.
  c What else could be done to manage the threats?
2 a List the organisations which have an interest in the Daintree.
  b Complete a conflict matrix for the organisations to show how far they are likely to agree or disagree.

  c How far are conflicts in the Daintree a case of economic versus environmental interests?
  d How far do you think these conflicts can be resolved, and how?

On your own

3 Research photographs and text about the Cairns area in terms of **(a)** what the area is like, **(b)** how it has developed, and **(c)** the kind of environment it provides for tourists.

Exam question: Evaluate the relative importance of global and local threats to one named global ecosystem. **(15)**

3.7 Mangroves

In this unit you'll learn about mangroves – what they are, what mangrove ecosystems are like, and mangroves and ecosystem services.

◄ *Devastation caused by the Boxing Day tsunami in Thailand*

The extent of the destruction caused by the Boxing Day tsunami was self-inflicted.

What do you think ?

Destroying Asia's coastal barrier

After the 2004 Boxing Day tsunami, many of the worst-affected countries admitted that the extent of the damage caused was partly self-inflicted. Over the last few decades, those countries have, in the name of development (particularly tourism and aquaculture), systematically destroyed one of their most effective barriers to ocean forces – their mangrove forests. Since 1961, more than 50% of Thailand's mangroves have been removed. Thailand suffered extreme damage from the tsunami, with over 5000 people killed – many of them in the new tourist resorts.

In Sri Lanka, Wanduruppa (which is located within degraded mangrove forests) was badly affected, with 5000-6000 of its people dying. Nearby Kapuhenwala is surrounded by 200 hectares of dense mangrove forest and lost only two people.

While tsunami are rare, all Asian coastlines are prone to extreme weather – such as cyclones and storm surges. Mangrove deforestation is thought to have significantly increased the damage from such natural disasters.

Why are mangroves special?

On first appearance, mangroves do not look too inviting. Full of sticky mud and biting insects, they are smelly and look like wastelands. But look more closely and mangrove forests provide an important habitat for a diverse range of marine and terrestrial flora and fauna as varied as manatees, crab-eating monkeys, fishing cats, monitor lizards, Royal Bengal tigers and mud skipper fish.

Mangroves are essential to marine, freshwater and terrestrial biodiversity, because they stabilise coastlines against erosion, collect sediments and provide a nursery for coastal fish.

▼ *The Sunda Shelf mangroves, Borneo*

Where are mangroves located?

Mangroves live in two worlds at once. They grow in **inter-tidal areas** (areas between high and low tides) and estuary mouths. Mangrove forests are found along the tropical and sub-tropical coasts of Africa, Australia, Asia and the Americas – as the map shows. The greatest diversity of mangrove species exists in South-East Asia.

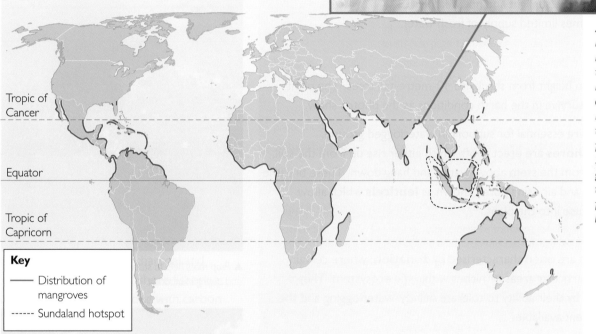

▲ The largest remaining tract of mangrove forest in the world is found in the Sundarbans on the edge of the Bay of Bengal. In this image, taken from space, the Sundarbans are dark green, surrounded to the north by farmland which is lighter green.

Key
— Distribution of mangroves
------ Sundaland hotspot

▲ The global distribution of mangroves

Sunda Shelf mangroves

The Sunda Shelf mangroves are found on the island of Borneo and the east coast of Sumatra. They cover an area of 37 400 km² (nearly twice as large as Wales). They fall within the Sundaland Wetlands hotspot. They are considered to be among the most biologically diverse mangroves in the world, although diversity in terms of endemics or species richness is not great.

- Five major mangrove types grow here, depending on differences in soils, salinity and tidal movements.
- More than 250 bird species are found in this **eco-region**, but many of them are transitory or migrant.
- The mangroves are home to the proboscis monkey, one of the few large mammals limited to mangrove and peat swamp forest habitats.

- An **eco-region** is a large area of land or water with geographically distinct natural communities, where the majority of species interact in ways essential for their long-term survival.

The proboscis monkey is threatened with extinction due to habitat loss and hunting. ▶

Ecosystem processes

Background

Ecosystems depend on two basic processes: the recycling of nutrients (see page 114) and the flow of energy.

Energy flows

Ecosystems are sustained by a flow of energy through them. An ecosystem's main source of energy is sunlight. Sunlight is absorbed by plants and converted into chemical energy by the process of **photosynthesis**. Energy then passes through the ecosystem in a food chain. Each link in the chain feeds on, and obtains energy from, the preceding link. Each link is known as a **trophic** or energy level.

Trophic level 1 Producers (convert energy by photosynthesis)	Trophic level 2 Consumer	Trophic level 3 Consumer	Trophic level 4 Consumer
grass	worm	blackbird	hawk
leaf	caterpillar	woodmouse	fox
phytoplankton	zooplankton	fish	human

▲ Examples of food chains and trophic levels

◄ Energy flows in the ecosystem

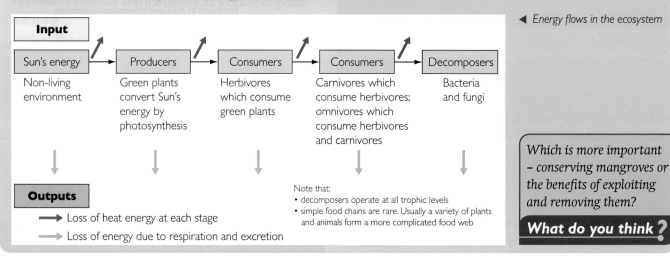

Input

Sun's energy	Producers	Consumers	Consumers	Decomposers
Non-living environment	Green plants convert Sun's energy by photosynthesis	Herbivores which consume green plants	Carnivores which consume herbivores; omnivores which consume herbivores and carnivores	Bacteria and fungi

Outputs

→ Loss of heat energy at each stage

→ Loss of energy due to respiration and excretion

Note that:
• decomposers operate at all trophic levels
• simple food chains are rare. Usually a variety of plants and animals form a more complicated food web

Which is more important – conserving mangroves or the benefits of exploiting and removing them?

What do you think ?

Over to you

1 a What is likely to affect nutrient levels in mangroves?
   b What effects could a change in nutrient levels have?
2 Draw a diagram to show the links between mangrove destruction and economic activity. Identify local and global factors on your diagram.
3 Complete the following table on the impacts of mangrove depletion.

Impacts of mangrove depletion	Explanation with examples
Declining fisheries	
Threatened migratory bird species	
Degraded water supplies	
Salinisation of coastal soils	
Increased erosion and land subsidence	
CO_2 released into the atmosphere	
Increased vulnerability of coastal populations	
Loss of biodiversity	

4 Look at the diagram at the bottom of page 122 showing the main drivers of change in coastal wetlands. Draw a diagram like that to show the main drivers of change in mangroves.

On your own

5 a Research species living in mangrove forests, so that you can draw:
   • a simple food chain
   • a more complex food web for a mangrove ecosystem.
   b Assess in detail how the mangrove destruction affects the ecosystem, using the food web you have drawn.
6 Write a 500-word report on a the threats to biodiversity in mangroves, b how far these are likely to change, and c whether these threats reflect people's attitudes to mangroves and biodiversity.

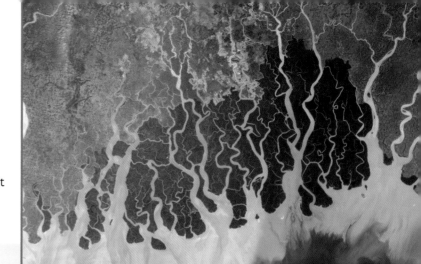

Where are mangroves located?

Mangroves live in two worlds at once. They grow in **inter-tidal areas** (areas between high and low tides) and estuary mouths. Mangrove forests are found along the tropical and sub-tropical coasts of Africa, Australia, Asia and the Americas – as the map shows. The greatest diversity of mangrove species exists in South-East Asia.

▲ *The largest remaining tract of mangrove forest in the world is found in the Sundarbans on the edge of the Bay of Bengal. In this image, taken from space, the Sundarbans are dark green, surrounded to the north by farmland which is lighter green.*

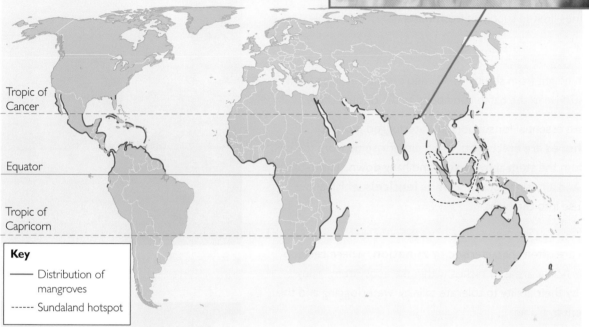

Tropic of Cancer

Equator

Tropic of Capricorn

Key

—— Distribution of mangroves

------ Sundaland hotspot

▲ *The global distribution of mangroves*

Sunda Shelf mangroves

The Sunda Shelf mangroves are found on the island of Borneo and the east coast of Sumatra. They cover an area of 37 400 km² (nearly twice as large as Wales). They fall within the Sundaland Wetlands hotspot. They are considered to be among the most biologically diverse mangroves in the world, although diversity in terms of endemics or species richness is not great.

- Five major mangrove types grow here, depending on differences in soils, salinity and tidal movements.
- More than 250 bird species are found in this **eco-region**, but many of them are transitory or migrant.
- The mangroves are home to the proboscis monkey, one of the few large mammals limited to mangrove and peat swamp forest habitats.

> ● An **eco-region** is a large area of land or water with geographically distinct natural communities, where the majority of species interact in ways essential for their long-term survival.

The proboscis monkey is threatened with extinction due to habitat loss and hunting. ▶

Mangrove ecosystems

Mangrove ecosystems are found:

- in the tropics and sub-tropics (where average temperatures are always above 20 °C)
- where shorelines are protected from the full force of oceanic waves, e.g. by coral reefs
- on depositional shorelines which have fine-grained sediments.

Mangroves grow in areas:

- with permanently waterlogged soil, which is also **anaerobic** (there is no oxygen due to waterlogging)
- with high salinity (from sea water)
- which are frequently flooded or inundated
- which have intense sunlight and hot weather
- with a sometimes limited supply of fresh water.

Vegetation

Mangroves vary in height from shrubs to 40-metre-tall trees. The plants have adapted to survive in the harsh conditions as the photos show.

- **Prop roots** are essential for support in waterlogged ground.
- **Pneumatophores** are erect roots which either rise up from the soil, or protrude from the stem above water, and hang down. These breathe in both water and air using pores known as **lenticels** which allow oxygen to diffuse into the plant.

Zonation

Mangrove forests are often characterised by **zonation**, where certain species occupy particular areas or niches within the ecosystem. They are distinguished by their ability to tolerate salinity, waterlogging and the amount of sediment available.

- **Red mangroves** are found closest to the coast, where they take the brunt of wave action, protect inland areas, and can survive permanent waterlogging.
- **Black mangroves** live further inland, protected behind the red coastal mangroves. They have pneumatophores with lenticels to obtain oxygen, but would die if permanently waterlogged.
- **White/grey mangroves** live furthest inland from the coast and are least able to survive waterlogging.

Productivity

Mangroves are highly productive habitats and can be almost as productive as tropical forests.

It is not just the flora and fauna that make the mangrove ecosystem so important. Mangrove muds play a critical role as a 'nursery' for the sea's fish, shrimp larvae, crabs and crustacea. A teaspoon of mud from a north Queensland mangrove forest in Australia contains more than 10 billion bacteria. These densities are among the highest in marine mud anywhere in the world, and are an indication of the incredibly high productivity of this habitat.

▲ Prop roots help to stabilise the trees in the soft mud and during inundation.

▲ Pneumatophores

Mangroves and ecosystem services

Ecosystem services are the benefits that people get from ecosystems. Look back at page 108 for definitions of provisioning, regulating, cultural and supporting services. Mangroves provide a range of both goods and services as the table shows.

The table shows only those goods and services which have a high or medium order of magnitude. Goods and services which are considered low (which includes cultural services) are not shown, although mangroves do provide opportunities for tourism, recreation and education. In addition, they provide some medicines, e.g. the ashes, or bark infusions of certain species can be used to treat skin disorders.

Putting a value on mangroves

It is possible to put an economic value on an ecosystem. Intact mangroves provide social benefits of timber, charcoal and other forest products, offshore fisheries and storm protection, along with carbon sequestration, etc. Adding up the contribution of all of the services which mangroves provide, gives intact mangroves an economic value of at least $1000 a hectare, and possibly as much as $36 000 a hectare.

Converting mangroves to **aquaculture** (see page 123) for the purposes of shrimp farming in Thailand might seem to make short-term economic sense, but the value of a hectare of mangrove cleared for aquaculture plummets to a mere $200 a hectare.

> ● **Aquaculture** is the production of aquatic organisms under controlled conditions. It is also called aqua farming, and includes shrimp farming.

The scale is: ● medium, ● high. Note that the scale shows the global average, but there will be local and regional differences in relative magnitude.

Services	Comments and examples	Mangroves
Provisioning		
Food	Production of fish, algae and invertebrates	●
Fibre, timber, fuel	Production of timber, fuelwood, charcoal, fibres and dyes, and construction materials	●
Regulating		
Climate regulation	Regulation of greenhouse gases and carbon dioxide sequestration	●
Biological regulation	Resistance of species invasions, regulating interactions between different trophic levels (see page 126), preserving diversity	●
Pollution control and detoxification	Retention, recovery and removal of excess nutrients and pollutants, including nitrates and phosphates	●
Erosion protection	Retention of soils and prevention of shoreline erosion	●
Natural hazards	Mangroves act as a buffer zone to protect coastlines from storm surges, floods, hurricanes, etc. and the resultant loss of life	●
Supporting		
Biodiversity	Habitats for resident or transient species – 75% of all tropical commercial fish species spend part of their lives in mangroves	●
Soil formation	Sediment retention and accumulation of organic matter	●
Nutrient cycling	Storage, recycling, processing and acquisition of nutrients	●

▲ *The relative magnitude of ecosystem services and goods for mangroves*

Over to you

1 Look at the table above. Consider the following people:
   - A Thai fisherman
   - A tourist visiting Thailand
   - A Thai owner of a shrimp farm
   - A Sri Lankan living in Wandaruppa
   a Which of the services and goods are valuable to each person?
   b Why might their attitudes to the value of biodiversity vary?
2 a Why does the economic value of cleared mangroves decline?
   b How far do the people who make decisions about mangroves benefit from ecosystem goods and services?
3 Which processes and factors influence biodiversity in mangroves?

On your own

4 Use these websites to find out more about the Sunda Shelf mangroves:
   www.nationalgeographic.com
   www.worldwildlife.org
   www.eoearth.org
   Find out about mangrove species, threats to mangroves and protected areas.

In this unit you'll find out about some of the threats that mangroves face, and about ecosystem processes (energy flows).

If there are no mangroves, the sea will have no meaning. It is like having a tree without roots – for the mangroves are the roots of the sea …
A Thai fisherman from the Andaman coast

Mangrove loss

Over 50% of the world's original mangrove forests have been lost – an area equivalent to 13 times the size of the UK. By 2007, less than 15 million hectares remained. According to the UN's Food and Agriculture Organisation (FAO), the current rate of mangrove loss is approximately 1% p.a. (roughly 150 000 hectares – the size of London). This alarming rate of loss has a major impact on biodiversity.

- Thailand has lost over 50% of its mangrove forests since 1961.
- In the Philippines, 210 500 hectares of mangrove (40% of the country's total mangrove cover) were lost to aquaculture between 1918 and 1988. By 1993, only 123 000 hectares were left – equivalent to a loss of 70% in 70 years.
- In Ecuador, estimates of mangrove loss range from 20% to nearly 50% of the country's once 362 000 hectares of mangrove-forested coastline.

Reasons for the mangrove loss vary:
- In Asia, over 50% of the mangrove loss has been due to increasing aquaculture (38% for shrimp and 14% for fish), about 25% has been due to deforestation, and another 11% to upstream freshwater diversion.
- In Latin America, mangrove destruction has occurred because of the expansion of agriculture and cattle rearing, the cutting of fuelwood and building materials, and the establishment of shrimp aquaculture.

▲ The coastal region of southern Thailand was once a landscape of rice paddies, but now there are shrimp ponds as far as the eye can see.

Drivers of change

The main drivers, or threats, to biodiversity are: habitat change, climate change, the introduction of invasive species, over-exploitation, and pollution. The diagram below shows the impacts that these drivers have had on coastal wetlands, which includes mangroves, over the last century – and their current trends.

▼ The main drivers of change on coastal wetlands, which include more than just mangroves

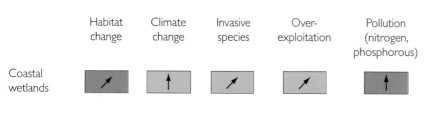

	Habitat change	Climate change	Invasive species	Over-exploitation	Pollution (nitrogen, phosphorous)
Coastal wetlands	↗	↑	↗	↗	↑

Key

Driver's impact on biodiversity over the last century
- ☐ Low
- ☐ Moderate
- ☐ High
- ☐ Very high

Driver's current trends
- ↘ Decreasing impact
- → Continuing impact
- ↗ Increasing impact
- ↑ Very rapid increasing impact

Shrimp farming in Thailand

Shrimp farming has been practised in Thailand for over 60 years. The traditional method of production was what is now called 'extensive production'. Shrimp 'fry' were trapped in salt beds and padi fields around estuaries, and harvested when mature.

However, between the 1970s and 1990s, Thailand's coastal shrimp aquaculture industry expanded dramatically. The Thai government encouraged shrimp farming and production became more intensive.

Asia's shrimp production in tonnes per year. 75% of farmed shrimp comes from Asia. China's production is nearly all consumed by its domestic market; most of Thailand's is exported. ▼

Dulah Kwankha used to be a rice farmer earning $400 a year, but now he earns six times that amount. He took a $12 000 bank loan to convert his rice padi into a shrimp pond, which produces three crops a year. Dulah has money to spend for the first time in his life, but he is haunted by the possibility of disaster. He still has to pay off his initial loan and must borrow more money each year to run the shrimp farm.

'With rice farming, I only made enough money to feed my family and it was very hard work,' he said. 'But I didn't worry much. Shrimp farming is easier, but I'm worrying a lot more – especially about debt.'

Country	1985	1990	1995	2000	2004
China	40 000	185 000	78 000	218 000	935 000
Thailand	10 000	115 000	259 000	309 000	389 000
Vietnam	8000	32 000	55 000	90 000	276 000
Indonesia	25 000	84 000	121 000	118 000	218 000
India	13 000	35 000	70 000	97 000	133 000
Bangladesh	11 000	19 000	32 000	59 000	58 000
Philippines	29 000	48 000	89 000	41 000	37 000
Burma	0	0	1000	5000	30 000
Taiwan	17 000	15 000	11 000	6000	12 000

Thailand's shrimp production helps to satisfy a huge global demand. The largest market for imported shrimp is the USA, which imported more than 500 000 tonnes in 2003. Japan imported about 250 000 tonnes; France, Spain, the UK and Italy together imported another 500 000 tonnes.

However, since the 1990s, the amount of shrimp consumed in Thailand itself has also increased, due to rising incomes as a result of improved economic conditions. An expanding tourist industry has helped to increase domestic demand.

Impacts of shrimp farming

Although shrimp farming has had some positive impacts – such as increased wealth, which has led to improved infrastructure in local areas and a reduction in migration to Bangkok – many of the impacts have been negative:

● Effluent from shrimp farms has to be removed before the next crop cycle begins. It is full of decaying food, shells, and chemicals (including antibiotics). If discharged into mangroves, it can have a harmful effect on their flora and fauna.

● Conflicts arise over the discharge of effluent.

● There is a constant threat of disease and the spread of infection.

● Rice fields and canals experience salinisation.

● There is a depletion of biodiversity in shrimp farms and their surrounding areas.

▲ *The shrimp-farming area of Thailand*

Toxic sludge at the bottom of a shrimp pond during harvesting ▼

The threats continue

Climate change

One in ten people worldwide lives less than 10 metres above sea level and near the coast (the at-risk zone). Asia contains about 75% of the people living in these highly vulnerable coastal areas. Global warming will lead to increased melting of the Arctic ice caps and rising sea levels; Greenpeace predicts rises of 15-95 cm in sea levels this century. A rise of 95 cm could cause shorelines to retreat by 100-200 metres.

Healthy coastal fringes of mangroves can:

- act as a buffer against, and lessen the dangers from, hurricanes and cyclones – which are expected to increase in intensity and frequency as a result of global warming
- protect the coast from erosion.

But rises in sea levels mean that mangroves are likely to be submerged and drowned and their protective qualities lost – putting people at increasing risk. Mangroves themselves need a buffer zone – an area they can retreat to in order to re-establish themselves above the rising low-tide mark.

Creating a buffer zone for mangroves could be difficult because of the massive amounts of development behind them, including roads, shrimp farms, industrial complexes and hotels. ▶

Background

Carbon sequestration

Mangroves take up (or **sequester**) approximately 1.5 metric tonnes/hectare/year of carbon. Current rates of mangrove destruction mean that almost 225 000 metric tonnes of carbon sequestration potential is lost each year. Not only that, but the layers of soil and peat which make up the mangrove substrate have a high carbon content of 10% or more. When mangroves are destroyed for shrimp farming, sediment is dug out and carbon released back into the atmosphere. According to Dr Jin Eong Ong of the Universiti Sams in Malaysia, clearing mangroves and excavating the substrate for shrimp ponds means a release of carbon which is 'some 50 times the sequestration rate. This means that, by converting a mere 2% of mangroves to shrimp aquaculture ponds, all of the advantages of mangroves as a sink of atmospheric carbon will be lost…'

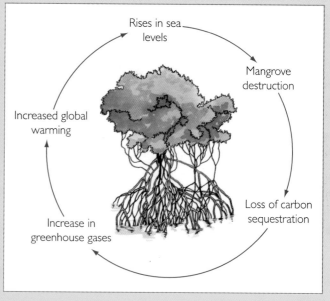

▲ *Mangroves and climate change*

Over-harvesting

Mangrove trees are used for many purposes, including fuelwood, construction material, wood chip and pulp production, charcoal production, and animal fodder. While harvesting has taken place for centuries, in some parts of the world it is no longer sustainable – threatening the future of the forests.

Over-exploitation of marine life

Unregulated fishing is depleting stocks of fish, prawns and other marine species. As stocks of one species become uneconomic, people target others. The result is a decline in biodiversity, an increase in effort (for fishermen) for lower returns, and a loss of jobs. In addition, over-fishing alters the balance of food chains and webs and mangrove fish communities can be altered as a result.

▲ Barbeque anyone? Prawns and charcoal – each produced in Asia and destroying mangroves.

Chao Mai, Trang Province, Thailand

Subsistence fishing is vital to the coastal village of Chao Mai in Trang Province, southern Thailand. However, starting in the late 1960s, some fishermen were drawn into the commercial fishing industry by being given loans to purchase larger boats and sophisticated equipment.

At the same time, the government opened up the coastal mangrove forests to cutting for charcoal production. The mangroves were depleted and, more recently, the rapid growth of shrimp farming has degraded the mangroves even further.

The depletion of the mangrove forests has meant a reduction in marine nurseries and the depletion of sea grass, which grows on silt washed down from the mangroves. The larger fishing boats, which use drag nets on the seabed, have also destroyed the sea grass that, like the mangroves, acted as marine nurseries.

This has had a severe impact on the livelihood of villagers with small boats who can only fish in shallow coastal waters. Conflicts have arisen between those villagers who depend on the mangroves and those who are destroying their livelihoods.

Unit 3.9 explains how this problem has been managed to improve the situation.

Sea grass – home to many marine species ▶

Other threats

- Tourism is spreading along mangrove-fringed coasts. Mangroves are frequently viewed as mosquito-infested muddy swamps, holding back progress and hindering development. They are being destroyed to build golf courses, cruise ship ports, marinas, hotels, apartments and restaurants.

- Oil exploration and development has many environmental impacts on mangroves, including deforestation to accommodate infrastructure, changes in soil pH, and the discharge of solid and liquid wastes.
- Environmental degradation results not only from tourism and oil development, but also from domestic and other industrial waste disposal, waste from aquaculture, and exploitation of minerals.

Ecosystem processes

Background

Ecosystems depend on two basic processes: the recycling of nutrients (see page 114) and the flow of energy.

Trophic level 1 Producers (convert energy by photosynthesis)	Trophic level 2 Consumer	Trophic level 3 Consumer	Trophic level 4 Consumer
grass	worm	blackbird	hawk
leaf	caterpillar	woodmouse	fox
phytoplankton	zooplankton	fish	human

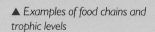

▲ Examples of food chains and trophic levels

Energy flows

Ecosystems are sustained by a flow of energy through them. An ecosystem's main source of energy is sunlight. Sunlight is absorbed by plants and converted into chemical energy by the process of **photosynthesis**. Energy then passes through the ecosystem in a food chain. Each link in the chain feeds on, and obtains energy from, the preceding link. Each link is known as a **trophic** or energy level.

◄ Energy flows in the ecosystem

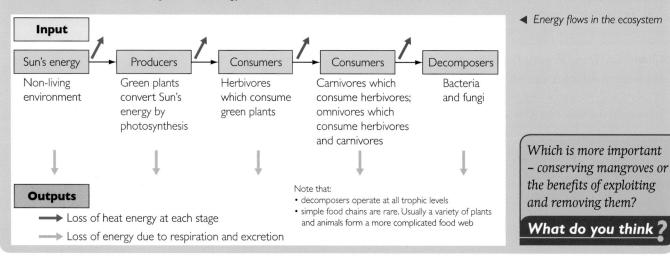

Input

Sun's energy	Producers	Consumers	Consumers	Decomposers
Non-living environment	Green plants convert Sun's energy by photosynthesis	Herbivores which consume green plants	Carnivores which consume herbivores; omnivores which consume herbivores and carnivores	Bacteria and fungi

Note that:
• decomposers operate at all trophic levels
• simple food chains are rare. Usually a variety of plants and animals form a more complicated food web

Outputs

➡ Loss of heat energy at each stage

➡ Loss of energy due to respiration and excretion

Which is more important – conserving mangroves or the benefits of exploiting and removing them?

What do you think?

Over to you

1 a What is likely to affect nutrient levels in mangroves?
  b What effects could a change in nutrient levels have?
2 Draw a diagram to show the links between mangrove destruction and economic activity. Identify local and global factors on your diagram.
3 Complete the following table on the impacts of mangrove depletion.

Impacts of mangrove depletion	Explanation with examples
Declining fisheries	
Threatened migratory bird species	
Degraded water supplies	
Salinisation of coastal soils	
Increased erosion and land subsidence	
CO$_2$ released into the atmosphere	
Increased vulnerability of coastal populations	
Loss of biodiversity	

4 Look at the diagram at the bottom of page 122 showing the main drivers of change in coastal wetlands. Draw a diagram like that to show the main drivers of change in mangroves.

On your own

5 a Research species living in mangrove forests, so that you can draw:
  • a simple food chain
  • a more complex food web for a mangrove ecosystem.
  b Assess in detail how the mangrove destruction affects the ecosystem, using the food web you have drawn.
6 Write a 500-word report on a the threats to biodiversity in mangroves, b how far these are likely to change, and c whether these threats reflect people's attitudes to mangroves and biodiversity.

Managing the threats to mangroves – 1

In this unit you'll explore a range of management options to halt the loss of biodiversity in mangroves.

Different players

There are many players who are trying to halt the loss of mangroves to protect and restore them. These include: community development organisations, such as Yadfon; regional organisations, such as TVE Asia Pacific; NGOs, like Wetlands International; EU-funded initiatives, such as the Coastal Biodiversity in Ranong project; intergovernmental agreements, like Ramsar Sites.

Community mangrove forests

Yadfon is a small organisation in Thailand, formed in 1985 to help fishing communities. Yadfon has worked with villagers in Chao Mai (see page 125) on environmental issues and securing their future livelihoods. Chao Mai has faced two problems:

- **Mangrove destruction**. Mangroves were being destroyed to produce charcoal, which affected the supply of seafood. Yadfon created an 80-hectare community managed mangrove forest – and did not allow shrimp farms within its boundaries.
- The **destruction of sea grass** by large fishing boats and drag nets. Sea grass is important as a nursery and habitat for commercially important species of fish and prawns. To protect the sea grass, a publicity campaign was launched among the fishing communities, which led to reduced destruction by boats.

Now the mangrove forest is managed by the community, and community managed forests like this have become the cornerstone of Yadfon's work with coastal villages in order to revive the coastal fisheries.

▲ Dugong, or sea cows, are large slow-moving marine mammals which feed on sea grass. The village of Chao Mai 'adopted' a dugong named 'Tone' as a symbol of their community's commitment to conservation.

Coastal Biodiversity in Ranong, Thailand

One of the most extensive areas of mangrove forest in Thailand (nearly 200 km²) is found in Ranong on southern Thailand's Andaman coast. The mangroves are found along the estuaries of rivers, such as the Kampaun, and fringing the more sheltered areas of off-shore islands.

Coastal Biodiversity in Ranong is a collaborative project between organisations including the Natural History Museum (London), Wildlife Fund Thailand and the Kampaun Fisheries Co-operative. It is supported by funding from the EU.

Its main objective is to assess biodiversity and enable long-term environmental monitoring. Equally important is to make information available to local communities, conservationists, government organisations and scientists. It has developed a number of activities:

- meetings with local communities
- open days
- a newsletter
- an education officer who liaises with local schools
- Marine Biodiversity Youth Camps to raise awareness.

Managing the threats to mangroves - 2

Wetlands International

Wetlands International is an independent, not-for-profit global NGO dedicated to the conservation and wise use of wetlands of all types. It has four long-term goals.

Global goal 1 All those who use and have impacts on wetlands should be well informed about their importance.

Global goal 2 The functions and values of wetlands are recognised and integrated into sustainable development.

Global goal 3 Conservation and **sustainable yield** of wetlands is achieved through effective uses of water resources and coasts.

Global goal 4 Large-scale strategic initiatives result in improved conservation status of species, habitats and ecological networks

Wetlands International is involved in numerous projects, including the ones on the right in South-East Asia.

> ● **Sustainable yield** means ways in which ecosystems can be productive, but at levels which can be maintained and are sustainable, without causing damage.

▲ *Replanting mangroves in Aceh, Indonesia*

Project	Aim	Methods
Green Coast; for nature and people after the tsunami	To support coastal communities affected by the December 2004 tsunami.	Provides capital, grants and technical advice to support local communities in restoring and managing mangroves, coral reefs and sand dune ecosystems.
Community based mangrove reforestation	To reduce poverty and vulnerability and increase sustainability in Aceh, Indonesia.	By enabling community groups to understand and replicate a mangrove-aquaculture ecosystem.
Mangrove replanting in Kuala Gula, Perak	To rehabilitate degraded mangrove areas	Involving private companies in funding community projects

Saving greenbelts

Asia's coastal ecosystems – including coral reefs, mangroves and sand dunes – are collectively called **greenbelts**. Their role in protecting coastal communities against natural disasters and the anticipated impacts of climate change was brought home by the 2004 Boxing Day tsunami.

TVE Asia Pacific produced a series of films called *The Greenbelt Reports*. Established in 1996, TVE Asia Pacific is a regional not-for-profit organisation that uses the media to raise awareness of environmental, developmental, and health and social justice issues.

Its CEO says: 'These stories are not about the tsunami itself, but they reflect a lesson from the mega-disaster'. For many Asian coastal areas, the message about greenbelts '... arrived too late. For decades, mangroves,

coral reefs and sand dunes have been degraded or destroyed by population pressures, poverty, and economic activities. The good news is that local communities are taking up the challenges without government help.'

Examples include:

● an old law which allows local communities in Tuntaset village in southern Thailand to manage their mangroves and transform an area devastated by the charcoal and shrimp-farming industries.

● the people of Jaring Halus in Indonesia who, for many years, managed their own mangrove forest using traditional methods. Now the government has asked them to co-manage mangroves in a wildlife sanctuary.

Had Chao Mai Marine National Park

Since signing the Ramsar Convention (see bottom right), Thailand has designated ten areas of wetland as 'Ramsar Sites'. They include the Had Chao Mai Marine National Park. This Ramsar Site is located in Trang Province and consists of three distinct wetland areas, with a range of wetland habitats that include mangroves, mudflats, sandy beaches and coral reefs. It is important in terms of its biodiversity:

- At least 212 bird species have been recorded in the area, including vulnerable and endangered species.
- It is home to at least 22 mammal species, including the endangered dugong.
- At least 75 fish species are found here – critical to local people's livelihoods.

Reliance on wetlands

- Chao Mai is home to about 10 000 people, many of whom make a living from fisheries either in canals, the Trang River or coastal areas (see photo).
- Local people rely on the wetlands as a source of water, for agriculture and aquaculture.
- At least 250 000 tourists visit the area each year, with local communities earning income from tourist activities.

Threats

Despite Chao Mai being listed as a Ramsar Site, biodiversity is still under threat from:
- habitat change (conversion of wetlands to aquaculture)
- over-fishing and destructive fishing activities
- a lack of pollution control.

Background

Ramsar Sites

The *Ramsar Convention on Wetlands* is an intergovernmental treaty established in 1971 in the Iranian city of Ramsar. Its mission is 'The conservation and wise use of all wetlands through local, regional and national actions and international co-operation, as a contribution towards achieving sustainable development throughout the world.' The Ramsar Convention recognises that wetlands are ecosystems which are highly important for biodiversity conservation and for human well-being.

The Ramsar Convention has 158 member states and there are 1720 wetland sites, totalling 159 million hectares (larger than the surface area of France, Germany, Spain and Switzerland combined) designated for inclusion in the Ramsar List of Wetlands of International Importance. These are known as 'Ramsar Sites'.

Over to you

1 Complete a table to show the advantages and disadvantages of the five approaches towards achieving sustainable yields in mangrove ecosystems.
   - community development organisations, e.g. Yadfon
   - Regional organisations, e.g. TVE Asia Pacific
   - NGOs, e.g. Wetlands International
   - International, e.g. Coastal Biodiversity in Ranong
   - Intergovernmental agreements, e.g. Ramsar Sites
2 In pairs, rank the projects in terms of:
   a costs and benefits
   b effectiveness
   Does one approach seem better than the others?

On your own

3 Use three outline maps of South-East Asia.
   a Identify the location of mangrove ecosystems and services on map 1.
   b Annotate the threats to mangroves on map 2.
   c Show how the loss of mangroves is being managed on map 3.
4 'Human well-being and sustainable yields are inter-linked.' Using the examples from mangroves in Units 3.7-3.9, discuss whether you agree.

Exam question: Assess the role played by different players in managing areas in which biodiversity is under threat. **(15)**

In this synoptic unit you'll explore a range of resources to assess the future of biodiversity.

Biodiversity and well-being

Biodiversity is a key resource. Its sustainable management is critical to our well-being now and in the future.

Biodiversity and the Millennium Development Goals

The protection and sustainable management of biodiversity – including genetic resources, species, and ecosystem services that support human development – is central to achieving the Millennium Development Goals (MDGs) adopted by world leaders at the UN Millennium Summit in September 2000 (see pages 214-217). Although one of the MDGs (MDG 7) deals most explicitly with biodiversity, wise use of biological resources is important for the full range of development priorities encompassed by all eight MDGs.

For example:

- **eradicating hunger** (MDG 1) depends on sustainable and productive agriculture, which in turn relies on conserving and maintaining soils, water, genetic resources, and ecological processes. The **capacity of fisheries** to supply hundreds of millions of the world's people with the bulk of their animal protein intake depends on the maintenance of ecosystems such as mangroves and coral reefs, which provide fish with habitat and sustenance.

- **Improving health and sanitation** (MDGs 4, 5 and 6) require healthy, functioning freshwater ecosystems to provide adequate supplies of clean water; and genetic resources for both modern and traditional medicines. The burden of **water and fuel collection** is lessened by keeping ecosystems intact and healthy.

- This contributes to MDG 3 on **gender equality** and empowerment of women, who are primarily responsible for these tasks.

Poverty and biodiversity are intimately linked. The poor, especially in rural areas, depend on biodiversity for food, fuel, shelter, medicines and livelihoods. Biodiversity also provides the critical ecosystem services on which development depends, including air and water purification, soil conservation, disease control, and reduced vulnerability to natural disasters such as floods, droughts and landslides. Biodiversity loss exacerbates poverty, and likewise, poverty is a major threat to biodiversity.
United Nations Development Programme

The Millennium Development Goals
1   Eradicate extreme poverty and hunger
2   Achieve universal education
3   Promote gender equality and empower women
4   Reduce child mortality
5   Improve maternal health
6   Combat HIV/AIDS, malaria and other diseases
7   Ensure environmental sustainability
8   Develop a global partnership for development

We will have time to reach the Millennium Development Goals – worldwide and in most, or even all countries – but only if we break with business as usual … So we must start now. And we must more than double global development assistance over the next few years. Nothing less will help to achieve the Goals.
United Nations Secretary General

▲ *Women and children collecting water from a new village pump in Burkina Faso*

The Convention on Biological Diversity

The Convention on Biological Diversity was signed by 150 government leaders at the 1992 Rio Earth Summit. Those who signed up, committed themselves 'to achieve by 2010 a significant reduction of the current rate of biodiversity loss at a global, regional and national level, as a contribution to poverty alleviation and to the benefit of all life on Earth.'

The Convention recognises that biological diversity is about more than plants, animals and micro-organisms and their ecosystems – it is about people and our need for food security, medicines, fresh air and water, shelter and a clean and healthy environment in which to live – in short it is about our well-being. But biodiversity is in decline at all levels and geographical scales. Targeted response options, e.g. through protected areas, resource management and pollution prevention programmes – can reverse this trend for specific species or habitats.

There have been some positive trends:
- Protected area coverage has doubled over the last 20 years; terrestrial protected areas now cover over 12% of the Earth's land surface.
- Water quality in rivers in Europe, North America, and Latin America and the Caribbean has improved since the 1980s.

> Failure to conserve and use biological diversity in a sustainable manner would result in degrading environments, new and more rampant illnesses deepening poverty, and a continued pattern of inequitable and untenable growth.
> **United Nations Secretary General**

▲ Ecotourists in Ranthambore National Park and tiger sanctuary in India. Protection of ecosystems is especially important in environments where biodiversity loss is sensitive to changes in key drivers.

What are the prospects for achieving the target?

According to the Millennium Ecosystem Assessment Biodiversity Synthesis 'an unprecedented effort would be needed to achieve, by 2010, a significant reduction in the rate of biodiversity loss at all levels'.

- Although rates of habitat loss are decreasing in temperate areas, they are projected to increase in tropical areas.
- Time lags of years or even decades can exist between taking action and observing the impact of that action on biodiversity and ecosystems.
- The food and agriculture sector needs to improve agricultural efficiency, plan for agricultural expansion, moderate the demand for meat, and halt destructive fishing practices, to reduce its negative impact on biodiversity.
- Trade policies have a big impact on economic development, including food and agricultural production. Measures to protect biodiversity should be included with any changes to trade arrangements.
- Strategies to reduce poverty need to include the conservation and sustainable use of biodiversity.

> Short-term goals and targets are not sufficient for the conservation and sustainable use of biodiversity and ecosystems. Given the characteristic response times for political, socio-economic and ecological systems, longer-term goals and targets (such as for 2050) are needed to guide policy and actions.
> **Millennium Ecosystem Assessment Biodiversity Synthesis**

Biodiversity – what is the future? - 2

The Millennium Ecosystem Assessment

The Millennium Ecosystem Assessment (MEA) assessed the consequences of ecosystem change for human well-being from 2001-2005. The MEA involved the work of more than 1360 experts worldwide. Their findings provided an up-to-the-minute scientific appraisal of the condition of and trends in the world's ecosystems and the services they provide, as well as the scientific basis for action to conserve and use them sustainably.

The MEA's view is that there would be less biodiversity today if NGOs, governments – and to a growing extent business and industry – had not taken action to conserve biodiversity, mitigate its loss and support its sustainable use. However, further progress will require initiatives to tackle all the ways in which biodiversity is being lost and ecosystems degraded. There is a range of possible options:

- **Conservation and protection**, e.g. conservation of species both in their native areas (in situ) and in species banks, such as the one at Kew Gardens in London (ex-situ), and protecting and restoring ecosystems.
- **Sustainable use**, e.g. incorporating conservation measures into farming, fishing and forestry, and promoting economic activities such as eco-tourism.
- Increasing the effectiveness of various international agreements, e.g. Ramsar Sites.
- Intensifying agricultural production using sustainable methods, and preventing human activities which deplete ecosystem resources and biodiversity.

Four MEA scenarios

The MEA has put together four possible scenarios for the future, focusing on alternative approaches to sustaining ecosystem services.

Scenario 1:
Global Orchestration

In this scenario, policies improve the well-being of those in poorer countries by removing trade barriers and subsidies. Environmental problems are dealt with in an ad-hoc way. Ecosystem services deteriorate.

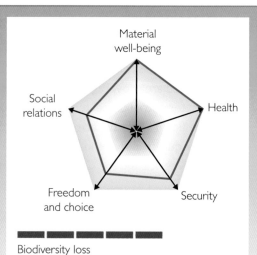

Biodiversity loss

Scenario 2:
Adapting Mosaic

In this scenario, local and regional management is the main approach to sustainability. Some regions will manage ecosystems well, but others will not.

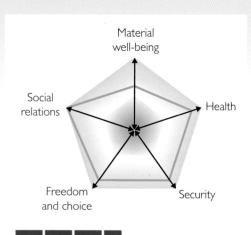

Biodiversity loss

Is it possible to reconcile the maintenance of a healthy biodiversity with a healthy economy?

What do you think?

Changes in biodiversity due to human activities were more rapid in the past 50 years than at any time in human history, and the drivers of change that cause biodiversity loss and lead to changes in ecosystem services are either steady, show no evidence of declining over time, or are increasing in intensity. Under the four plausible future scenarios developed by the MEA, these rates of change in biodiversity are projected to continue, or to accelerate.
Millennium Ecosystem Assessment Biodiversity Synthesis

Synoptic question

In order to answer this question fully, you need to choose either the Daintree rainforest (Units 3.4-3.6), or mangroves in South-East Asia (Units 3.7-3.9). Use examples from these units in your answer.

a Assess the importance of biodiversity in the Millennium Development Goals **(12)**

b How far is it possible to reconcile the desire for development with the need to manage biodiversity? **(14)**

c Analyse the four MEA scenarios, and assess which you think is most likely to be achieved **(14)**.

Science can help to ensure that decisions are made with the best available information, but ultimately the future of biodiversity will be determined by society.

What do you think ?

Scenario 3:
Order from Strength

In this scenario, the rich protect their borders by attempting to confine poverty, conflict, environmental degradation and destruction of ecosystem services to areas outside their borders.

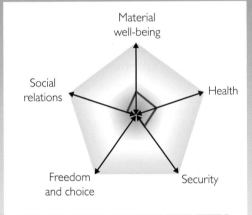

Biodiversity loss

Scenario 4:
The TechnoGarden

In this scenario, people push ecosystems to their limits of producing the optimum amount of ecosystem services for humans through the use of technology. However, the success in increased production of ecosystem services can undercut the ability of ecosystems to support themselves.

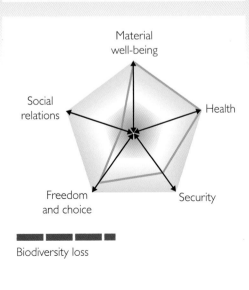

Biodiversity loss

Key to the diagrams

Each arrow in the star diagrams represents one component of human well-being. The area marked by the lines between the arrows represents well-being as a whole. The 0 line represents the status of each of these components today. If the coloured line moves more towards the centre of the pentagon, this component deteriorates between today and 2050; if it moves towards the outer edges of the pentagon, it improves.

But there is a trade-off. Loss of biodiversity is least in the two scenarios that feature a proactive approach to environmental management (TechnoGarden and Adapting Mosaic) while the Global Orchestration scenario does most to promote human well-being and achieves the fastest progress toward the MDG of eradicating extreme poverty. The Order from Strength scenario performs badly on both sets of objectives.

Chapter summary

What key words do I have to know?

There is no set list of words in the Specification that you must know. However, examiners will use some or all of the following words in the examinations, and would expect you to know them and use them in your answers. These words and phrases are explained either in the Glossary on pages 330-333, or within this chapter.

alien species
anaerobic
aquaculture
biodiversity
biodiversity hotspots
biomass
biome
buy-back
carbon sequestration
conservation
coral bleaching
cultural services (in terms of biodiversity)
eco-region
ecosystem diversity

endemic
endemism
eutrophication
genetic diversity
Gersmehl's nutrient cycles
global orchestration
inter-tidal area
leaching
lenticels
mangrove
net primary productivity
organic productivity
permafrost
photosynthesis

pivotal areas
pneumatophores
prop roots
provisioning services (in terms of biodiversity)
Ramsar Sites
red, black and white / grey mangroves
regulating services (in terms of biodiversity)
species diversity
supporting services (in terms of biodiversity)
sustainable use
sustainable yield
trophic
zonation

Films, books and music on this theme

Music to listen to

'Big Yellow Taxi' – Pick your version, either by Counting Crows (2003), Amy Grant (1994), or the original by the song's writer – Joni Mitchell (1970)
'Earth Song' (1995) by Michael Jackson

Books to read

Six Degrees: Our Future on a Hotter Planet (2008) by Mark Lynas
The Global Casino (4th edition 2008) by Nick Middleton

Films to see

'State of the Planet' – a series made by the BBC in 2001 and presented by David Attenborough

Try these exam questions

1 Explain the pattern of alien species invasions, and suggest the possible impacts of alien species on ecosystems. **(10)**

2 Referring to examples, discuss the threats to biodiversity hotspots and why these threats could prove critical. **(15)**

What do I have to know?

This chapter is about how some nations and players have a disproportionate influence over regional and global decision-making. The geography of power develops over time; some nations gain power and influence, while others lose it. The nature of power varies, from direct control to indirect, e.g. trade, culture, and resources. The Specification has three parts, shown below, with the examples used in this book. (You will also find it useful to refer to Chapter 5 to help you understand further the impacts of some of the world's superpowers.)

1 Superpower geographies

What you need to learn	Examples in this book
• What 'superpower' means. • How international influence develops, e.g. military, economic, cultural. • Theories of superpower emergence. • Power changes over time; some decline while others rise. • The influence of superpowers varies regionally and globally.	• Defining 'superpower' by size, population, resources, military and economic indicators and influence. • UK colonialism – the development of trade between the UK and its colonies; cultural impacts, e.g. cricket, English language. • Modernism, Mackinder's heartland theory, capitalism, evangelical Christianity, social Darwinism; dependency and development theory. • The emergence of the USA and USSR as superpowers; China as an industrial giant; the EU. • The collapse of colonialism and communism. • Economic influence – modernisation theory versus communism; Global organisations with US influence – IMF, World Bank.

2 The role of superpowers

What you need to learn	Examples in this book
• Power can be maintained directly (e.g. colonial rule) or indirectly.	• European colonialism; legacies of colonialism (Ghana). • Dependency theory and the influence of trade (e.g. neo-colonialism, trade, aid, debt).
• Superpowers play a key role in international decision-making through direct and indirect processes.	• Superpower influence through economic blocs (e.g. the EU), or international groups (e.g. G8, IMF, NATO, UN, WTO, World Bank). • How these maintain power, e.g. investment, military force.
• How superpowers control trade, power and global influence.	• Energy and resource influence – Russia as an energy superpower, China as a consumer of resources (e.g. Australian iron ore).
• Superpowers exert cultural influence.	• Global culture – brand names, film, TV and music, publishing. US culture and companies (e.g. News Corporation), India and Bollywood.

3 Superpower futures

What you need to learn	Examples in this book
• Economic superpowers and their demands on energy, water and land. • Power is shifting globally, which has implications for the majority of the world.	• The changing world order – the BRICs. • Growing resource demands by superpowers, e.g. China on Australia's iron ore reserves. • The environmental impacts of economic growth, e.g. air pollution in Beijing. • The rise of China; Russia's re-emergence as a superpower.
• Shifting power can cause tensions or conflict between global cultures.	• Russia's re-emergence; its political and economic influence as an energy provider – relations with its immediate neighbours and Western Europe.

4.1 Changing world order

In this unit you'll learn how the balance between superpowers can change over time.

Die Mauer ist gefallen!

The year 1989 was an important year for European and world history. Europe was divided between the capitalist West and the communist East (controlled by the Soviet Union or USSR). Germany was also divided into West Germany and East Germany, with the city of Berlin split in two by the Berlin Wall. Throughout 1989, there were demonstrations and protests in Eastern Europe against the communist governments there – beginning in Poland. The demonstrators were demanding more personal freedom, and expressing hatred of the secret police and the restrictions of living under communist rule.

On 9 November 1989, queues started to build up at the East Berlin border checkpoints through which people had to pass to get to West Berlin. That evening, the rules were relaxed and people began to pass through the checkpoints at will. Watched by the border guards, some of the crowd climbed on to the Berlin Wall itself and started attacking it with pickaxes and sledgehammers. 28 years of enforced division in the city had ended.

▲ East German border guards begin to remove the Berlin Wall on 12 November 1989, watched by crowds of West Berliners

All change!

You do not always recognise it when you are living through a revolution, but everyone knew in 1989 that big changes were afoot in the communist countries of Eastern Europe.

For those witnessing the dismantling of the Wall, this was more than just a single event. It marked not only the end of the Wall, but also of the East German government and a divided Germany. Three weeks earlier, the East German leader, Erich Honecker, had been forced to resign after mass demonstrations from crowds demanding freedom. Within four months, free elections – the first for over 50 years – took place in East Germany. Within a year, West and East Germany were reunified for the first time since 1945. Government functions in the West German capital, Bonn, were gradually relocated to the pre-war German capital, Berlin.

For anyone who didn't experience the Wall, it will be hard to imagine what an overwhelming feeling of relief, of joy, of unreality filled one that this monster was dead, and that people had conquered it. The years of degrading searches at border crossings, the loved ones who were walled in on the eastern side. The East German restrictions, regulations, bureaucratic border formalities, the dogs and soldiers with machine guns, the main streets cut off suddenly by this cold, hard cement wall – all of this was suddenly defeated, and one could dance on its dead body or chip off a piece of it with a hammer. Just days before, either action might have meant arrest or even being shot. I tell you, it was a giddy, delirious feeling – even for someone completely sober.
Richard Pinard's experience of the Berlin Wall, quoted on the BBC website

In less than two years, the communist bloc based on the USSR (the world's second superpower after the USA) had been dismantled. The communist countries of Eastern Europe, such as Poland and Hungary, became independent. At the end of 1991, the USSR itself broke up into 15 independent republics – the largest of which is Russia. Throughout the region:

● different political parties replaced decades of single-party communist rule
● free elections were scheduled
● free-market economies replaced State control, and promised new wealth and jobs.

Which do superpowers consider most important – economic or military strength?

What do you think ?

A new generation of superpowers?

The past two decades have brought enormous political and economic changes.

● The former USSR used to be the world's third largest economy. Russia, the USSR's largest component, is now re-emerging as a major economic power in its own right – and is a member of the G8.

● China's economic influence is growing rapidly, following intensive industrialisation.

● India and Brazil are emerging as major economies.

● The EU, which now includes former communist countries, has expanded and become the world's largest economic trading bloc.

The expansion of the EU to 27 member states means that it now competes with the USA for global economic dominance. It exceeds the USA in terms of economic power (see the table), although its GDP per capita, productivity per employee and other indicators – such as Internet usage and car ownership – lag behind.

▲ Australian wagons of iron ore heading for port and then China. In 2007, China absorbed as much iron ore as Australia could provide.

	USA	EU
GDP total	1.62 trillion	2 trillion
Growth per year	2.2%	2.6%

Comparing economic data between the USA and the EU (the data are in US dollars)

Now, a new era of superpowers is beginning. One global writer, Richard Scase, predicts a rapid rise in the economies of the BRICs – Brazil, Russia, India and China. He predicts that, on current trends:

● India will overtake the UK's economy in 2022 and Japan's in 2032. By 2050, India's economy will almost rival that of the USA.

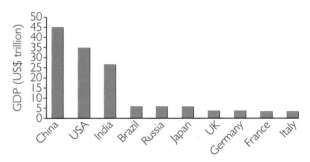

▲ The world's economic superpowers by 2050

● By 2050, the BRICs will produce US$84 trillion (see graph) – compared to the US$35 trillion of the USA.

Who are the superpowers? – 1

In this unit you'll learn about the meaning of 'superpower', and how to decide whether a country is a superpower or not.

What does 'superpower' mean?

Type 'superpower' into Google and you'll find a range of definitions. One, by a writer called Alice Lyman Miller, defines superpower as a country with 'the capacity to project dominating **power** and **influence** anywhere in the world, sometimes in more than one region of the globe at a time.' By gaining power, a country may become a **hegemon**, or a supreme power over others. The country most closely linked to this definition now is the USA.

The use of the word 'superpower' is recent; it was used towards the end of the Second World War to describe the global influence held by the British Empire, the USA and the Soviet Union (USSR). Over 60 years later, the world has changed:

- The British Empire disappeared after the war and became the Commonwealth, as its members, such as India, Pakistan, Kenya and Jamaica, gained independence.
- The Soviet Union collapsed in 1991, breaking up into 15 independent states.
- Only the USA remains as a major world power.

In another sixty years the world is likely to be different again. Many economists now talk about the BRICs – Brazil, Russia, India and China – becoming superpowers in their own rights within 60 years.

But how does a country gain power and influence? A number of factors help to assess who might rival the USA now or in the future:

- Physical size and geographical position
- Resources
- Population size
- Economic power and influence
- Military force
- Dominant belief systems

Physical size and geographical position

Russia is the world's largest country (see the table), covering a vast area from the Baltic to the Pacific. Physical size and position are important, because each one

Country	Approx. area (km²)	Time zones	Land neighbours it potentially influences
1 Russia	17 million	11	Azerbaijan, Belarus, China, Estonia, Finland, Georgia, Kazakhstan, Latvia, Lithuania, Mongolia, North Korea, Norway, Poland, Ukraine
2 Canada	10 million	6	USA
3 USA	9.8 million	4	Canada, Mexico, and only 75 km from Russia across the frozen Bering Strait
4 China	9.6 million	5	Afghanistan, Bhutan, Burma, India, Kazakhstan, Kyrgyzstan, Laos, Mongolia, Nepal, North Korea, Pakistan, Russia, Tajikistan, Vietnam
5 Brazil	8.5 million	4	Argentina, Bolivia, Colombia, French Guiana, Guyana, Paraguay, Peru, Suriname, Uruguay, Venezuela
6 Australia	7.7 million	2-3	None

▲ The world's six largest countries

determines the area over which a country has potential influence. Larger countries also tend to have greater resources.

- The future of the Arctic, and the resources that lie beneath it, lies mainly in the hands of Russia and Canada, because they are the largest countries bordering on the ocean.
- Russia's relations with its neighbours are complex, because there are 14 of them and they vary hugely between, for instance, Norway (a liberal-thinking democracy) and China (an authoritarian communist one-party state). Russia can exert its influence over them in different ways.

◀ A map showing the world's iron ore exporting countries. Each country is drawn in proportion to its volume of exports of iron ore. It shows the iron ore 'superpowers', e.g. Chile and Australia.

▼ The world's ten most populous countries

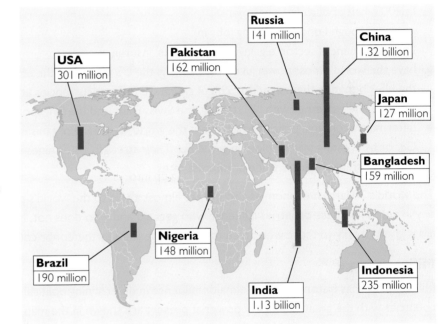

Russia 141 million

China 1.32 billion

Pakistan 162 million

USA 301 million

Japan 127 million

Bangladesh 159 million

Nigeria 148 million

Brazil 190 million

Indonesia 235 million

India 1.13 billion

Resources

Some resources are critical to economic development, e.g. oil or metals. Iron ore is the basis of a steel industry, while energy sources such as oil, gas and coal are used to generate most electricity or provide fuel for industrial processing. However, large oil or iron ore reserves do not guarantee economic development. For instance:

● many reserves are in the hands of TNCs from wealthier overseas countries, e.g. Shell and BP earn profits from Nigeria's oilfields.

● Australia has huge reserves of iron ore, but it exports almost all of it and gains little of the value added by manufacturing.

Only when countries withhold production and create shortages do their resources provide a means of influence. For instance:

● in the past, countries in the Middle East have threatened to withhold oil supplies unless the price per barrel increases

● Russia has recently threatened to cut supplies of gas to Ukraine and Europe from its huge Siberian reserves.

Population size

Some countries see population as a key to economic success. Economic growth can only be maintained with a sufficient labour force. Since the 1990s, the UK has used immigration as a means of supplementing the domestic labour force in areas with shortages, such as doctors

and agricultural labourers. China and India each use their population as a source of cheap labour in manufacturing.

A large population also acts as a spur to economic growth, because it provides a large market. Countries in the EU have access to 550 million people – much larger than their individual populations.

Yet a large population is not crucial to economic growth. Singapore's population is about half that of London, yet it controls much of South-East Asia's economy. Its low tax and status as an Export Processing Zone (where no duties or tariffs are payable) attract business and make it a key player in Asia's economy.

Who are the superpowers? – 2

Economic power and influence

The map opposite shows the world's top 12 economies. These countries influence much of what happens in the global economy. Between them, they:

- earn 68% of global GDP. They not only earn the greatest amounts, but spend it too. The four highest-spending nations are the USA, Germany, the UK and Japan – not surprising, given that between them they earn nearly half of global GDP.

- control investment. Most investment is targeted in these 12 countries because companies are more likely to make profits. In 2005, half of all global investment passed through London, creating jobs in London's financial sector, e.g. investment banks, insurance.

- have the world's most powerful currencies, e.g. the US dollar and the euro. The EU and the USA between them produce 59% of global GDP. Their currencies are therefore seen as 'safe', together with UK sterling, the Swiss franc and the Singapore dollar.

- determine economic policies which affect the world, by joining organisations such as the G8, or by creating trading blocs, e.g. the EU. Their strength determines what happens to – for instance – global trade or interest rates.

The world's 20 largest economies also donate almost all aid to the world's poor. Although tiny in amounts, these countries influence who gets aid and who does not. In some cases, aid is granted only to those who agree to policies specified by the donor countries.

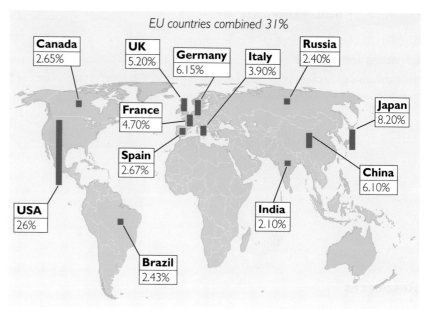

EU countries combined 31%

Canada	2.65%
UK	5.20%
Germany	6.15%
Italy	3.90%
Russia	2.40%
France	4.70%
Japan	8.20%
Spain	2.67%
China	6.10%
USA	26%
India	2.10%
Brazil	2.43%

▲ *The world's 12 largest economies in 2007. The data show the percentage of global GDP produced by each country.*

Military force

Military force has historically been a major influence in determining political power. The countries with the greatest military power at present are shown in the map below. Some organisations try to balance this influence, e.g. the UN, where the Security Council is responsible for world peace. The UN Security Council consists of five permanent members (the USA, UK, Russia, China and France) and 10 non-permanent members, who are elected for two years. It approves military intervention only when it considers it to be justified. The permanent membership is based on those countries which possessed nuclear weapons in the 1970s. Other countries now possess nuclear weapons (India, Pakistan, and - according to many sources - Israel) but they are not part of the permanent Security Council.

Countries with the largest military in 2002. Like the map of iron ore, the countries are shown in proportion to the size of their military forces. ▼

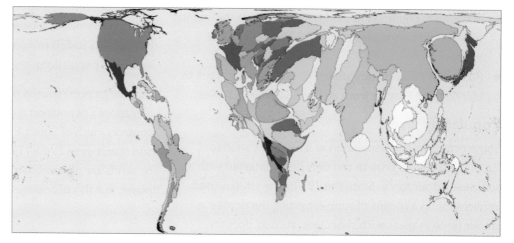

Dominant belief systems

Religion has been a source of conflict throughout history, e.g. the Crusades between Christians and Muslims in the Middle Ages. More often, though, religions seek to extend their influence through their teachings. European Christian states, such as Spain and Britain, used their empires from the sixteenth century onwards to expand the influence of Christianity throughout the South American and African continents. Now, Christianity and other belief systems vary in influence throughout the world, as the map shows. Religions gain influence over countries by influencing their politicians in particular (e.g. their policies on abortion, birth control or stem cell research).

▼ *The world's dominant religions*

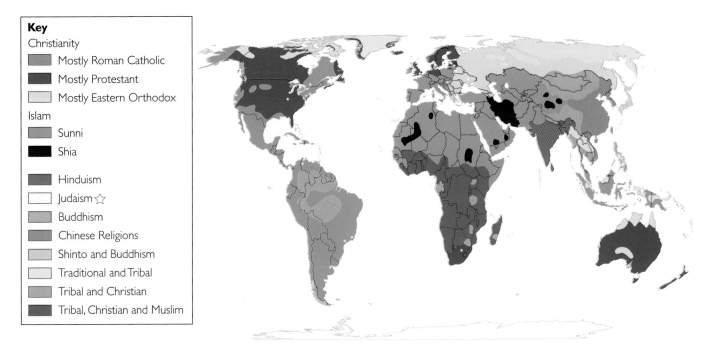

Key

Christianity
- Mostly Roman Catholic
- Mostly Protestant
- Mostly Eastern Orthodox

Islam
- Sunni
- Shia

- Hinduism
- Judaism ☆
- Buddhism
- Chinese Religions
- Shinto and Buddhism
- Traditional and Tribal
- Tribal and Christian
- Tribal, Christian and Muslim

However, religion is not the only form of belief. Global capitalism is promoted by the media, especially via global news channels like CNN. These channels are available on Sky, a part of American-based News Corporation, which has global coverage through Fox TV (North America), Sky (UK and Europe), and Star TV (Asia). While the media alone does not make the USA a superpower, it does help to maintain the USA's position by using its influence politically. For instance, Fox TV has been a strong supporter of the Iraq war and the 'war on terror' in the USA.

Over to you

1. In pairs, discuss and then rank which criteria are most important in creating the world's superpowers: physical size and geographical position, resources, population size, economic power and influence, military force, or dominant belief systems.

2. Using those criteria, use the data in this unit to select, with reasons, the top three world superpowers from this list: the USA, Russia, China, Brazil, the UK, France, Germany, Japan, India (plus one country not in this list but which you think deserves to be considered).

On your own

3. Research the 'worldmapper' website (www.worldmapper.org) and justify three **other** criteria to help decide which are the world's superpowers.

4. Write 400 words on 'Is the EU really the world's major superpower?'.

Are superpowers essential for keeping world peace?

What do you think?

In this unit you'll learn, in particular, how the British Empire became the world's first major superpower in modern times, and also theories behind colonialism.

Rule Britannia!

Imagine sitting in a geography class a century ago and reading the following passages from *A new geography comparative* – a textbook published in 1894 by Professor J.D. Meiklejohn:

'When the first European settlers visited [Australia], they found … not the smallest trace … of what is called civilisation.'

'The native Australian … is one of the most degraded of savages, with … no religion.'

Or this passage from another textbook, written in 1881 by Charlotte M. Mason:

'The aborigines, who are a miserable race of savages … do not seem able to learn the ways of civilised men.'

The three extracts demonstrate how these two Victorian writers saw the world. No mention was made in their textbooks about the skills with which Australia's aboriginal groups lived in the Australian desert environment; or that the aboriginal beliefs about spirits, land, and culture were as complex as Christianity.

At the beginning of the twentieth century, the British Empire was the biggest global superpower. The map shows the extent of British **colonial** rule in the 1920s. This rule did not happen by accident, but was part of a system by which British people believed that it was right for them to obtain colonies and rule over much of the world, both:

- politically, because land gave Britain power, and
- economically, because the colonies were seen as a source of cheap raw materials and a market for British manufactured goods.

However, the indigenous people of the colonies were looked down on, and racist language was often used to describe them. Colonialism was supported by a sense of moral superiority, fed by Christian teachings that God was on Britain's side. Christian missionaries helped to maintain control during colonial rule.

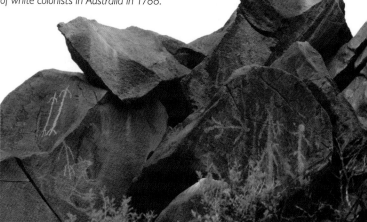

▼ *The Australian desert environment provided food, water and home to aboriginal peoples. These paintings are several hundred years old, well before the arrival of white colonists in Australia in 1788.*

- **Colonialism** is a system by which an external nation takes control of a territory in another part of the world, often by force. It then reinforces this control by settling the new colony with its own people.

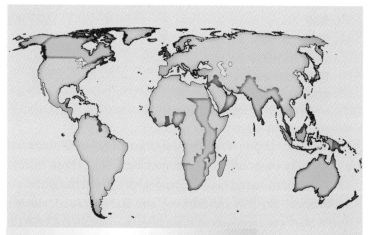

Rule Britannia!
When Britain first at Heav'n's command
Arose from out the azure main;
Arose, arose from out the azure main;
This was the charter, the charter of the land,
And guardian angels sang this strain:
Rule, Britannia! Britannia, rule the waves!
Britons never, never, never shall be slaves!

▲ *The extent of the British Empire in the 1920s. The areas shaded in pink were controlled by Britain. The lyrics of 'Rule Britannia' show that many Britons believed that they were fulfilling 'Heav'n's command'.*

How did colonialism develop?

It is easy to think of Africa as always having been a poor continent. This is far from the truth. Before colonialism, West Africa was home to many wealthy tribal kingdoms, which traded across the Sahara with the Arabic states of North Africa. Spices and metals were traded, together with salt. One city – Timbuktu – was a centre of learning, with a large university. Books were printed there long before printing reached Britain in 1476. Timbuktu is now a World Heritage Site, because of its fine buildings pre-dating the French colonial invaders.

▲ Timbuktu, in Mali, with buildings from the fifteenth century which help to give it World Heritage status today

Colonialism, particularly in Africa and the Americas, took place between the fifteenth and nineteenth centuries. European countries sought land overseas to expand their political control – exploring, invading, and taking control of large areas of the world.

- Many parts of Africa were colonised by the British and other Europeans (see the map). Trade in salt, spices and gold was established first, but was soon followed by the capture of millions of slaves. Some slave trading was done with the co-operation of tribal leaders, but much of it was done using force. The loss of generations of healthy adults impoverished many formerly wealthy African empires. Between 1550 and 1850, nearly 8 million West Africans were taken to Brazil alone to work as slaves.

> **Should colonisers say 'sorry' for slavery and any other misdeeds of colonialism?**
>
> **What do you think?**

▼ European control over Africa by 1914

- British manufacturers saw colonialism as a way of obtaining cheap raw materials and markets for their products. Industrialists were fiercely supportive of colonial government policy.
- Spain conquered Central and South America, together with the Portuguese, British, Dutch, and French. By 1560, Spain had overthrown the Inca and Aztec Empires and gained land and precious metals, especially gold and silver. The indigenous people were forced to work in mines and tobacco and sugar plantations, and millions died from European diseases like smallpox and syphilis.
- In the Caribbean, the new British and French rulers created sugar plantations by taking over land previously used by the local people to grow food. When most of the indigenous people had died through European diseases, slaves were brought in from West Africa to work in their place.

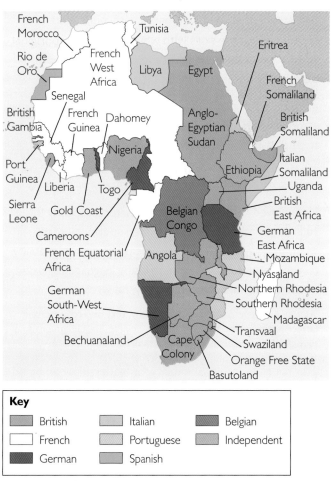

Maintaining colonial rule depended on military power. For instance, the British navy of 100 000 men – the world's largest – existed largely to protect Britain's trade routes and project its power overseas. The surprising thing about the British Empire was how cheap it was to defend. The defence budget for 1898 was £40 million, or 2.5% of GDP – little higher than the UK's defence budget now, and far less than was spent during the Cold War (see page 146).

Theories of colonialism

Colonialism arose from an **ideology**, or set of beliefs. Industrialists, for example, believed in free trade and the freedom to operate without interference – known as laisser-faire. In their eyes, they felt free to travel and conquer the world in the name of trade. However, the desire to colonise and dominate the world was also driven by philosophies, such as modernism, Mackinder's heartland theory, evangelical Christianity and social Darwinism.

Modernism

Modernism was a philosophy of ideas from the late nineteenth century about 'modern society'. Modernists believed that the world could be **improved** by human intervention and achievement – that Europe could improve the world by colonising. According to modernists, Europe was the continent most naturally able to give the rest of the world a lead. Through their conquest and knowledge of 'foreign' peoples, Europeans could portray themselves as modern, civilised, superior and progressive when compared with indigenous peoples, like the Australian aboriginals, who they believed were none of these things.

Early modernists also believed that the British were naturally intelligent. Lionel Lyde, Professor of Geography at University College, London, wrote a geography school textbook in 1895 in which he said:

'If we were not a clever industrious people, able to make things which other nations are glad to have instead of corn, we should be badly off.'

He believed that the British had a natural capacity for ruling over others. Believing in the superiority of the west, he assessed peoples overseas for their own ability to contribute to development. Using language regarded as completely unacceptable today, he wrote:

'As a race, Negroes are strong and healthy; they are really excellent farmers, being ... more intelligent than Chinese. The Chinaman, if not so useful as the Negro, is more useful than the Malay.'

This natural sense of suitability for leadership, combined with Mackinder's theory, led many modernists in Britain to believe that Britannia not only could, but **should** rule the waves.

Mackinder's heartland theory

In 1904, Mackinder, a British geographer, developed theories about global power and the continents. He believed that whoever controlled Europe and Asia – the biggest landmass – would control the world. He believed in a **heartland** extending from Eastern Europe into Russia, at the centre of which was a 'pivot' (see the map). Three things determined power in this region:

- Whoever ruled the pivot would command the **heartland**.
- Whoever ruled the heartland would command the **world island**, i.e. Europe, Russia, and into Asia.
- Whoever controlled the world island would rule the **world**.

In theory, the further away from the heartland a country was, the less influence it would have. On the map these are referred to as 'Inner or Marginal Crescent' and 'Outer or Insular Crescent'. Mackinder believed that Russia **ought** to be the world's global power, because its location and resources gave it natural advantages. He also believed that Britain **should** lie outside the heartland. However, he also believed that Russia had 3 disadvantages that stopped it being too powerful:

- Its many borders meant that it could be attacked from many directions.
- It had few all-year ports, because much of its coast is frozen in winter.
- It had a weak government in the early 20th century.

Mackinder believed that Britain disturbed the theoretical balance, and that its industrialisation and efficient government had shifted the heartland westwards. Two factors caused the heartland to shift geographically towards Britain:

- Its industrial revolution gave it **economic power**.
- Its naval strength gave it **sea power**, which did not form part of Mackinder's original theory.

Through these two factors, Mackinder believed that the UK could dominate everywhere from Western Europe to the Pacific, the Eurasian land mass and potentially the world – and become the new pivot.

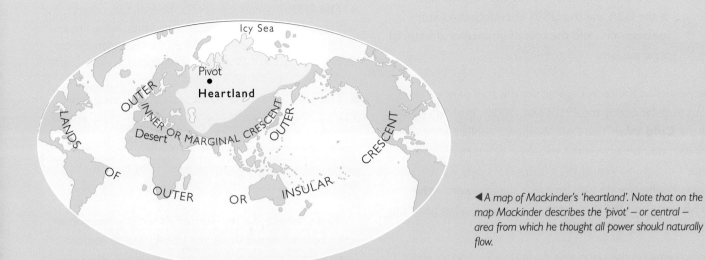

◀ *A map of Mackinder's 'heartland'. Note that on the map Mackinder describes the 'pivot' – or central – area from which he thought all power should naturally flow.*

Evangelical Christianity & social Darwinism

Part of the thinking behind colonialism was to spread 'the Christian word' through churches and schools – known as **evangelism**. The British sent missionaries to their African colonies with this aim in mind. In 1500, less than 5% of Africa's population was Christian. The proportion now is about 50%. However, evangelism brought with it two myths about colonial occupation:

- That the people already living in the new colonies were racially inferior.
- That colonialism was part of a divine call to civilise the inferior peoples.

This was based on **social Darwinism** – a theory put forward by Herbert Spencer. An elitist, he believed in natural superiority on the basis of 'might makes right'. His ideas then gained a boost from Darwin's theories about adaptation and natural selection – that the strong should survive over the weak, and therefore improve the gene pool. In this way, even liberal thinkers believed that colonialism brought benefits to the colonised.

▲ *A nun teaching a class in a missionary school in the Congo, Africa*

Over to you

1 In pairs, research and prepare a PowerPoint presentation on one area of the world that was colonised by European settlers. Divide the areas of the world amongst the class, e.g. some pairs could cover the Inca and Aztec Empires, and others the various African kingdoms. Focus on **a** extent of territory, **b** how they were treated by European settlers, **c** issues faced by them then and now.

2 Divide the class into six groups. Three should prepare a one-minute presentation in favour of **a** modernism, **b** Mackinder's heartland theory, **c** evangelical Christianity and social Darwinism. Three should prepare a one-minute statement against. Present your ideas and be prepared to challenge and question each other.

3 Finally, debate the question 'Were early colonialists racist, or just looking for economic opportunities?'.

On your own

4 Research Australian aboriginal peoples and colonialism. Focus on **a** lands occupied by aboriginal groups, **b** how they were treated by European settlers, **c** the later 'White Australia' policy, and **d** the issue of the 'stolen generations'.

Cold War and superpower rivalries

In this unit you'll learn about the development of the USA and the USSR as twentieth-century superpowers, and the role of international financial organisations.

Secret police and secret agents are the stuff of novels and films popular during the 1950s and 1960s – in an era known as the **Cold War**. The Cold War, i.e. a conflict without fighting, was set against the backdrop of international rivalry after the Second World War. It raged for over 40 years between the world's two post-war superpowers – **the USA** and **the USSR** – and their respective allies (until the break up of the USSR in 1991).

From being on the same side during the Second World War, relations between the two new superpowers froze in the late 1940s. The closest they came to a 'hot' war was in 1962, when the Cuban Missile Crisis brought them to the brink of nuclear war.

During the Cold War, relations between the USA and the USSR were hostile. The USA, believing in **capitalism**, was opposed to the Soviet **communist** government (see page 148). These different ideologies often prevented mutual understanding on major issues, such as:

- the Soviet policy toward Eastern Europe after 1945. Faced with growing hostility from the West towards

- **The USA** is a federation of 50 states and one district. It was first recognised as an independent nation in the Treaty of Paris (1783), which ended its War of Independence from Britain. It began with 13 states and continued to expand until the last two joined in 1959 (Alaska and Hawaii).
- **The USSR** (Union of Soviet Socialist Republics) was formed when four Soviet Socialist Republics merged in 1922 after the communist Russian Revolution of 1917 and the following Russian Civil War (1918-21). Also known as the Soviet Union, it is sometimes referred to by the name of its largest and most powerful constituent republic, Russia.

communism (and after Germany's terrible invasion in the Second World War), the USSR tried to remove any further threat of direct attack by occupying the countries of Eastern Europe, such as Poland and Hungary, as 'buffer states' – establishing a 'friendly' communist government in each one. These countries were heavily defended along their western borders, which became known as the 'Iron Curtain'.

- the use of nuclear weapons by the USA against Japan in 1945, followed by nuclear testing by the USSR in 1949. Over 40 years of nuclear expansion (or proliferation) followed, so that – by 1990 – the USA and USSR had 27 000 nuclear weapons between them. Even now, Russia and the USA control most of the world's nuclear weapons.

◄ The USSR before it broke up into 15 separate states in 1991. On its western border were the Eastern European 'buffer states' behind the Iron Curtain, which were communist for 44 years (1945-1989).

Estonia
Latvia
Lithuania
Belarus
Ukraine
Moldova
Georgia
Armenia
Azerbaijan
Uzbekistan
Turkmenistan
Tajikistan
Kyrgyzstan
Kazakhstan
Russia

UNION OF SOVIET SOCIALIST REPUBLICS

Key	
1	Poland
2	East Germany
3	Czechoslovakia
4	Hungary
5	Romania
6	Bulgaria
—	Iron Curtain

0   1000
Km

The USA as an economic superpower

By 1900, the USA had taken over from Britain as the world's largest economy, for the reasons shown on the map. It has managed to maintain that position, because:

- by 1940, it was the world's largest manufacturer of industrial and consumer goods.
- it produced the goods and finance to help rebuild Europe after the devastation of the Second World War.
- the US dollar was the world's major currency by 1950; 60% of all global bank reserves were held in dollars.
- the USA now produces over 25% of global GDP.
- the US military dominates global arms and defence spending.

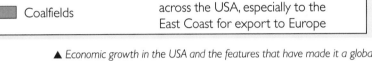

Key

▨ Steel production region	☐ Great Plains - one of the world's most productive grain/arable areas
■ Oil fields	◀── East-West rail network - moves goods across the USA, especially to the East Coast for export to Europe
■ Coalfields	

▲ Economic growth in the USA and the features that have made it a global economic superpower

The USSR becomes a superpower

By 1921 (the year before the formation of the USSR), Russia had been wrecked by the effects of the First World War, the 1917 Revolution and the following Civil War. The economy was devastated and industrial and farm production had been decimated (see the table). Millions of Russians had been killed and injured since 1914 and the outbreak of the First World War – and the continuing political unrest had caused production to collapse. People were starving in the cities.

Yet twenty years later, in 1941 – when Germany invaded – the USSR had the industrial strength to take on and defeat Hitler, and it also had the world's second largest economy – having overtaken Britain.

Stalin, the Soviet leader from 1922-53, was aware of anti-communist hostility across Europe, and he realised that the USSR's only hope of survival was military and industrial strength. Under a harsh authoritarian regime, a series of five-year plans dramatically increased economic production – all under State ownership. During Stalin's rule:

- industrial production multiplied several times over. Coal and iron ore mines fed steel industries located far away from national borders and potential attack.

- the State took control of farm production, to supply food to the cities.
- the armaments industry was expanded hugely.

	1913	1921	1937
Grain	80.0	37.6	(1934) 67.6
Coal	29.0	9.0	128.0
Oil	9.2	3.8	29.0
Iron	4.2	0.1	15.0
Steel	4.3	0.2	18.0

▲ Economic production in million tons in Russia (1913 and 1921) and the USSR (1937, at the end of the second five-year plan). Although the USSR was a much larger economic area than Russia, the growth in output between 1921 and 1937 was due mostly to the first two five-year plans.

After 1945, the USSR competed with the USA – matching its research in technology and testing and producing nuclear weapons by 1949. In 1961, the USSR beat the USA to put the first man into space (Yuri Gagarin). By the 1960s, it was the world's second economic superpower, although its GDP was less than half that of the USA.

Capitalism versus communism

Nowadays most countries are **capitalist**. Capitalism believes in the **market economy**, where people are free to:

- set up their own businesses (free enterprise)
- keep profits (subject to taxes)
- compete (prices of goods and services are determined by supply and demand)
- work for whoever they like (for a wage or salary).

Under this uncontrolled system, the poor may become poorer and the wealthy richer. Employers argue that keeping wages down keeps costs down.

By contrast, **communism** believes in communal wealth. It is based on the ideas of Karl Marx (1818-1883) of a classless society, communal ownership of property, and equal wealth distribution. The State owns land, banks, resources, industry, transport, and the media; it controls what will be made and how much. Capitalists argue that this stifles business enterprise; communists argue that all people are provided for equally.

There are now few examples of truly communist government; most communist nations, like China, are encouraging enterprise. Vietnam, Laos, North Korea and Cuba retain their communist governments. Kerala, in southern India (see photo), has had a communist state government since 1957.

Many countries operate a combination of free market and State control – promoting enterprise, but intervening through taxation, planning controls and subsidies at home, and through financial and trade regulations internationally.

▼ *Kerala in southern India. Although one of India's poorest states, its communist state government has invested heavily in education and health care; and life expectancy, medical care and education are at levels associated with much richer countries.*

The influence of the USSR

The balance of power between the USA and the USSR affected world geography. The USSR was keen to influence countries and promote communism. During the 1970s, there was considerable Soviet influence within Africa (see opposite). Communist China and Cuba were also influential in unstable African countries where there were active left-wing groups. The influence was of three kinds:

- Military assistance, with aid and equipment to left-wing groups, such as the MPLA in Angola.
- Attempts to destabilise countries such as South Africa and Zimbabwe (then called Rhodesia), where white minority governments survived.
- Financial aid. Several countries, including Guinea, used Soviet investment to develop their raw materials.

Soviet influence was not restricted to Africa; political, financial and military support was also given to countries in Central and South America. Meanwhile, China played a major role in supporting the communist Viet Cong forces during the Vietnam War.

◄ Soviet and other communist influence in Africa in the mid-1970s

- The USSR gave military assistance to 17 sub-Saharan countries. Two thousand Soviet military advisors were based in Africa. Military hardware worth more than $340 million was delivered between 1974 and 1976.
- China provided military assistance to 14 African nations. One thousand Chinese military advisors were based in sub-Saharan countries. It delivered military hardware worth US$28 million between 1974 and 1976.
- Cuba also had 7900 advisors and troops in Africa, mainly in Angola.

Guinea

Guinea's bauxite reserves were opened up by the USSR's largest economic aid programme in the region. The USSR was granted military privileges in return, and was allowed two naval reconnaissance aircraft over the region.

Somalia

Somalia received the largest Soviet support in East Africa – military equipment worth US$165 million. Numbers of Soviet military advisors were increased to 1000 in 1974. Soviet naval reconnaissance aircraft, operating from Somali airfields, gave Moscow the potential to cover the entire Indian Ocean.

Angola

The USSR supported the Marxist MPLA with arms and equipment in the civil war, including civil and military aircraft.

Uganda and Tanzania

The USSR provided military assistance to both countries.

Republic of Congo

The USSR provided military assistance. In return, it was given access to natural resources and was allowed to establish a military base in the country, which was seen as a danger to the white minority governments in Rhodesia and South Africa. This base also allowed reconnaissance into the South Atlantic.

The influence of the USA

The spread of communism throughout Eastern Europe, China and Korea in the mid-to-late 1940s caused shockwaves in the USA. By the mid-1950s, American money was pouring into India and South-East Asia, because the USA feared further communist expansion. The fight against communism dominated US foreign policy until the collapse of the USSR in 1991. The American promotion of economic development through **modernisation theory** was to dominate global economic development. Two organisations, established after the Second World War, helped the USA to achieve this influence – the International Monetary Fund (IMF) and the World Bank (see pages 150–151).

Background

Modernisation theory and the USA

The USA aimed to prevent communism by increasing economic development, using modernisation theory. Modernisation theorists aimed to explain that:

- poverty was a trap or cycle
- traditional family values in poorer countries had held their economies back, preventing, for instance, geographical mobility
- capitalism was THE solution to poverty
- investment and development loans to countries closest to China and the USSR could prevent the spread of communism.

Their aim was to help those countries close to communist areas, where war damage had brought economic collapse. Investment was poured into Japan, India, Singapore, South Korea, Taiwan, Thailand, and the Philippines in South-East Asia – as well as Mexico. Many of these countries were ex-colonies struggling in their first years of independence. All are now major economies.

Modernisation policies remain important now. With the collapse of communism in the USSR in 1991, investment and aid packages were made a priority for ex-communist countries.

International financial organisations

The International Monetary Fund (IMF)

The IMF was formed in 1944 at Bretton Woods, USA, to stabilise currencies after the Depression of the 1930s and the devastation caused by the Second World War. Forty-four governments initially joined, to create a fund (mostly paid for by the wealthiest countries) that could be loaned out to help those countries in debt – thus helping to stabilise their currencies and economies. The intention was that poverty – and communism – would therefore be prevented.

The IMF now has 185 members. However, the members are not equal – voting rights are proportional to the amounts invested. The world's leading 21 economies invest most of the money and therefore control over 70% of the votes between them (see the table). Of all the votes:

- the USA has nearly 17%, because it has the largest economy
- the EU countries combined have 25.7%
- the Western developed countries between them have 54.5%
- the industrialising economies of Brazil, Russia, India and China (the BRICs) have only 9.7%
- most of Africa's poorest countries have less than 1% between them.

Therefore, the IMF reflected American and European interests; it loaned money to stabilise those countries threatened by communism. It also imposed conditions on those borrowing money, which were decided by its wealthy members. For example, in the 1980s, the right-wing governments in the USA and the UK created IMF conditions which forced developing countries to cut spending on health and education in return for re-financing their debts. You can read more about this later (see pages 186 and 203 on Structural Adjustment). The IMF's main role now remains to stabilise countries faced with debt, but its impacts can be harsh.

The World Bank

The World Bank was also formed at Bretton Woods in 1944. Like the IMF, its role is to be a bank to finance development. Its first loan was to France for post-war reconstruction. It also focuses on natural disasters and humanitarian emergencies.

By the 1950s, it began to finance the development of ex-colonies. It gained a bad reputation in the 1970s and 1980s for financing projects which were either environmentally damaging (e.g. developing the Amazon rainforest), or beyond the ability of developing countries to repay. By the 1990s, its focus was on debt. Now it aims to achieve the Millennium Development Goals to eliminate poverty and implement sustainable development (see pages 214-217). Its decision-making structure is like that of the IMF; the USA has 16% of the votes, and power lies in the hands of the wealthiest economies.

IMF member	% of total votes
1. The USA	16.8
2. Japan	6.0
3. Germany	5.9
4. France	4.9
5. The UK	4.9
6. China	3.7
7. Italy	3.2
8. Saudi Arabia	3.2
9. Canada	2.9
10. Russia	2.7
11. The Netherlands	2.3
12. Belgium	2.1
13. India	1.9
14. Switzerland	1.6
15. Australia	1.5
16. Mexico	1.4
17. Spain	1.4
18. Brazil	1.4
19. South Korea	1.3
20. Venezuela	1.2
21. Sweden	1.1
The remaining 164 countries	**28.6**

▲ The voting rights of IMF members. The votes are proportional to the amounts invested by each country in the IMF.

How good a solution to poverty is capitalism?

What do you think?

The IMF and the World Bank in action

1 Singapore

Singapore's fall to Japan in 1942, and its occupation until the Allies retook control in 1945, led to huge damage. Fearing communism, the USA (supported by the IMF) encouraged rebuilding through loans and investment. Singapore's economic recovery was rapid. It became one of the 'Asian Tigers', the economies of South-East Asia which grew rapidly from the 1970s onwards (see below).

1979	5.9 billion
1989	27.5 billion
1999	98.0 billion
2007	153.0 billion

◀ Singapore's GDP (in US$)

▲ Singapore – South-East Asia's economic leader

2 Japan

By the end of the Second World War, in August 1945, Japan had been devastated (see the photo); cities, industries and transport networks had been severely damaged. The Allies wanted to encourage Japan's recovery, because of fears that the USSR might invade, and they occupied the country between August 1945 and 1952. The new Japanese government:

- gained substantial loans and grants from the World Bank and IMF.
- handed back territory acquired after 1894.
- created universal suffrage (the right to vote), giving power to Parliament instead of the Emperor when the Allies left in 1952.
- forbade Japan ever to lead a war overseas or maintain an army.
- broke up powerful interest groups, e.g. companies, landowners, government officials and police.

In spite of social unrest, investment in Japan's economy by the IMF and/or World Bank led to a rapid recovery, with rising living standards and political stability. By the 1970s, Japan had overtaken the USSR as the world's second largest economy.

▲ Hiroshima – Japan's most extreme destruction, following the dropping of the first atomic bomb by the USA on 6 August 1945

Over to you

1 Compare the USA and the USSR in terms of **a** how they became superpowers, **b** dates during which they were superpowers, **c** how they maintained their influence.
2 In pairs, compare the strengths and weaknesses of communism and capitalism as theories.
3 In pairs, research one communist country (e.g. North Korea), and one where the World Bank invested in development (e.g. South Korea). Compare their **a** economy, **b** quality of life, **c** education and welfare systems, **d** environment, **e** personal freedoms. Feed back your findings in class.

4 Discuss in class: 'Should donors retain greater power in the IMF and the World Bank than borrowers?'.

On your own

5 Research material about Japan's post-war occupation, and the recent Iraq conflict. Compare the US occupation of Japan with that of Iraq under **a** motives, **b** methods used, and **c** successes and failures.

In this unit you'll learn whether gaining independence from British colonial rule has given Ghana true freedom.

It is Ghana's 50th birthday

The year: 2007. The event: a discussion among Ghanaians, aged 20-30, entitled 'Ghana@50: success or failure?' – to celebrate 50 years of Ghana's independence from British rule.

Frank Agyekum, speaking for the Ghanaian government, is having a rough time. 'Ghana has done well', he says. The audience groans. 'Ghana is a thriving democracy with a stable economy', he claims; 'kids go to school and are fed'.

The crowd do not agree. Samuel Ablakwa, a graduate, explains. 'So many kids are still on the streets hawking,' he says. 'You have to pay for uniforms and textbooks.' He questions Ghana's prosperity. 'I'm still carrying buckets [of water] on my head – at my age! Maybe our economists are cooking the figures.'

Freedom from colonialism?

In 1957, Ghana gained independence from British colonial rule – the first colony in sub-Saharan Africa to do so. Fifty years later, some aspects of that rule still survive (for example, English is Ghana's official language). British colonial rule did bring some benefits to Ghana, such as schools and hospitals. However:

- the best-paid and managerial jobs were held by the British

	1987	2007
1 Economic indicators		
GNP total	US$5.7 billion	US$14.9 billion
GNP per capita	US$385	US$650
Annual growth rate	5.3%	6.2%
Annual Inflation rate	39%	11%
Unemployment	26%	11%
Exports	US$863 million (60% from cocoa. Also: timber, gold, bauxite, aluminum and tuna)	US$4.2 billion (from gold, cocoa, timber, tuna, bauxite, aluminum, manganese ore, diamonds, horticulture)
Main export markets	EU countries 25%, USA 23%	EU countries 40%, USA 7%
Imports	US$783 million (petroleum 16%, consumer goods, foods, capital equipment)	US$8.1 billion (capital equipment, petroleum, foods)
Main import sources	USA 10%, EU, Japan, South Korea	Nigeria 17%, China 13%, EU, USA, South Africa
Debt	US$3.3 billion	US$3.4 billion
2 Social indicators		
Population growth rate	2.9% per year	1.97% per year
Birth rate	42 per 1000	29.9 per 1000
Death rate	10 per 1000	9.6 per 1000
Infant mortality	68 per 1000 live births	53.6 per 1000 live births
Life expectancy at birth	59.5 years	59.1 years
Fertility rate	5.6 children per woman	3.9 children per woman
Literacy of adults	30%	58%

▲ *Comparing Ghana's indicators of development in 1987 and 2007. Were the young Ghanaians right – that little progress has been made? (Note that aid from overseas countries, such as the UK, which is used among other things to help reduce Ghanaian debt, is not included.)*

- **Neo-colonialism** means 'new' colonialism, where countries remain under control from overseas – even though they are supposedly independent.

- Ghana's crops and minerals were exported to Britain in a raw state, so Britain – not Ghana – processed and added value to them.

In 1957, Ghana's GDP per capita was equal to that of South Korea. But South Korea has developed rapidly since then, while a third of Ghana's people live on less than a dollar a day. Ghana still depends on decisions made by the world's wealthy nations – a situation known as **neo-colonialism**. Unlike colonialism, neo-colonialism is often hidden; it is difficult to identify where the power lies.

The rest of this unit examines the effects of neo-colonialism on Ghana.

Who is really in charge?

The cocoa trade has always been important to Ghana. During colonial times, it was the largest producer of cocoa in the world. The British colonial government used to dictate the price that the Ghanaian cocoa farmers received. Did things change after independence? The short answer is no! Today, three external influences control the price of cocoa: **commodity** traders, overseas tariffs, and the World Trade Organisation (WTO).

1 Commodity traders

The price of cocoa is not decided in Ghana, but in the **commodity trading exchanges** in London and New York. Buyers seek supplies of cocoa for large companies like Cadbury or Nestle, where continuous supply is essential. To guarantee supplies, and to get the best prices, buyers trade in the **futures market**, i.e. they buy supplies now to ensure delivery in 3-6 months. Cocoa prices depend on global supply and demand, which may vary. Other countries produce cocoa besides Ghana; Ivory Coast is now the world's largest producer. It means that if prices from Ghana are too high, dealers will purchase cocoa from other, lower-priced countries. Therefore, there is downward pressure on prices.

As a result of these factors, the price of cocoa on the world market is volatile (see the graphs).

- From 1991 to 1995 the price of cocoa changed 60 times.
- From 1996 to 2002 it changed 90 times.
- From January 1991 to December 1993 it increased by 112%.
- From June 1998 to December 2000 it dropped by 32.5%.

When prices fluctuate like this, it impacts badly on exporting countries and their workers. Unstable prices result in irregular income for workers and uncertain tax returns for the government. In 2000, government representatives from Ivory Coast, Ghana, Nigeria and Cameroon agreed to destroy part of their cocoa crop in order to reduce supply and force the global price up. This was a response to a theory known as **Dependency theory**, where developing countries remain dependent on wealthier developed nations for their trade and income (see the next page).

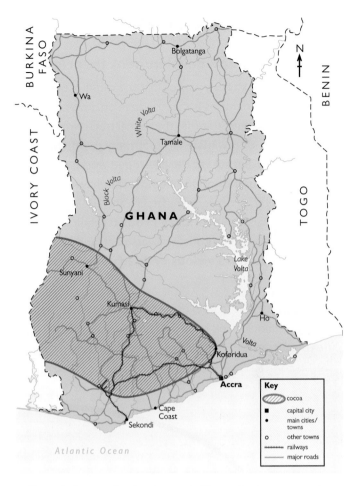

▲ The main features of and settlements in Ghana, plus the cocoa growing region

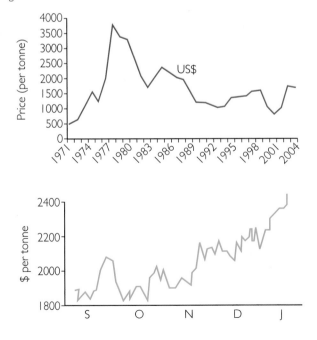

▲ A long-term graph (1971 – 2004) and a short-term graph (September 2007 – January 2008) to show trends in global cocoa prices. The fluctuations shown make planning very difficult; Ghana never knows how much income it will receive from cocoa.

Colonialism – gone but not forgotten? – 2

Dependency and development theories

Dependency theory argues that the cause of poverty in developing countries is their reliance on developed economies. It maintains that traditional trade, where surplus crops or raw materials (**primary products**) are produced and sold to the developed world, keeps developing countries poor – because they have no opportunity to process, manufacture and add value to these materials (as **secondary products**) before selling them. This leaves no profit for investment, and countries become trapped in a **vicious cycle of development** or poverty (see below).

The theory states that, in order to develop, these countries should seek to adopt a **virtuous cycle of development**. In other words, they should keep their surplus production and raw materials and invest in their own processing and manufacturing industries – which would add value to their primary goods.

Andre Frank proposed an alternative, known as **development theory**. He was more specific about the poverty trap – arguing that colonialism alone was responsible for its persistence. Colonialists, he claimed, exploited their colonies for raw materials and kept them poor by preventing investment in domestic manufacturing – and by using trade tariffs to prevent cheap imports from the colonies undercutting their own workers. This process continues today with the ways in which the world's richer nations maintain economic control over poorer countries (see below).

▼ *The vicious cycle of development (or poverty)*

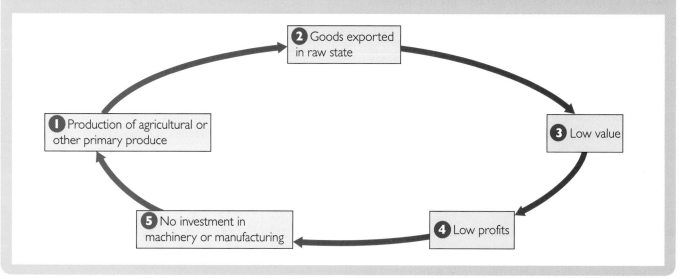

1. Production of agricultural or other primary produce
2. Goods exported in raw state
3. Low value
4. Low profits
5. No investment in machinery or manufacturing

2 Overseas tariffs

Why are cocoa-exporting countries, like Ghana, unable to develop their own processing industries – turning raw cocoa beans into finished goods, to create a virtuous cycle of development? The **value-added** would really boost their economies.

Most processing and packaging of cocoa is done in Europe. EU import **tariffs** (duties on imported goods) are much higher for processed cocoa than for raw cocoa beans. For example, in 2007, the EU charged a 7.7% import tariff on cocoa powder, 15% on chocolates containing cocoa butter, but no tariff at all on raw cocoa beans. This is a process known as **tariff escalation**. Similarly, Japan and the USA have no import tariff on unprocessed cocoa beans, but charge significant tariffs of up to 65% on cocoa chocolate imports.

Therefore, Ghana is forced to export raw cocoa beans, and lose out on the value added by processing them. As a result, people like William Korampong (see below) have little opportunity to earn higher incomes.

Cocoa is seasonal, so we have long periods of poverty. After paying off my debts, there is no money left to send the children to school or pay for food. Get the government to process the cocoa here and not abroad – then there would be more money in our pockets.
Cocoa farmer, William Korampong

Information about cocoa production

Cocoa supplies can be irregular in the short-term, because cocoa is:

- mainly grown by smallholders with low average yields (they grow other cash crops alongside the cocoa to protect themselves against price falls)
- prone to pests and diseases.

In the longer-term, supplies could be threatened by:

- a lack of planting – it takes 8 years for a cocoa plant to mature enough to produce pods
- young workers leaving rural cocoa-growing areas for better jobs in the cities
- low prices, which lead to reduced investment and lower productivity on many farms.

Information about cocoa consumption

- There was a sustained growth in demand throughout the twentieth century.
- Western Europe and North America are the most important markets.
- Asia, Eastern Europe and Latin America are all rapidly growing markets.

- Consumers react badly to high prices or supply shortages – buying alternatives to chocolate, or chocolate products with a low cocoa content.
- Consumers in Western Europe and North America respond well to brands like Fair Trade – offering farmers a fair income – but not to foreign name brands they do not know. So it is hard for Ghana to produce and market its own chocolate.

▲ Cocoa – one of these trees can take 8 years to produce any cocoa pods

3 The World Trade Organisation

Ghana joined the World Trade Organisation (WTO) - see next page - in 1995, in an attempt to increase its global trade. Until then, the Ghanaian government had subsidised its farmers to encourage them to stay on the land and grow food for Ghana's growing cities. But the WTO imposed the joining condition that Ghanaian farmers should no longer be subsidised, even though other countries subsidise their own farmers.

As a result, farmers in Ghana now suffer greatly from imports of heavily subsidised foreign food. For example, Ghanaian farmers in a tomato-growing area in the Upper East District are finding it hard to sell their own produce, because imported EU tomatoes are cheaper. Tomato canning factories have closed and there is no other market for the tomatoes. Ghana's rice growers have also been overwhelmed by cheap imported rice, much of it from the USA.

▲ Tomato farmers in Ghana now find themselves unable to sell their own crop, because cheap subsidised imports from the EU have flooded the market

Do countries such as Ghana gain or lose from membership of the WTO?

What do you think?

155

The World Trade Organisation (WTO)

The WTO belongs to the same family of organisations as the World Bank and the IMF, which were established after 1945 to promote international economic cooperation. It deals with the rules of global trade, with the aim of easing trade and getting rid of anything hindering it. The WTO negotiates new trade agreements and makes sure that its members keep to its rules.

Unlike the IMF, the WTO operates on a one-country, one-vote system, which, in theory, should be fairer to developing countries. However, votes have never actually occurred! Decisions are generally taken by 'mutual' consent – with the biggest markets usually deciding the outcome. The poorest countries, with the smallest markets, feel that the WTO is a rich man's club and that bargaining favours the EU and the USA.

Subsidies

The WTO is currently focusing on helping the world's poor by reducing and removing farm subsidies – grants paid to farmers to encourage production. However, its rules are difficult to apply.

In theory, the WTO should be against subsidies for two reasons:

- It believes in free trade, without subsidies or tariffs.

- Only the rich countries can afford to give substantial subsidies – the poor cannot. (Although Ghana did subsidise its farmers, the amounts involved were very small.)

However, the WTO's Agreement on Agriculture (AoA) allows 'domestic support' (i.e. subsidies) for producers. There are three categories: Green, Amber and Blue.

- Green allows subsidies to producers for environmental reasons, e.g. if they reduce crop output and plant woodland instead.
- Amber allows subsidies that governments have agreed to reduce in amount but not cut altogether.
- Blue allows subsidies to be given, so long as production is reduced in the long-term.

This system allows the EU and the USA to spend nearly $400 billion annually on farm subsidies – arguing that subsidies protect small farmers whose costs are greater for every item produced. However, 50% of EU subsidies go to the largest 1% of producers, and 70% of American subsidies go to the largest 10%. The effect of these farm subsidies allows large farmers to go on producing huge volumes, which the EU and American governments then buy, and 'dump' on developing countries in the form of aid – which reduces prices there and undercuts local producers (such as Ghanaian tomato and rice farmers).

◀ Agri-business on English farms – large landholders, such as insurance companies, receive half of all EU subsidies. Although some countries, such as the UK, have tried to get rid of EU subsidies, EU representatives from France, in particular, have resisted this.

Where next?

To gain some control, Ghanaian cocoa farmers have begun to form cooperatives. Kuapa Kokoo ('good cocoa farmers') began in 1993, and now has 40 000 members in 650 villages – producing 1% of the global crop. With stronger bargaining power than individuals, it pays farmers more per bag of cocoa by selling to Fair Trade organisations in Europe. In 1998, it joined a UK company to create the Day Chocolate Company. It makes 'Divine' chocolate, a Fair Trade chocolate aimed at the UK's mainstream market. Kuapa Kokoo is managed by a farmer's trust and provides its members with credit. Profits have paid for improved drinking water in rural areas and health insurance for farmers.

▼ *Processing cocoa butter for the Kuapa Kokoo cooperative in Ghana – which now sells under Fair Trade labels in the UK via its website – www.divinechocolate.com*

Over to you

1 Study the table on page 152 comparing Ghana's development indicators in 1987 and 2007. In what ways has Ghana **a** made progress, **b** fallen behind?

2 In pairs, draw a spider diagram on a large sheet of paper, with Ghana at the centre. On it, draw links to 'Commodity markets in London and New York', 'Overseas tariffs', and 'World Trade Organisation'. Write down ways in which Ghana is affected by each of these influences, and classify them as positive or negative.

3 How far do these influences work for Ghana's benefit? Explain.

4 Summarise the arguments for and against subsidies for EU and American farmers. Should the WTO prevent further subsidies to farmers in the EU and the USA?

On your own

5 Research the Kuapa Kokoo cooperative through www.divinechocolate.com. Find out **a** who owns it, **b** how it is organised, **c** how it promotes itself in the UK compared to the mass producers, **d** how communities in Ghana benefit.

6 Prepare a report 'Ghana – 50 years on'. In 750 words, assess whether life in Ghana is improving, and what issues it faces for the future.

Exam question: Using examples, assess the view that the relationship between the developed and the developing world is a neo-colonial one. **(15)**

Superpower geographies

In this unit you'll learn how China is gaining influence as a new economic superpower.

Karratha – boom town!

Karratha is a medium-sized town of about 16 000 people, in the north-western part of Western Australia. It is an isolated and remote community – nearly 1000 km or two hours' flying time away from Perth, the state capital. Yet the news stories shown on the right, from 2006-7, show that this area is in the midst of a major boom. Labour is in short supply and there is a high demand for housing.

The reason for the boom in Karratha is the local iron ore. Karratha is part of an area of Western Australia known as the Pilbara, which contains among the world's largest and highest-quality iron ore reserves.

Most of the mined ore is sold to China to fuel its economic expansion (see pages 17-21). Every day, several trains – each 2.2 km in length and carrying 30 000 tonnes of ore – run from the mining areas around the town of Tom Price (300 km inland in the Hammersley Ranges), to the port of Dampier, Karratha's nearby port. Iron ore bulk carriers – each carrying up to 250 000 tonnes – are then loaded automatically and depart fully laden for China.

Feeding China's demand

China's demand for raw materials to feed its growing economy is so great that:

- it has accounted for 90% of the global increase in sea traffic so far this century.
- iron ore imports to feed the Chinese steel industry grew 23% in the first *4 months* of 2007!
- it now makes 500 million tonnes of steel a year – 33% of the world's output – and has a steel industry four times larger than that of the USA.
- it is now, by far, the largest producer and consumer of steel in the world – having overtaken the USA in 2001.
- most of China's steel production is used for domestic consumption.
- the construction boom in its cities, driven by the rapid economic growth, has led to soaring demand for both steel and cement. China is the largest consumer of cement in the world, and is expected to use 50% of global cement production by 2010.

News stories from boom town Karratha ▶

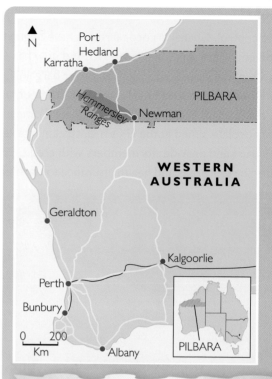

Qantas starts flights between Brisbane and the Pilbara

Qantas has commenced weekly direct flights between Brisbane and Karratha in Western Australia.

10 October 2007

McDonald's in Karratha has begun recruiting workers from the Philippines to beat what it says are labour shortages in the remote town.

ABC News, 2 August 2006

Global demand fuels Australian resources boom

Karratha's house prices now rival Sydney's, and workers are flocking in from all over the country to cash in on the boom. But these people are straining the facilities of this town – sucking Karratha dry of accommodation.

ABC News, 20 December 2006

'Motel' planned to house 1000 workers at Karratha

The mining company, Woodside, will generate new housing accommodation for mine workers at Karratha in Western Australia. A short-term, motel-style facility to house 1000 workers will be built at Woodside's Gap Ridge construction camp.

Hotspotting, 10 September 2007

The impact of mining on Australia is enormous – controlled by the large companies that extract the iron ore. The demand from China for metal ores has been so great that the world's three largest mining companies (Vale, Rio Tinto and BHP Billiton), who between them control 80% of the iron ore market, decided to raise the global price of iron ore by 70% in early 2008. Their control of the supply of ore is so great that they are able to control prices and can restrict supply when demand is high.

China's going shopping

Faced with the increasing costs of raw materials, like iron ore, China has recently tried to guarantee supplies by buying up companies which supply metals such as iron ore, copper and aluminium (or its raw form, bauxite). In this way, it hopes to control the future supplies and prices of raw materials:

- In 2008, Chinalco, the State-owned Chinese aluminium company, teamed up with the American aluminium company, Alcoa, to buy a 12% stake in Rio Tinto.
- In 2007, Chinalco bought Peru Copper Inc.
- Anshan Iron & Steel Group agreed to take part in joint ventures (one stage short of a merger) with Australian iron ore suppliers.
- In 2007, Industrial and Commercial Bank of China Ltd. bought a 20% stake in Standard Bank Group Ltd. (South Africa's largest bank and a major influence on investment in South African mineral reserves).

China is also aiming to guarantee its fuel supplies:

- In 2006, Cnocc Ltd., a Chinese oil corporation, spent US$2.7 billion buying Nigerian oil fields.
- Until 2007, a large portion of Kazakhstan's oil and gas fields belonged to a Canadian company. But then, China's State oil company, CNPC, became the main shareholder of PetroKazakhstan. The purchase includes oil refineries, petrol stations and a 1000-km pipeline overland into China. Kazakhstan has enormous oil reserves, and, by 2012, will be one of the world's major producers. The country shares a 1700-km border with China. Therefore it makes sense for China to secure its own supplies.

▲ Shanghai's construction boom – part of the cause of the demand for iron ore

▼ A Chinalco aluminium smelting plant in China. In 2008, Chinalco bought 12% of Rio Tinto. In future, Chinese companies like this may take over more overseas companies to try to control supplies of raw materials.

How far has China become a superpower?

Undoubtedly, China's economic growth has had major impacts – not only on the demand for fuel and raw materials, but also on the cost of living for everyone else in the world. Its cheap labour rates and drive for economic growth – free of any environmental restrictions or planning applications – have allowed relentless industrialisation.

Nevertheless, China has some way to go when compared with the USA and the EU. In 2005, only two of the world's biggest 200 companies were Chinese (see below). However, until recently, almost all of China's wealth was used for domestic re-investment – to create a better infrastructure (e.g. the new airport in Beijing) and to expand the industries producing consumer goods for the rest of the world. Its trading position is very strong – in 2007, it earned US$360 billion **more** through exports than it spent on imports. Now some of the surplus wealth is available for takeovers and mergers with overseas companies. By the end of 2007, over 5000 Chinese companies had invested in 172 countries and regions around the world, so the map below is likely to change in the next few years.

▼ *The distribution of the world's 200 largest companies (based on valuation) in 2005*

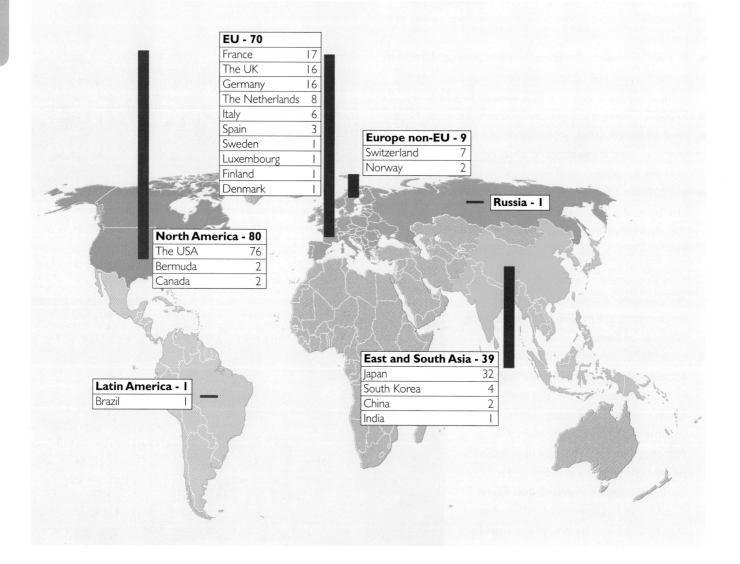

EU - 70	
France	17
The UK	16
Germany	16
The Netherlands	8
Italy	6
Spain	3
Sweden	1
Luxembourg	1
Finland	1
Denmark	1

Europe non-EU - 9	
Switzerland	7
Norway	2

Russia - 1	

North America - 80	
The USA	76
Bermuda	2
Canada	2

Latin America - 1	
Brazil	1

East and South Asia - 39	
Japan	32
South Korea	4
China	2
India	1

China's rise to superpower challenger

China has undergone massive change. Since the early 1980s, its economy has doubled in size every eight years. China has the largest sustained GDP growth in history, and is no longer a low-income country. Public spending on health and education over 50 years has provided a healthy, literate and skilled workforce. But how has this change happened?

After it seized power in 1949, China's communist government kept the country separate and disconnected from the rest of the world. The economy was planned centrally, goods were produced for the consumption of China's own people, and no private wealth was permitted. Government income was spent on improving health, education and life expectancy.

However, from 1986, China's government developed an 'Open-Door Policy' to overseas investment. In the 1990s, it transformed into a more capitalist economy – allowing individuals to accumulate wealth by producing goods and services, without State interference. But it is still not a pure free-market economy – most of China's largest companies are either totally or in part State-owned, so any profits are reinvested or ploughed back into State spending on health and education. The government is also very authoritarian and voters' choices are between members of the Communist Party. As a country, its political and social organisation is tightly controlled.

Since 2000, China has been the largest recipient of overseas investment, and now foreign-owned companies,

▲ China has formed several partnerships with overseas companies, and this factory in Shenyang – a partnership with BMW – has already starting producing cars. In 2008, BMW increased its Chinese sales to over 60 000 cars.

or those in partnership with Chinese companies (like BMW), produce half of its exports. 60% of the increase in world trade since 2004 has been as a result of China's industrialisation. With 20% of the world's population, and increasing wealth, it has now overtaken the USA as the world's largest consumer.

- In 2004, China consumed 382 million tons of wheat (USA 278 million tonnes).
- In 2006, China consumed 67 million tonnes of meat (USA 39 million tonnes).
- In 2006, China consumed 258 million tonnes of steel (USA 104 million tonnes).

By 2050, China will consume annually more oil and paper than the world currently produces!

The environmental costs of growth

Rapid economic growth in China has been achieved at a high environmental cost (also see pages 88-89):

- China now has 16 of the top 20 most air-polluted cities in the world. The polluted air was blamed for over 400 000 premature deaths in 2003.
- 30% of China suffers from acid rain, caused by emissions from coal-fired power stations. In June 2007, it was estimated that China was opening new coal-fired power stations at the rate of two per week (with a goal of 550 new power stations in total). China's CO_2 emissions for 2006 were estimated at more than 6.2 billion tonnes,

an increase of 9% (the USA produced 5.8 billion tonnes in the same period, an increase of 1.4%). However, *per person*, the USA still produces between 5 and 6 times more CO_2 than China.

- 70% of China's rivers and lakes are polluted. Water quality in 207 of the Yangtze River's tributaries is not fit for spraying on farmland, yet 1 in 12 of the Earth's population depends on it for drinking water. Beijing has poor-quality tap water. The water is safe when it emerges from the water treatment plants, but it becomes contaminated by old, leaking pipes.

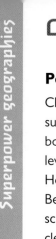

China - economic superpower? - 3

Pollution in Beijing

China is now the world's largest emitter of CO_2. Beijing, in particular, suffers from terrible air pollution (see photo). Vehicle emissions and booming car sales have made the problem worse. The city's pollution levels are 2-3 times higher than the levels considered safe by the World Health Organisation. In 2007, the Chinese government claimed that Beijing had clean air for over 240 days that year, but only using its own scale of pollution severity! On the usual international scale, even Beijing's cleanest days would still be considered as highly polluted. Before the 2008 Olympics, Beijing's poor air quality was worrying those due to take part in endurance sports and distance and track events. Australia's Olympic Committee decided that its athletes would not arrive in Beijing until just before the 2008 Games began – to avoid respiratory problems.

Social concerns & China's economic growth

China's economic growth has also been achieved at a social cost:

- Its rural population has yet to see evidence of China's economic boom.
- 20% of China's population live on less than US$1 a day, and most of those working on the construction of Beijing's Olympic facilities earned just US$4 a day.
- Child labour was used in some factories manufacturing Olympic souvenirs and toys (see right).
- Housing in some of Beijing's old, narrow streets (known as hutongs), was demolished to make way for Olympic facilities. One human rights group estimated that 300 000 people were evicted (see next page).

Meanwhile, human rights groups are concerned about China's record of abuse. Beijing's city authorities banned non-residents from being in the city during the 2008 Olympics, including vagrants, beggars, and those with mental illness. The pressure group Amnesty International believes that China deserves criticism because of its foreign policies, lack of democracy, and treatment of Tibet. It is difficult to get accurate news reports from China, although there are fewer restrictions now on what foreign journalists can or cannot report. But Amnesty claims that freedom of thought and communication is highly restricted – citing the huge numbers of political prisoners, the lack of freedom to access the Internet, and radio jamming.

How China's human rights issues have been reported ▶

▲ *Beijing's poor air quality in 2007*

Chinese factories are churning out licensed bags, caps and stationery for the 2008 Beijing Olympics using child labour, and paying workers less than half the minimum wage, a report says. The report – 'No Medal for the Olympics' – finds evidence of children as young as 12 producing Olympic merchandise.

The Playfair Alliance, represented in Britain by the Trades Union Congress (TUC) and Labour Behind the Label, researched working conditions at four factories making 2008 Olympic bags, headgear, stationery and other products. 'It also reveals that factory owners are falsifying employment records and forcing workers to lie about wages and conditions,' the TUC said. Researchers found adults earning half the legal minimum wage in China, and employees made to work 15 hours per day, seven days a week.

Fears about the use of child labour, as reported in the *Sydney Morning Herald*, 11 June 2007

International media should have unrestricted access to all areas of Chinese life during the Olympics, including human rights and China's criminal justice system.
Nine Television, Australia

In 2000, Amnesty International knew of 213 people still in prison for protesting in 1989 against the lack of democracy, a fraction of the total unfairly tried and sentenced.
Amnesty International, 2000

One man was sentenced to four years' imprisonment in 2004 for opposing the forced evictions in Beijing for construction work for the 2008 Beijing Olympics. He was tortured, reportedly suspended from the ceiling by the arms and beaten repeatedly by police.

Amnesty International, 2007

A traditional street in Beijing, known as a 'hutong' ▶

China as a military superpower?

At present, China has limited conventional military reach and therefore influence; it has no aircraft carriers, and its planes cannot be refuelled while in flight - but it does possess nuclear weapons. The US Pentagon estimated in 2008 that China's annual budget on defence could be US$90 billion, making it the world's third largest defence spender – and the largest in Asia. Will China's new economic strength lead to it becoming a military superpower?

Can a country be a superpower if it denies human rights to its people?

What do you think?

Over to you

1 Compile a factfile of 15-20 economic, social and environmental indicators about China from the CIA Factbook (type 'CIA Factbook' into Google).

2 Using an atlas, identify and mark on a world map **a** the world's largest sources of iron ore, copper and bauxite, **b** the world's major sources of oil.

3 Using the information under the heading 'China is going shopping', annotate your map to show how the Chinese are buying companies elsewhere in the world. Research examples of Chinese take-overs on the BBC News website (news.bbc.co.uk) and add them as well.

4 In pairs, take **one** environmental problem from the list below. Consider **a** what problems it causes, and **b** how easy it would be to solve. Present your ideas in class.

- China has 16 of the top 20 most air-polluted cities in the world.
- 70% of China's rivers and lakes are polluted.
- Beijing's tap water is unsafe and its air quality poor.

5 In pairs, decide what strengths and weaknesses China has in becoming an economic superpower.

6 Draw a spider diagram to show what future threats could occur to China's continued economic growth.

On your own

7 In 2008, there were protests against China's human rights record during the lead-up to the Olympics. Research whether things have changed in China since the Olympics.

8 Write 500 words on the title 'Is child labour ever justifiable?'

Exam question: Assess the view that economic development is not possible without causing environmental degradation. **(15)**

Russia – the re-emerging superpower? – 1

In this unit you'll learn about the collapse of the USSR, and the re-emergence of its largest republic, Russia, as a global player.

The collapse of the USSR

The airport runway of Tashkent International Airport looks fine, but all around it are the decaying relics of old aircraft that used to be owned by Aeroflot – the State airline of the former USSR. They happened to be located in Tashkent on 31 August 1991 – the day that Uzbekistan declared its independence from the USSR. The new state kept what it believed was its share of the USSR's equipment and resources, and used the planes for its new national airline – Uzbek Airways.

By the end of December 1991, the USSR had completely collapsed. It was virtually dead anyway – most of its 15 constituent states had, like Uzbekistan, already declared themselves independent during the course of 1990-1991.

The former USSR's largest member, Russia, is now fighting back as a future superpower, while some of the USSR's smaller members – like Uzbekistan – are struggling. This unit looks at the fall of the USSR and the rise of a powerful independent Russia.

The relaxation of communism

In the 1980s, the USSR was one of the two global superpowers (the other being the USA), and it had a **command** style of government. There were few personal freedoms and the State decided all production targets. However, despite these restrictions, the USSR did have universal literacy, a high proportion of educated professionals, and a large research workforce – so it looked like it would continue to be a major power in the 21st century.

However, in 1987, its communist leader, Mikhail Gorbachev, introduced a number of changes to combat political and social discontent, and economic stagnation. After 70 years of communist rule, two of these changes became significant – *perestroika* (meaning 'restructuring'), and *glasnost* (meaning 'openness').

- **Perestroika** relaxed economic controls, allowing businesses to sell their products privately. For instance, Russian farmers were allowed to sell their surplus produce and keep the proceeds (see photo). By 1992, the economy had changed significantly to free market capitalism.

- While communism did bring benefits to Russia, such as rapid industrialisation, it was often brutal – especially under Stalin (1922-53). **Glasnost** allowed greater political freedom, such as freedom of speech and of the Press, and the establishment of rival political parties. Under *glasnost*, the Communist Party, unopposed for 70 years, had to compete for votes with **nationalist** parties and a growing call for independence among the constituent states of the USSR – particularly the Baltic States of Estonia, Latvia, and Lithuania.

▲ *A collective farm in Russia in 1992. On farms like this, production targets were formally determined by the State. Under perestroika, surplus farm produce could be sold privately.*

Communism falls, new rulers emerge

The changes introduced by Gorbachev led ultimately to the collapse of the USSR in 1991. His economic reforms allowed trade, private investment, and personal wealth. However, the new political freedoms exposed State corruption and other embarrassing problems, such as pollution and alcoholism, and led people to question why the State should control so many other aspects of their lives. A power struggle took place in which a new leader, Boris Yeltsin, overthrew the Soviet state – demanding faster economic changes.

Under communism, power had always rested with an elite, the *nomenklatura*. Under Yeltsin, the *nomenklatura* **privatised** the State's vast resources. This took place not – as in the UK – by floating shares which anyone could buy, but by State decree from Yeltsin. By 1995, his government was in severe financial difficulties and it auctioned off Russia's mineral and oil reserves at knock-down prices.

Fighting off a threat from resurgent communists in elections, an alleged loans-for-shares scandal took place, in which the government sold shares in State assets in return for loans to help win its re-election.

To avoid a backlash from any future government, the new owners closed some production facilities and sold off the assets – usually to their own subsidiaries! In this way, some insiders made massive personal profits. Major companies like Norilsk Nickel (one of the world's richest suppliers of metals) and Gazprom (Russia's largest energy company) owe their origins to this sell-off period. The State still owns 50% of Gazprom, but the company's other large shareholders have made huge personal fortunes from it.

The economic impacts of change

As the State-run economy collapsed, there were massive economic and environmental impacts. An 80% decline in investment in Russia's economy during the 1990s caused old factories to grind to a halt and economic production to fall by 50% – causing mass unemployment.

The environmental legacy of communism

Contaminated land and polluted rivers followed decades of rapid industrialisation in the Soviet Union, where little regard was paid to any environmental impacts. The USSR had few environmental controls and left a huge environmental legacy after 1991. Russia had to undertake massive clean-up operations in order to qualify for loans from the IMF (see page 150).

▲ Rusting chemical barrels on the shore of Lake Dzerzhinsk in 1997. According to Greenpeace, this lake is the most poisonous in the world and this area, east of Moscow, has the worst chemical pollution in Russia.

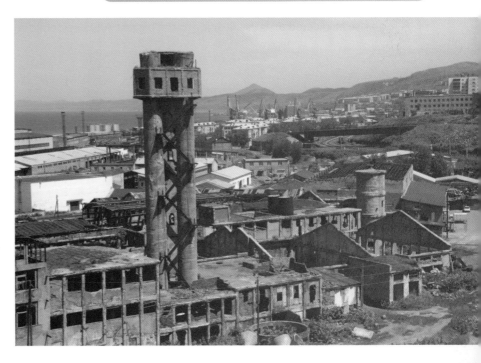

A derelict factory after the 1990s economic collapse ▶

Economic fallout

The new Russian economy produced little in the 1990s.

- Once Government protection was removed, Russians were exposed to higher prices – causing inflation. Farmers benefited from higher food prices, but many people, especially in the cities, were unable to afford basic foods.
- People on fixed incomes (e.g. teachers and pensioners) faced a sharp drop in living standard.
- As factories were sold off and stripped, there was mass unemployment.
- A **disparity** grew between the new super-wealthy and the poor (resembling other industrialised countries – see the table).

	Lowest 10%	Highest 10%
Australia	2.0%	25.4%
China	1.6%	34.9%
France	3.0%	24.8%
The Netherlands	2.5%	22.9%
Russia	1.9%	30.4%
Sweden	3.6%	22.2%
The UK	2.1%	28.5%
The USA	2.0%	30.0%

▲ *Income distribution in Russia and seven other industrialised countries. This table shows the percentage of total national income received by the lowest 10% and the highest 10% of income earners, by household.*

The social impacts of change

The negative economic consequences of change in Russia led to a number of social impacts:

- The emigration of young Russians seeking employment overseas.
- A fall in birth rates, caused by that emigration. The Russian population has declined steadily since 1995.
- A sharp decline in Government revenue, causing cutbacks in welfare provision. From free health care under communism, medical care in Russia is now amongst the worst in the industrialised world. A World Health Organisation report in 2000 ranked Russia's health care system 130th out of 191 countries.
- A fall in male life expectancy to 59, caused mainly by increased alcoholism. In 2006, research in the UK medical journal The Lancet showed that excessive drinking now causes 50% of all deaths in working-age men in Russia.
- A sharp rise in the spread of infectious diseases, such as tuberculosis (TB).
- A huge rise in injected drug use, leading to the biggest AIDS epidemic in Europe.

Which is more effective in the current world – economic or military muscle?

What do you think ?

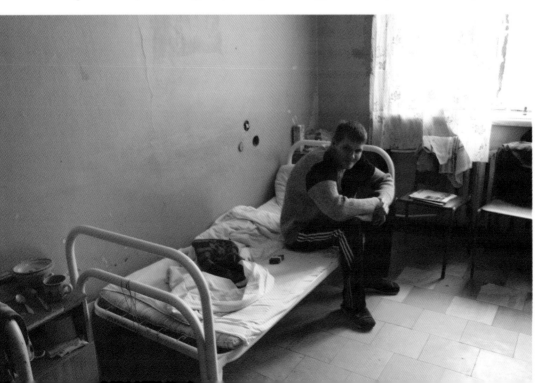

◄ *Russian health care in the twenty-first century - an HIV/AIDS patient in a St Petersburg hospital in 2008. Now, although State health care exists, many patients find that they have to supplement hospital charges with their own money – typically US$90 in 2005, when the average monthly income was about US$500.*

Russia's re-emergence as a global power

Russia has the world's largest known reserves of natural gas, and is the second largest oil producer after Saudi Arabia. These two commodities accounted for 60% of Russian exports in 2007, and have helped to fund Russia's economic recovery since 1999 (see the graph). Other factors include:

- massive loans from the IMF, which helped to prevent total economic collapse during the 1990s.
- an economic crisis in 1998, which led to a big **devaluation** of Russia's currency (the rouble) – making Russian goods cheaper overseas.
- the rapid growth of the Chinese and Indian economies, which has led to huge demand for energy and increased global energy prices. This has made Russia's massive energy reserves very valuable indeed. The Russian Government is benefiting from this, because it kept a 50% stake in Gazprom (Russia's largest energy company). Gazprom controls a third of the world's natural gas reserves, and its profits have increased Government revenue and enabled Russia to repay its debts to the IMF.

The growth in global oil prices since 2004, and the inflow of 'petro-roubles', have altered the balance of power between Russia and its neighbours. Russia now sees itself as an energy superpower, and energy security will be central to its relations in the future (see pages 22-27 and 170-171).

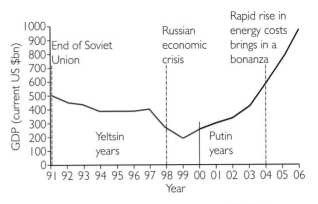

▲ *The changing fortunes of the Russian economy 1991-2006*

Russia's military influence

There was a decline in Russian military spending in the 1990s, due to the economic situation. So far, Russia's new prosperity has had little impact on defence – overall military spending is up, but not as a percentage of GDP (see right). Russia's current national security depends largely on outdated Cold War-era weapons, with only a few new additions. Because of conscription, the Russian army is still large, but it has relatively few professional soldiers. However, even with its elderly weapons and conscript army, Russia does have one of the world's largest military forces.

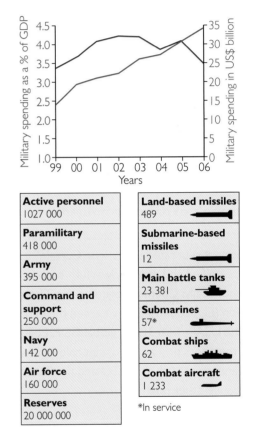

Active personnel 1027 000	Land-based missiles 489
Paramilitary 418 000	Submarine-based missiles 12
Army 395 000	Main battle tanks 23 381
Command and support 250 000	Submarines 57*
Navy 142 000	Combat ships 62
Air force 160 000	Combat aircraft 1 233
Reserves 20 000 000	*In service

▲ *Russian military spending and forces in 2006*

Over to you

1 As a class, divide up and research the 15 countries of the former USSR, using the CIA's Factbook (type 'CIA Factbook' into Google). For each one, prepare a report on **a** its current economy, **b** its assets, **c** its social indicators. Present a summary of each report in class.
2 In pairs, classify the economic, social and environmental impacts of changes in Russia since 1991.
3 In pairs, decide whether Russia now deserves the title 'superpower'.

On your own

4 Research the state of Russia's environment by typing 'Russia environmental issues' into Google. What are the key problems? How are they being tackled?
5 In 1991, a poll showed that 76% of Russians wanted to keep the USSR. Does your work show that they were right or wrong? Explain in 750 words.

Superpower influences over nation states

In this unit you'll learn how superpowers play a key role in international decision-making

The Eurovision Song Contest keeps on growing!

Billed as the longest-surviving music competition in the world, the Eurovision Song Contest reflects a changing world! In the first contest, in 1956, just seven countries competed. By the time Sandi Shaw sang the UK's first winner in 1967, 17 countries were entered. Rapid expansion in the early years of the twenty-first century raised the number of competing countries to 43 in 2008. For a contest which activates more sweepstakes than musical quality, it seems that everyone wants to be a part of it!

Of course, Eurovision is about another agenda altogether – the desire of many countries to 'be' European. Europe has no clear geographical definition; along the Russian border, it becomes 'Eurasia', which at some stage becomes Asia. This is, therefore, more about identity and aspiration than a desire to compete in a bizarre musical contest!

Europe's expanding influence

Europe has changed dramatically since 1989; from a divided continent separated by an 'Iron Curtain', it has grown more unified and more powerful – both politically and economically. Counted together, the 27 current European Union (EU) countries now exceed the economic weight of the USA – with 31% of global GDP. Politically, several EU members play an important role in international organisations such as NATO and the G8, and increasingly influence other powers.

The EU and its influence

The EU was set up by the Treaty of Rome in 1957 to achieve economic and political co-operation following the Second World War. Originally with six members, its aim was to bring economic co-operation through a common European market free of tariffs. Since 1957 it has become much more, with:

- a permanent staff (the European Commission)
- a European Parliament
- a common currency (the euro) used by 13 of its members
- common laws: economically (e.g. consumer law), socially (human rights), and environmentally (pollution).

▲ The 43 entrants for the Eurovision Song Contest in 2008. Note that Italy and Austria withdrew under protest about voting in 2007. The original seven entrants from 1956 are shown in hatched shading.

▲ The EU member states in 2008. Now expanded to 27 members from the original six, the list of would-be members (see the panel on page x) could see this number increase considerably in the future.

Three factors affect the future influence of the EU:

- On the one hand, further **EU enlargement** is likely, with a long list of would-be members. If all the countries listed on the right were to join, EU membership could total 51!
- Significant enlargement of the level of government could also take place. Some members support the creation of a European defence force, to which all member countries would contribute. A major obstacle to this is funding. While all member states currently contribute to the EU, the new defence force would mean a big increase in cost. Would defence be paid for by taxation paid directly to the EU? To create such a military force would also lead to a loss of sovereignty on taxation and defence – over-riding national governments and a big issue in the UK. Some members see this as a chance to move away from American-dominated NATO. Others, such as the UK, see it as a threat to transatlantic relations.
- On the other hand, nationalist movements are strengthening, e.g. in Scotland, where support for devolution is growing. Some (like Kosovo) support EU membership, while others (like the UK Independence Party, UKIP) want to withdraw the UK from membership of the EU.

North Atlantic Treaty Organisation (NATO)

Since 1949, when NATO was created as a reaction to Cold War tensions between Western Europe and North America on the one hand and Eastern Europe and the USSR on the other, its membership has grown from 12 to 26 members (by 2008) – again with a growing queue of potential members (see below). All of the new members since 1999 are former communist countries from Eastern Europe – old NATO enemies – so NATO has changed considerably since the end of the Cold War. Although NATO's main role is to defend member states and the North Atlantic, it now also has a key role in sensitive areas of the world, such as Afghanistan.

> **Who would like to be in the EU, then?**
> **Current candidates**: Croatia, Former Yugoslav Republic of Macedonia, Turkey
>
> **Potential candidates expressing interest**: Albania, Bosnia and Herzegovina, Kosovo, Montenegro, Serbia
>
> **The following countries are either considering EU membership or have expressed a wish for closer links**:
>
> - *The European Free Trade Association*: Switzerland, Norway, Iceland, Liechtenstein
> - *Former Soviet republics*: Ukraine, Moldova, Belarus, Georgia, Armenia, Azerbaijan, Russia, Kazakhstan
> - *Microstates*: San Marino, Andorra, the Vatican, Monaco
> - *Dependencies of EU member states*: Greenland, Faroe Islands, plus all European dependencies
> - Northern Cyprus
> - *Non-European states*: Morocco, Israel, Cape Verde Islands

▼ NATO members and the years in which they joined, plus potential new members

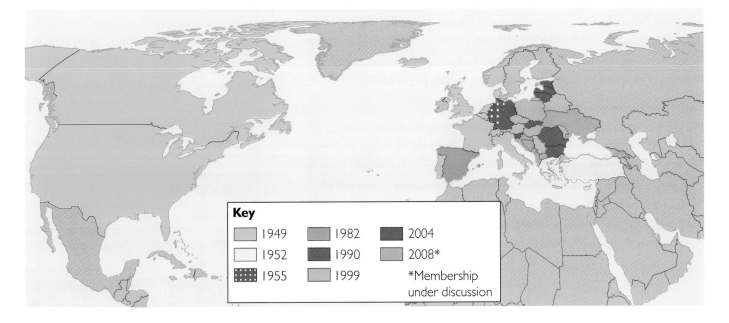

Key

1949	1982	2004
1952	1990	2008*
1955	1999	*Membership under discussion

Russia and its influence

The recent absorption of former communist countries into both the EU and NATO has increased tensions with Russia. In spite of its rapid growth since 2001, Russia's economy is still smaller than those of the other G8 nations, and is dwarfed by that of the expanded EU. However, its **direct influence** is much greater over former regions of the USSR, such as Ukraine and the South Caucasus.

Ukraine

Russia's relations with Ukraine soured in the winters of 2005-6 and 2007-8, when Gazprom threatened to cut Ukraine's gas supplies unless it paid a 400% price rise. Ukrainian winters are very severe, as the climate graph shows, so Ukraine must have gas for heating. Russia claimed that Ukraine's new gas price would be the same as for Russia's other customers.

Three factors help to explain the conflict over gas between Ukraine and Russia:

- The states of the former USSR received cheap gas under deals signed during the Soviet years. Once the USSR collapsed, Russia felt under no obligation to continue to offer cheap gas to its former partners.
- The Soviet gas infrastructure was developed by Ukraine in the 1920s, using its own pipelines, engineers and workers (who later developed Russia's giant Siberian gas fields in the 1960s). Ukraine therefore felt that it deserved gas at preferential rates.
- In November 2004, Ukraine elected a pro-European, anti-Russian government, which has increasingly sought membership of NATO and the EU. Therefore, Russia has questioned why Ukrainians should have the same access to cheap gas as Russians. So, it now expects Ukraine to pay for its gas at commercial rates.

Europe's worries about the Ukrainian gas dispute

Wider European involvement in the gas dispute has arisen because Europe depends on Russia for 25% of its gas. Three major gas pipelines bringing gas from Russia to Europe pass through Ukraine. All of the supply valves for this gas also lie in Ukraine, so, if Russia threatens to cut Ukraine's gas supplies, Europe's supplies could also be reduced. As a result, there is European support for new gas supply routes (see pages 26-27).

Russia's worries about the expansion of NATO

Russia has serious misgivings about Ukraine's proposed membership of NATO. If, as a NATO member, Ukraine allowed NATO military bases or missile defence systems on its territory, Russia says that it would have to respond by targeting its own missiles on Ukraine. Russia is concerned that NATO has absorbed so many former communist countries, and it fears that it is being encircled – a feeling strengthened because Georgia in the Caucasus is also seeking NATO membership.

> ● **Direct influence** is where countries gain influence through direct contact with each other.

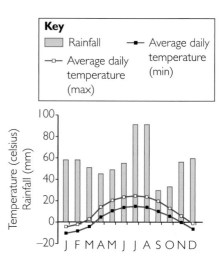

▲ Climate in Kiev, Ukraine. Note how low temperatures reach, December-March, and hence Ukraine's problem if Russia cuts off its gas supplies.

Gazprom cuts Ukraine's gas supply

Gazprom, Russia's gas monopoly, has cut supplies to Ukraine. State-owned Gazprom said it would cut shipments to Ukraine by 25%, but Ukraine's national gas firm, Naftogas, has claimed that the reduction is 35%. The dispute centres on a US$1.5 billion (£770 million) debt that Gazprom says it is owed, and Ukrainian officials say has been paid.

Gazprom has offered assurances that supply to the rest of Europe will not be affected. A previous row saw Russia cut gas to Ukraine in 2006, also hitting exports to Western Europe.

Adapted from a BBC news article on 3 March 2008

> *Should Russia be free to charge whatever it likes for its gas?*
>
> **What do you think ?**

The South Caucasus

The South Caucasus region lies between the Black Sea and the Caspian Sea (see the map below). Once part of the USSR, it now consists of the independent countries of Azerbaijan, Georgia and Armenia. The area is strategically significant, because it is situated on the boundary between:

- Christian countries to the north and Muslim to the south
- the Middle East and Eurasia
- NATO and former Soviet territories.

Now the region has gained greater significance, with the discovery in 1999 of natural gas in the Shah Deniz gas field in Azerbaijan (under the southern Caspian Sea). The South Caucasus pipeline (see the map) was opened in 2006 to bring this gas to Europe via Turkey – thus avoiding Russia. The gas field exploration and the pipeline were developed as a partnership between BP, American, Iranian and European companies.

The USA and the EU are keen to develop good relations with Azerbaijan, Georgia and Turkey – the countries through which the gas will flow. The new South Caucasus pipeline will weaken Russia's influence over the region by:

- reducing the three countries' economic ties to Russia
- providing an alternative gas route into Europe instead of the Ukrainian pipelines.

With Hungarian, Romanian and Bulgarian support, there is now a proposal to pipe the gas further into Europe, using the Nabucco pipeline to Austria and Italy.

▼ *The South Caucasus region, showing the South Caucasus gas pipeline route from Baku (capital of Azerbaijan) westwards to Turkey at Erzurum. Opened in 2006, it could eventually link up with a proposed trans-Caspian gas pipeline from the east in Turkmenistan.*

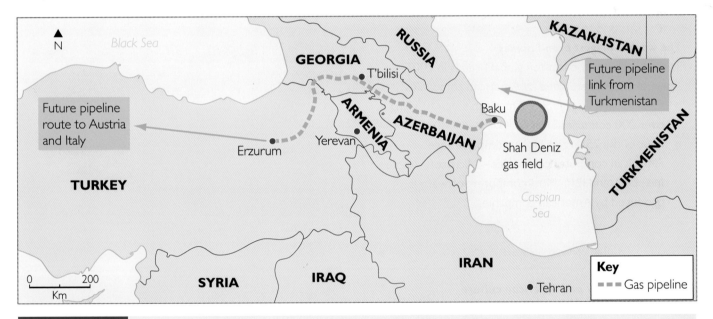

Over to you

1. In groups, list the benefits of joining **a** the EU and **b** NATO, as perceived by those countries wishing to join.
2. What do NATO and the EU have to **a** gain, and **b** lose by admitting more countries?
3. Explain the benefits and risks in developing pipelines such as the South Caucasus.
4. In class, research the benefits to and attitudes of the following countries to the South Caucasus pipeline: Azerbaijan, Georgia, Turkey, Hungary, Romania and Bulgaria. The BBC News website should help (bbc.co.uk/news).

On your own

5. Research how Russian-Ukrainian relations have affected energy supplies to Ukraine and Europe. The BBC News website should help again.
6. In 500 words, explain whether superpowers like the EU can become too large.

In this synoptic unit you'll learn how superpower influence extends to 'global culture', which helps to draw together this chapter.

The British abroad

Warm weather. Neat, superbly mown lawns. The bowler bowls. Restrained applause as another four is scored. Drinks in the Club lounge. Membership fees of up to £20 000. While this sounds like cricket in an affluent part of the UK on a summer's day, the venue is actually Singapore. The cricket club in the photo is surrounded by some of the world's most valuable property – occupied by banks that manage and control much of South-East Asia's investment. Yet the cricket club remains – a hangover of British rule – brought by them to their colonies all over the world.

Is the world now becoming Americanised?

Nowadays, the dominant global culture is American – which spreads by broadcasting its TV, films and music throughout the world. Many American programmes are broadcast through subsidiary companies, e.g. HBO Asia, Fox Channel and CNN International. Nearly all content on these channels is American. Use the following resources to assess the impacts of cultural Americanisation and whether it is beneficial (also see pages 287-294).

▲ *Singapore Cricket Club, surrounded by some of southern Asia's most valuable real estate*

The world's biggest brand names

The table shows the top-ten global brands in 2007 – seven of which are owned by American companies.

- Coca Cola is a symbol of Americanisation – selling in over 200 countries.
- Four of the 25 largest global companies are American computer companies – Microsoft, Apple Inc., Dell, and IBM. Most computer software is produced by American companies.

The top-ten global brands in 2007, by value of sales ▶

Rank	Brand name	Value (US$ million)	Country
1	Coca-Cola	65 324	USA
2	Microsoft	58 709	USA
3	IBM	57 091	USA
4	General Electric	51 569	USA
5	Nokia	33 696	Finland
6	Toyota	32 070	Japan
7	Intel	30 954	USA
8	McDonald's	29 398	USA
9	Disney	29 210	USA
10	Mercedes-Benz	23 568	Germany

Resource 1

The film industry and American culture

Each of the ten largest-grossing films of all time has some kind of American input – either produced with US finance, or cast or made there.

- Without adjusting for inflation, *Titanic* is the largest-grossing film worldwide.
- Adjusting for inflation, the highest grossing film is *Gone with the Wind*.

▶ *The top-ten largest-grossing films of all time, as of mid-2008 (note that these figures are not adjusted for inflation, so generally list more recent films)*

Rank	Film title	Global box office takings (US$ million)
1	Titanic (1997)	1800
2	The Lord of the Rings: The Return of the King (2003)	1100
3	Pirates of the Caribbean: Dead Man's Chest (2006)	1000
4	Harry Potter and the Sorcerer's Stone (2001)	970
5	Pirates of the Caribbean: At World's End (2007)	960
6	Harry Potter and the Order of the Phoenix (2007)	937
7	Star Wars: Episode I – The Phantom Menace (1999)	922
8	The Lord of the Rings: The Two Towers (2002)	921
9	Jurassic Park (1993)	920
10	Harry Potter and the Goblet of Fire (2005)	892

Resource 2

Global TV viewing

Data collection methods vary when assessing the number of programme viewers, so it is difficult to compare TV programme audiences. However, the following data give some idea of global viewing patterns:

- Beijing's 2008 Olympics were broadcast in every country in the world; they attracted 4.7 billion viewers - 70% of the world's population.
- The FIFA World Cup was reportedly the most-watched sporting event in 2006 – averaging 1.2 billion viewers per match.
- Wimbledon men's final regularly attracts up to 1 billion viewers worldwide.
- British TV show *Top Gear* is probably the most watched TV show in the world – broadcast in 117 countries, with about one billion viewers.
- Princess Diana's funeral was reportedly watched by 2.5 billion people simultaneously across the world.
- *Follow Me!*, a BBC beginner's English programme, was broadcast in China during the early 1980s, and was estimated to have attracted 500 million viewers each night.

Rank	Programme title
1	CSI: Miami
2	Lost
3	Desperate Housewives
4	Te Voy a Ensenar a Querer (A Colombian series produced by a US-based TV network)
5	The Simpsons
6	CSI: Crime Scene Investigation
7	Without a Trace
8	Inocente de Ti (A Mexican soap produced by a US-Spanish television network)
9	Anita, No Te Rajes! (A US programme for Hispanic speakers in the USA)
10	The Adventures of Jimmy Neutron: Boy Genius

The top-ten most watched television programmes in 2006, across 20 countries in America and Europe (note that this research was based only in 20 countries, and so may not match other estimates of global audiences) ▲

Culture and the music industry

Rank	Artist	Album
1	Michael Jackson (American)	Thriller
2	AC/DC (Australian)	Back in Black
3	Whitney Houston / Various artists (American)	The Bodyguard (soundtrack)
4	Eagles (American)	Their Greatest Hits (1971–1975)
5	Backstreet Boys (American)	Millennium
6	Bee Gees / Various artists (British)	Saturday Night Fever (soundtrack)
7	Pink Floyd (British)	Dark Side of the Moon
8	Meatloaf (American)	Bat out of Hell
9	The Beatles (British)	Sgt. Pepper's Lonely Hearts Club Band
10	Celine Dion (Canadian)	Falling into You

Artist	Album
Arcade Fire (Canadian)	Neon
Arctic Monkeys (British)	Whatever People Say I am That's What I'm Not
Bloc Party (British)	A Weekend In The City
Bright Eyes (American)	Cassadaga
Cold War Kids (American)	Robbers & Cowards
Ciara (American)	The Evolution
Jamie T (British)	Panic Prevention
Joss Stone (British)	Introducing
Kings of Leon (American)	Because Of The Times
Klaxons (British)	Myths of the Near Future
LCD Soundsystem (American)	Sound of Silver
Mark Ronson (UK-born, educated in the USA)	Version
Mika (Lebanese-born, UK career)	Life In Cartoon Motion
Nine Inch Nails (American)	Year Zero
Rakes (British)	Ten New Messages
The Good, The Bad & the Queen (British)	The Good, The Bad & the Queen
Timbaland (American)	Shock Value
Kaiser Chiefs (British)	Yours Truly, Angry Mob
Shins (American)	Wincing The Night Away
View (British)	Hats Off To The Buskers

▲ *The top-ten largest-selling albums of all time (based on data from different record companies, so the list may not be fully accurate)*

▶ *The 20 largest-selling albums in the UK, January–March 2007 (in alphabetical order by artist, not sales order)*

Cultural superpowers - 2

The film industry and Indian culture

If you have watched films starring actors such as Aamir Khan, Hritik Roshan, Shah Rukh Khan, Amitabh Bachchan, Kiron Kher, or Shiny Ahuja, it is probably because you are a fan of India's Bollywood – the world's largest film industry, both in terms of numbers of employees and the number of cinema goers.

The countries where people watch the most films at the cinema. Of 7.6 billion viewings at cinemas annually, almost 3 billion are in India. ▶

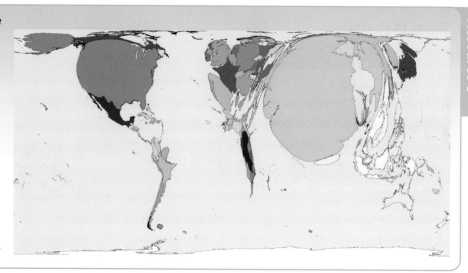

Background

Cultural imperialism

Cultural imperialism refers to *either*:

- an enforced spread of its culture by a larger power (e.g. the British Empire)

or

- the voluntary adoption of a foreign culture by other people (e.g. the adoption of American 'street' culture across the world).

For those on the receiving end of this, cultural influences from elsewhere can either enrich or threaten the local culture.

- One argument says that the 'receiving' culture absorbs a foreign culture passively by purchasing or using its goods and services. This is not enforced, like colonialism, so it has been described as 'banal imperialism'. It implies that one culture is better than another, and that those on the receiving end are in some way deficient because their own culture is put in second place.
- The other argument says that all cultures are a blend – that there is no such thing as 'pure' culture – and that the addition of other cultures is enriching.

The expanding global economy has led to a global culture, spread by information technology. **Electronic colonialism** has emerged – spread via multi-media TNCs, such as Time-Warner, Disney, News Corporation (see Resource 6), or Sony (most are American or Japanese). These companies operate a **hegemony**,

in that they dominate global culture. They decide which films and TV shows people will see, and which radio stations and music they will listen to. Most UK commercial radio stations are owned by a few large companies, and four large record companies (Sony BMG, EMI, Universal and Warner) dominate UK record production.

Some people oppose 'cultural imperialism', because they believe it suppresses cultural diversity. Like ecological diversity, cultural diversity (such as the preservation of language, musical styles or literature) is said to be valuable – and prevents the view that one culture is always 'right'.

▲ *A Hong Kong street with clearly visible global brands – cultural imperialism is alive and kicking*

News Corporation

If you read the *Times* or *The Sun,* watch *The Simpsons,* phone a friend on Sky, and log onto MySpace, you are providing income for News Corporation. A US-based company, originating in Australia, it owns film companies, newspapers, satellite TV, and is highly influential:

- Since 1979, every winning political party in UK General Elections has been supported by *The Sun.*
- *The Sun* campaigns against the EU.
- News Corporation actively promotes Christian programmes through its publishing, cable, DVD and television networks.
- Fox News in the USA has been strongly supportive of 'The War on Terror'.

▼ *Companies owned by News Corporation and their activities*

Publishing companies
- HarperCollins and Zondervan (a Christian book publisher)

Newspapers and magazines
- Australia: 101 newspapers, from national (*The Australian*) to suburban.
- UK: 4 national papers (*The Sun, News of the World, Times* and *Sunday Times*).
- USA: *The New York Post,* financial papers, e.g. *The Wall Street Journal.* Eight daily and 15 weekly regional newspapers, plus New York suburban papers.
- Russia: a 33% share in *Vedomosti* (Russia's leading financial newspaper).

Music and radio
- MySpace Records
- India: commercial FM stations in 20 cities.
- Russia: a 50% share in two radio stations

Television
- Studios: Fox is the main brand name, e.g. 20th Century Fox, Fox Television Studios in France, India, Australia, and New Zealand.
- Networks: Fox TV (USA and satellite), My Network TV, plus channels in Eastern Europe, Uruguay, Israel, Indonesia, and New Zealand.

Satellite television
- Owns or part-owns: BSkyB (UK), Foxtel (Australia), SKY (New Zealand, Italy, Germany), STAR TV (Asia, including parts of China, and India).

Internet
- Websites include Foxsports.com, MySpace

Sports clubs
- Owns 50% of the National Rugby League in Australia; majority or full ownership of Brisbane Broncos, Melbourne Storm and North Queensland Cowboys.

Should trans-national media corporations be prevented from trying to influence how people vote?

What do you think?

Synoptic question

a How far is the USA the world's 'cultural superpower'? **(12)**

b Using evidence, assess the extent to which 'cultural imperialism' can be beneficial. **(16)**

c Referring to examples, assess how healthy it is for the production of most music, television and film to be in the hands of so few. **(12)**

Chapter summary

What key words do I have to know?

There is no set list of words in the Specification that you must know. However, examiners will use some or all of the following words in the examinations, and would expect you to know them, and use them in your answers. These words and phrases are explained either in the Glossary on pages 330-333, or within this chapter.

capitalism
Cold War
colonialism
command government
commodity
commodity trading exchanges
communism
cultural imperialism
dependency theory
devaluation

development theory
direct influence
disparity
electronic colonialism
EU enlargement
evangelism
futures market
glasnost
hegemony
ideology
International Monetary Fund (IMF)

Mackinder's heartland theory
market economy
modernisation theory
modernism
NATO
neo-colonialism
perestroika
primary products
privatisation
secondary products
social Darwinism

superpower
tariff
tariff escalation
USSR
value-added
vicious cycle of development (or poverty)
virtuous cycle of development
World Bank
World Trade Organisation

Films, books and music on this theme

Music to listen to

'Two Tribes' (1984) by Frankie Goes To Hollywood
'The High Dam' (1970) by Abdel Halim Hafez – about colonialism in Egypt; available online via Google

Books to read

Tinker, Tailor, Soldier, Spy (1974) and Smiley's People (1979) by John Le Carre. Also watch the BBC's film series of these books, available on DVD.
The new rulers of the world (2003) by John Pilger

Films to see

'The Hunt for Red October' (1990)
'When the Wind Blows' (1986) – a chilling cartoon based on Raymond Briggs' book
And if you have never seen a Bollywood film before, 'Bride and Prejudice' (2004) is a modern classic!

Websites to see

If you Google 'Protect and Survive', you will reach the Protect and Survive website. This is an archive of material from the 1950s through to the 1980s about civil defence in the UK, and protection against nuclear attack.

Try these exam questions

1 Explain how membership of Intergovernmental Organisations gives some countries political and economic power. **(10)**

2 Referring to examples, discuss the factors that cause power to shift between superpowers over time. **(15)**

5 Bridging the development gap

What do I have to know?

This chapter is about ways in which the wealth of Western economies contrasts with
the continuing poverty of some peoples and nations in the developing world. The 'gap' is
generally increasing. In this chapter you will learn about this gap, how it occurred, and how
it might be reduced. The Specification has three parts, shown below, with the examples
used in this book. (You will also find it useful to refer to Chapter 4 to help you understand
further how and why the development gap arose.)

1 The causes of the 'development gap'

What you need to learn	Examples in this book
• The global development gap and how it is measured.	• Measuring development; indicators of development (e.g. economic, HDI).
• Theories used to explain the development gap.	• Core/periphery, dependency theory, cycle of poverty – vicious and virtuous cycles
	• Colonialism and neo-colonialism.
• Global players and organisations.	• The emergence of the World Bank, IMF, WTO (see also Chapter 4 on this).
	• The G8 and The Millennium Development Goals.
• Trade and investment and its role in the development gap.	• Colonialism and dependency theory; the development of neo-colonialism; Uganda, Ghana.

2 The consequences of the 'development gap'

What you need to learn	Examples in this book
• The consequences of the development gap for people in the most disadvantaged countries	• Uganda – standard of living and life chances.
	• The growth of debt in the 1970s/80s.
	• Structural Adjustment Packages; IMF conditions and the drive for privatisation.
	• The role of the World Bank / IMF in the development gap.
	• Resolving the debt crisis – debt cancellation in Uganda and Africa.
• The development gap in the developing world's megacities.	• Bangalore – contrasts in development and emerging costs, e.g. traffic, housing, poverty.
• Ethnic and/or religious dimensions to the development gap.	• Ethnic disparities in South Africa and apartheid.
	• The caste system in India.
• The positive and negative consequences of development.	• Benefits and problems brought by Bangalore's 'new economy', e.g. urban sprawl, unequal wealth among different castes.

3 Reducing the 'development gap'

What you need to learn	Examples in this book
• The theories that attempt to reduce the development gap.	• Neo-liberalism (e.g. 'economic man') and 'free market' development; water privatisation in Ghana.
• Strategies to reduce the development gap, e.g. aid and investment.	• Modernisation and investment – big development projects, e.g. Malaysia's Pergau Dam, Ghana's Akosombo Dam and Tema aluminium smelter.
	• NGO bottom-up approaches in Uganda (Barlonyo, Equatorial College School) and Moldova (Gura Bi Cului)
	• Populism – land redistribution in Zimbabwe.
• The future of the development gap.	• The Millennium Development Goals and how they are being met in Uganda and Bangladesh.

Food in crisis

In this unit you'll learn about how the global food crisis affects some people and countries more than others.

All good things come to an end …

The extracts on the right were adapted from *The Times* of 17 June 2008, when food, wine, and fuel prices were all rising. A 15-year period of economic growth and low inflation in the world's developed countries had stalled. During this period, life in those developed countries had been good for most people, because:

- the costs of imported manufactured goods had fallen – due to cheap labour in rapidly industrialising countries such as China
- food and wine prices had also fallen, because of food surpluses around the world
- low global interest rates had made purchasing property cheap. In spite of rocketing house prices, people spent a relatively low percentage of their incomes on servicing mortgages, and more on luxuries

… so what changed?

Food prices rose sharply worldwide after 2006. The UN's Food and Agriculture Organisation (FAO) found that, globally, rice for example rose in price by 70% between 2007 and 2008. These price increases were the result of global grain shortages, caused by:

- poor harvests after extreme weather in Europe and North America during 2007, which meant that world grain stocks were at a 26-year low
- the growth of biofuels, which reduced the amount of land under food cultivation
- rising demand in some rice-exporting countries, such as India and China, where rising incomes and food consumption reduced the amount available for export to global markets
- rising demand for cattle feed in Saudi Arabia, one of the world's biggest barley importers, caused by increasing milk demand there
- hoarding of food by traders expecting further price increases

While people in developed countries complained about the rising cost of food, paying for food in these countries accounts for only about a third of average incomes. In developing countries, almost all of some families' incomes can be used up just paying for food and water, so rapidly increasing food prices badly affect vulnerable families like these. The devastating effects on these families are often caused by decisions made elsewhere in the world. In 2008, these decisions/factors included:

- rising oil prices, which increased the costs of growing and distributing food (about 10% of food production costs come from energy) – costs which are then passed on to the consumer as higher prices
- speculators trading crops in the 'food futures' market, i.e. trying to guess how far crops might increase in value – in order to make a big profit.

Inflation bites

Shoppers set to count the cost of bananas

Bananas have joined the ranks of dairy, meat and wheat products among foods which are set to surge in price because of the sharp rise in fuel costs. Chiquita, one of the world's biggest banana groups, said yesterday that the price of Britain's most popular fruit had risen by 36% last month against the same period a year ago.

Inflation seeps into cellar as Majestic says price of wine will have to go up

Faced with prices rising on all fronts, from petrol pumps to the supermarket checkout, those tempted to drown their sorrows were handed a further case of bad news yesterday. Inflation, Majestic Wine announced, is creeping into the wine cellar and prices are about to jump by 10% over the coming weeks.

Oil prices surge

Oil prices surged to a new record high yesterday.

The food crisis illustrates the **development gap** between rich and poor worldwide. Just as there are low-income earners in the UK who struggle when food and energy prices rise sharply, the world's developing countries also have poorer **food security**, caused by high import costs placing some foodstuffs beyond the economic reach of their poorest citizens. For low-income, food-importing countries, such as Bangladesh, the Philippines and Afghanistan, times have become very hard. Even in India, which produces more than enough rice for its population, prices soared by 35% during 2008, following 20% increases in 2007 (see pages 280-282). Many governments of rice-producing countries banned the export of rice in order to guarantee supplies for their own people.

Stories from the Philippines

In Manate, a poor suburb of Manila (capital of the Philippines), most people earn little, and buying food – mainly rice – consumes most of their income. In 2007, rice prices rocketed in line with global increases. In 2008, the Government used troops to deliver subsidised rice to poor districts, following protests (see right). It also urged restaurants to offer customers just half the rice they would normally get, and warned rice hoarders that they would be imprisoned if found guilty.

Bizarrely, until the 1990s, the Philippines was self-sufficient in rice but, by 2008, it was the world's biggest importer. In 20 years, the country lost its best farmland to urban development (reducing the rice supply) and at the same time its population grew rapidly (increasing demand). While industries grew, farming received little investment. (See page 285 for more information about the situation in the Philippines.)

Should farmland be protected from future urban development?

What do you think ?

● The **development gap** is the social and economic disparity between the wealthy and the poor. It occurs globally (i.e. between countries) but also locally (i.e. within cities or local regions).
● **Food security** is the extent to which a country can feed its own population, or can afford to purchase food on global markets when shortfalls occur.

▲ *Farming in Afghanistan – made more difficult by years of conflict and war*

▲ *Subsidised Government rice being sold to the poor in Manila under army supervision*

Over to you

1 Classify the FAO's causes of rising grain prices into **a** physical and **b** human. Which is more significant?
2 Draw a spider diagram to show how each cause makes food security worse.
3 In pairs, discuss what actions could prevent **a** food hoarding, **b** future food shortages. Discuss your ideas in class.

On your own

4 Research the FAO website (fao.org) to find out **a** which countries are most affected by rising food prices and **b** how any one country is coping.
5 How far does your research show that **a** developing countries are less food secure, **b** the development gap is a wide one?

Identifying the development gap - 1

In this unit you'll learn how indicators are used to highlight the development gap.

The North-South divide

In 1981, a landmark report was published about global **development**. Its author was Willy Brandt, the former West German Chancellor. The report became known as the Brandt Report, and the disparity that it identified was called the development gap.

> ● **Development** means 'change', and implies that change is for the better. It usually means 'economic' change, which improves people's standard of living.

The report showed a divided world (see right) – a wealthy 'North', possessing 80% of the world's wealth, and a poorer 'South' with just 20%. The report's main concern was the poverty affecting two-thirds of the world's population, compared with the remaining third who enjoyed a higher quality of life.

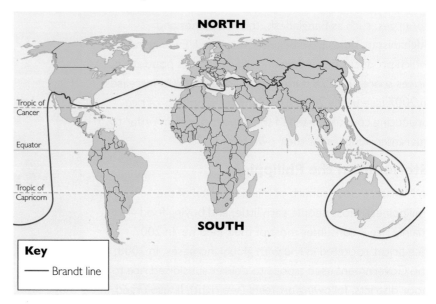

Key

—— Brandt line

▲ The North-South divide, as shown by the Brandt Commission in 1981

Measuring the development gap

Measuring development is difficult, because it requires data – but data are based on the **formal economy**, and ignore informal, subsistence, or unpaid work. Nonetheless, whether economic or social, all data show the same gap between the world's wealthiest and poorest countries.

*Street traders in Bangkok are part of the **informal economy**, where no record is kept of trading or income. So, traders like this do not figure in most development data about countries. ▼*

Measuring economic development

Two indicators are commonly used – Gross Domestic Product (GDP) and Gross National Income (GNI). Each is slightly different.

● **Gross Domestic Product (GDP)** means the value of the goods and services produced in a country over a year. It is calculated by combining the value of all goods produced with the value of services provided, such as tourism or banking. It is then usually divided by the total population to give a **per capita** (per person) value, which is converted to US$ to enable comparisons to be made between countries.

● **Gross National Income (GNI)** is like GDP, but it also includes income from overseas investments, such as shares and earnings/profits from overseas subsidiary companies and branches. As such, it is a better measure of a country's wealth than GDP – but more so for wealthier countries, which earn a lot of their wealth in this way. Like GDP, it is usually expressed as GNI per capita. GNI is shown in US$, using a conversion known as the Atlas method, which uses exchange rates averaged out over three years to smooth out fluctuations in currency values.

However, neither GDP nor GNI shows what per capita income is actually worth in terms of spending power (in other words, taking into account the cost of living in a country). To improve on this, the UN introduced **Purchasing Power Parity (PPP)**, to show what per capita income will purchase when the cost of living is taken into consideration. For instance, because of the low cost of living in China, US$100 will buy far more there than in the USA – China's GDP per capita in 2006 was US$2000, but its PPP$ (when the cost of living is allowed for) was US$7660. A PPP$ value which exceeds GDP per capita means that prices in a country are cheaper than average, and vice-versa.

▼ *The global distribution of GDP per capita in 2006, using PPP$ to allow comparisons to be made between countries. This map uses ten different bands, and provides a greater level of detail than the Brandt map.*

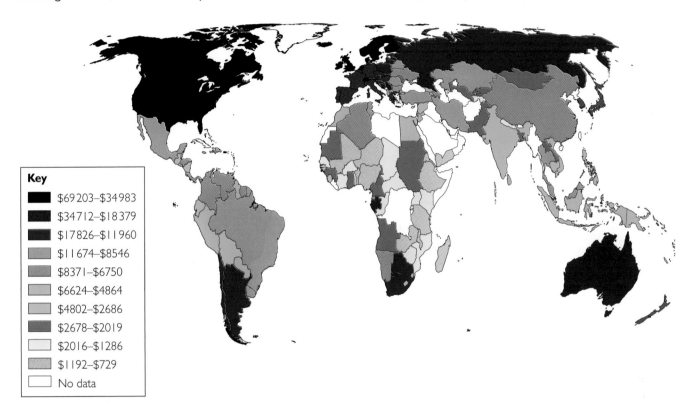

Key

■	$69 203–$34 983
■	$34 712–$18 379
■	$17 826–$11 960
▨	$11 674–$8546
▨	$8371–$6750
▨	$6624–$4864
▨	$4802–$2686
▨	$2678–$2019
□	$2016–$1286
▨	$1192–$729
□	No data

Measuring human development – the HDI

Economic growth is only of human value if quality of life increases too, and education and health care are central to human development. The **Human Development Index (HDI)** was devised by the United Nations to provide a measure of life expectancy, education and GDP for every country in the world. It is used widely because, by tying GDP to education and health, it shows how far people are benefiting from economic growth.

The HDI uses four indicators:
- Life expectancy
- Education (using the two indicators of literacy and average length of schooling)
- GDP per capita (using PPP$)

It averages these indicators out for each country, converts them into index figures – the value of which ranges from 0 (low) to 1 (high), and then combines the four figures into a single result – the HDI. Its calculation is complex but is basically as follows:

- The world's longest average life expectancy is currently 82 (Japan), and the shortest 32 (Zambia). In the HDI, values are converted to a number between 0 (worst) and 1 (best). Therefore, 32 years is converted to 0 and 82 years to 1. So, a country with an average life expectancy of 57 would be 0.5 on the index (i.e. half way between 32 and 82).

(Continued on the next page.)

Identifying the development gap - 2

- Education uses adult literacy rates to get one index figure and average years of schooling to get another. The lowest and highest values are expressed as a figure between 0 and 1. Therefore, 75% adult literacy would be 0.75 on the index. The same is done for schooling. If pupils spend six years on average (out of 12) in school, this is expressed as 0.5 on the index (i.e. 50% of a 12-year maximum).

- GDP per capita in PPP$ (which measures standard of living) uses a scale from the lowest (in 2006, this was Burundi with PPP$729) to the highest (in 2006, this was Luxembourg with PPP$69 203). The HDI index value is then calculated by placing all countries on a logarithmic scale between 0 and 1.

The four indicators are then averaged out to give a single figure. In 2005, Iceland (the top) had an HDI of 0.968, and Sierra Leone (the bottom) 0.336. The UK was sixteenth with 0.946. High GDP per capita does not necessarily convert into high HDI scores, or vice-versa. It is a matter of how committed a government is to converting the benefits of economic growth into improved education and health care.

Final HDI values are classified as follows:
- *A value of less than 0.5 represents a low level of development. All 22 countries in that category are in sub-Saharan Africa. Even the highest-scoring sub-Saharan countries (Gabon and South Africa) are only ranked 119th and 121st in the world, respectively.*
- *A value of 0.8 or more represents a high level of development. As well as the developed economies of North America, Europe, Oceania, and eastern Asia, it includes some in Eastern Europe, Central and South America, South-East Asia, the Caribbean, and the oil states of the Arabian Peninsula.*

▼ *The global distribution of the HDI in 2007*

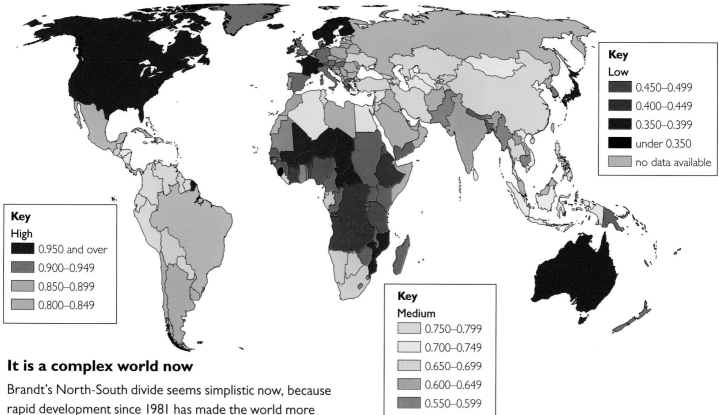

Key
Low
- 0.450–0.499
- 0.400–0.449
- 0.350–0.399
- under 0.350
- no data available

Key
High
- 0.950 and over
- 0.900–0.949
- 0.850–0.899
- 0.800–0.849

Key
Medium
- 0.750–0.799
- 0.700–0.749
- 0.650–0.699
- 0.600–0.649
- 0.550–0.599
- 0.500–0.549

It is a complex world now

Brandt's North-South divide seems simplistic now, because rapid development since 1981 has made the world more complex:

- Some countries in Latin America (e.g. Brazil) were already developing in the 1970s, which even in 1981 made Brandt's classification out of date.

- Economic development since 1981 has produced a complex pattern, so that some countries (e.g. Singapore) or parts of countries (e.g. Dubai, Hong Kong) are highly developed. Others in South-East Asia (e.g. Thailand, Malaysia) grew rapidly in the 1980s and 1990s. Now, China and India are growing fast. These are referred to as 'Newly Industrialising Countries' (NICs).

- While some countries have developed rapidly with rising incomes, others have barely changed. The gap between rich and poor has widened substantially.

The United Nations now uses four levels of income – 'High', 'Upper middle', 'Lower middle' and 'Low' to classify the world's 206 nations (see below). Even then, there are huge differences within each division; low-income countries include India – one of the world's largest and fastest-growing economies – and Ethiopia, one of its smallest and slowest.

However, one conclusion is clear. 26 of the 30 poorest countries in 2006 were in sub-Saharan Africa. Growth since 1981 has enabled some countries to develop rapidly – but Africa is being left further behind.

▼ *The classification of countries in 2006, using four levels of income – as defined by the World Bank and the United Nations*

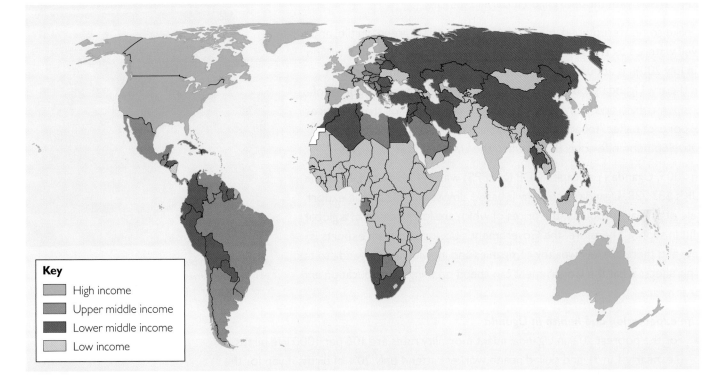

Key
- High income
- Upper middle income
- Lower middle income
- Low income

Over to you

1 Compare the map showing Brandt's North-South divide with those showing **a** global GDP, and **b** global HDI. In pairs, make a detailed list of:
   - which continents and countries have changed since 1981?
   - which have not changed since 1981?
2 Draw a table to show the advantages and disadvantages of using **a** GDP, **b** GNI, **c** HDI as indicators of development.

On your own

3 Use the worldmapper website (worldmapper.org) to research life expectation, infant mortality and child malnutrition in the regions on the right. Copy and complete the table, using 'high', medium', 'low', or 'varies between …' to describe each indicator.
4 How far is there a development gap **a** of the type that Brandt portrayed in 1981, **b** of a different type now? Explain your answer.

Region	Life expectation	Infant mortality	Child malnutrition
North America			
Western Europe			
Eurasia (Eastern Europe and the CIS)			
Latin America and the Caribbean			
The Middle East and North Africa			
Sub-Saharan Africa			
Eastern Asia (China)			
South-East Asia and the Pacific Rim			
South Asia (Indian sub-continent)			
Australasia			

Living on the wrong side of the gap

In this unit you'll find out what the development gap means for those affected by it.

The example of Uganda

The photo on the right is of a newborn baby in Uganda. This unit is about that baby and its probable life chances, and what living on the wrong side of the development gap is like. This new baby's life chances make a good comparison with those of children born in the UK.

Unlike some stereotyped images of Africa, Uganda is green, fertile, and well watered by two rainy seasons each year. Its population of 31 million (in July 2008) is about half that of the UK – within the same land area. In theory, Uganda should be a wealthy country, with good resources of copper, cobalt, and hydro-electric power – and fertile soil that produces exports of coffee, tobacco, sugar cane and tea. Why then is its Human Development Index only 0.505, when the UK's is 0.946?

In 2005, Uganda's per capita GDP (in PPP$) was $1454 – just 4.4% of the UK's $33 238. Uganda's economy is based almost entirely on the export sale of primary products, the prices of which are low because of a global glut in supply. Therefore, the Government's tax income from exports is low and there are few wealthy companies and individuals in Uganda to tax. This affects what the Government can spend on things like education and health care.

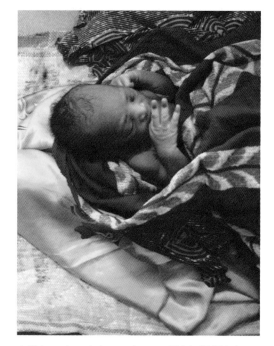

▲ This newborn baby was born on 10 July 2007 in Lacor Hospital, northern Uganda. It has very different life chances from a child born in the developed world.

Life expectation and health in Uganda

- For the poorest 20% in Uganda, infant mortality rates are 106 per 1000 live births (i.e. nearly 1 in 9), and skilled health workers attend only 20% of births. Even for the wealthiest 20%, health worker attendance is only 77% and infant mortality rates are 20 per 1000 live births – four times that of the UK.
- The new baby will probably not be well fed. 24% of Ugandan families are undernourished.

▼ A maternity ward in Uganda's capital, Kampala. Few children are born in hospitals like this – most are born at home or in a small clinic. The majority of children born to poor families have no health workers in attendance at all.

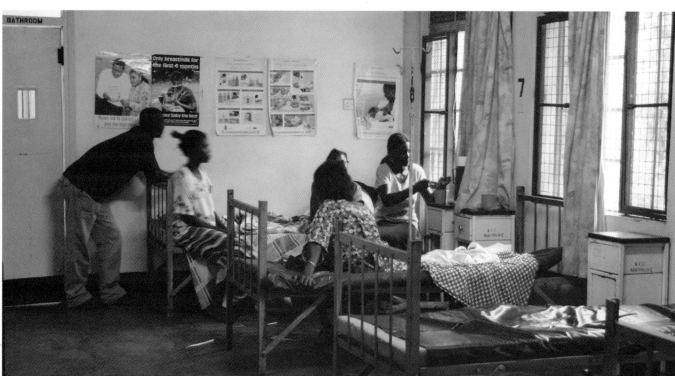

- Because of HIV/AIDS, the average life expectancy in Uganda in 2005 was only 49.7 years, compared to the UK's 79. Assuming that the new baby does not contract HIV/AIDS, it still has only a 62% chance of reaching the age of 40 (in the UK it is 98%), and only a 40% chance of reaching 60 (in the UK it is 91%).
- Malaria and cholera are also big killers. Living conditions are poor for many, with only 60% of people having access to safe water, and 43% to an improved sanitation system.
- Nonetheless, the new baby has at least a 94% chance of never contracting HIV/AIDS. Uganda's Government was the first in Africa to attract international aid to spearhead HIV/AIDS education programmes (see right). In the early 1990s, 20% of Uganda's population was HIV-positive; now it is only 6%.

Education in Uganda

- What education will the new baby get? Although primary education is free and universal, secondary school is different; only 17% of girls attend. There are few Government schools and most families have to pay fees. Even if a family earned a good yearly income of £200 from growing coffee, there would be little money left over to pay for education. Secondary school fees (£20 per term) are too expensive for most families. A family is more likely to pay for boys to receive secondary education than girls.
- The new baby is unlikely to go to university. The Ugandan Government funds only four universities, about a quarter of the total. To share resources, each district is allocated a number of university places, but that means one place for every 30 000 school students.

After a limited education, most women in rural Uganda marry by the age of 15 and become mothers soon after. They are then assets for their families because, when they marry, they attract a dowry. They may then have many children, because fertility rates are high (6.8 children per woman).

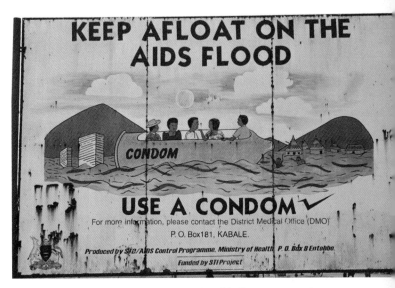

▲ An HIV/AIDS warning poster in Uganda. Uganda's Government made AIDS awareness and sex education (often a taboo subject in many African countries) a priority in the 1990s.

▲ Makerere University, Kampala, Uganda. There are few places for Ugandan students unable to pay fees – only those on scholarships or those whose families can afford the fees can attend.

The debt crisis - 1

Who is in control?

Like most people's, the life of the new baby on page 184 is influenced by wider systems and factors impacting on it. The biggest factor until recently has been Uganda's debt burden, which dates back to the 1970s and 1980s.

When it became independent in 1962, Uganda inherited a successful economy – it was self-sufficient in food and had a farming industry dominated by small farmers who provided the main exports of coffee and cotton. The economy was stable, debt low, and trade, manufacturing and finance thriving – thanks to an enterprising and well-established Asian immigrant community.

In 1971, Idi Amin came to power by overthrowing the democratic Ugandan Government and replacing it with a military regime. A period of violence resulted in civil war and the need for more weapons. But Amin's new government had expelled the wealthy Asian community and taken its assets (largely for personal gain), which led to the collapse of the Government's tax revenue. Large-scale borrowing was necessary to maintain Government spending and to finance the purchase of weapons. This situation persisted until Amin was himself overthrown in 1979.

During the 1980s, Uganda's debt burden increased massively for the reasons outlined below. By 1992, Uganda's debt was US$1.9 billion. Its **debt service** payments grew considerably during this time, so it was unable to repay its loans. During the early 1990s, its annual debt repayments exceeded its export earnings.

However, Uganda has benefited from a package of debt write-offs in recent years. In 2000, the IMF and World Bank cancelled most of Uganda's debt through the Highly Indebted Poor Countries (HIPC) debt relief initiative, worth nearly $1.5 billion (see opposite).

> ● **Debt service** means the payments of interest, plus a proportion of the original loan, which are required in order to pay back a debt over a given period of time.

How Uganda and other developing countries got into debt

1 In the 1970s, **OPEC** (see pages 32-33) raised the price of oil twice – massively increasing its members' earnings.

2 The OPEC countries banked these increased earnings in Western banks.

3 The banks then lent this money out to developing countries – sometimes responsibly for huge projects, e.g. dams, power stations – but at other times to finance conflict and keep regimes like that of Idi Amin in power.

4 By the 1980s, global interest rates had more than doubled – massively increasing the repayments needed to service loans taken out in the 1970s.

5 Uganda found itself unable to meet its debt repayments, so the unpaid interest was added to the original loan amounts. Every year, Uganda's overall debt burden grew as levels of unpaid interest mounted.

6 Most developing countries found themselves in similar situations. To prevent a collapse of the world's banking system, the IMF (see page 150) constructed a solution called **Structural Adjustment Packages (SAPs)**.

7 SAPs involved re-scheduling loans to make them more affordable – but only in return for cuts which the IMF imposed on Government budgets and spending (also see page 203).

8 Without IMF approval, no country would get further credit. SAPs therefore became compulsory.

9 The biggest Government budget items in Uganda were health and education. The imposed IMF cutbacks therefore affected both – with their greatest impacts being on the poor during the late 1980s and early 1990s.

The Highly Indebted Poor Countries initiative

The Highly Indebted Poor Countries (HIPC) are a group of 38 of the least developed countries, with the greatest poverty and debt (32 of which are in sub-Saharan Africa). Because of their situation, they are considered eligible for help from the International Monetary Fund and the World Bank. The HIPC initiative began in 1996, following pressure from Non-Governmental Organisations (NGOs – charities and voluntary organisations) to reduce debt burdens.

Further debt reductions came in 2005, when the UK held the presidency of the G8 – the world's richest nations (consisting of the USA, Canada, Britain, France, Germany, Italy, Japan and Russia). Under an agreement made at the G8 meeting in Gleneagles, the World Bank, the IMF and the African Development Bank cancelled all loans (worth US$40 billion) owed to them by 18 HIPC countries (see the map). This action alone saved those countries $1.5 billion annually in debt repayments.

However, there were two conditions:

- The government of each country had to demonstrate good financial management and a lack of corruption.
- The money saved had to be spent on poverty reduction, education and health care.

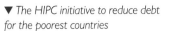

▼ The HIPC initiative to reduce debt for the poorest countries

Key

HIPC countries that qualified for debt relief in 2005

1	Benin
2	Bolivia
3	Burkina Faso
4	Ethiopia
5	Ghana
6	Guyana
7	Honduras
8	Madagascar
9	Mali
10	Mauritania
11	Mozambique
12	Nicaragua
13	Niger
14	Rwanda
15	Senegal
16	Tanzania
17	Uganda
18	Zambia

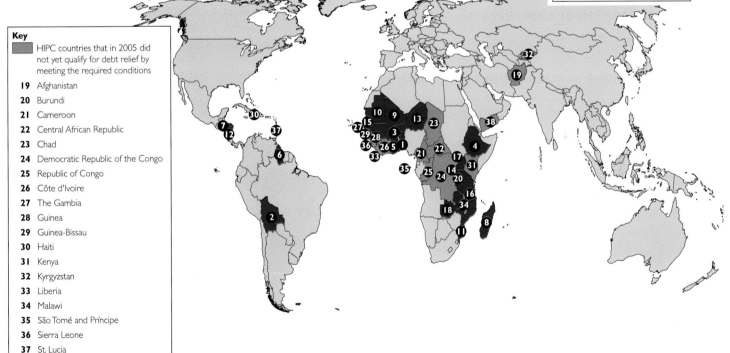

Key

HIPC countries that in 2005 did not yet qualify for debt relief by meeting the required conditions

19	Afghanistan
20	Burundi
21	Cameroon
22	Central African Republic
23	Chad
24	Democratic Republic of the Congo
25	Republic of Congo
26	Côte d'Ivoire
27	The Gambia
28	Guinea
29	Guinea-Bissau
30	Haiti
31	Kenya
32	Kyrgyzstan
33	Liberia
34	Malawi
35	São Tomé and Príncipe
36	Sierra Leone
37	St. Lucia
38	Yemen

By 2008, 27 of the 38 HIPC countries had met the conditions for debt relief, and had received a total of US$85 billion in aid. But the campaign is not over. African countries still owe US$300 billion, and there is at present little chance that any of them will be able to repay those debts. For countries affected by war (e.g. Somalia, Sudan), or natural disasters, new loans may be needed just as old debt is being cut. Other countries possess governments which have kept unreliable records or have come to power through force rather than democracy, and which the IMF is unwilling to support. Debt campaigners believe that an additional 24 countries should be added to the original 38, including Nigeria, India, and Indonesia.

The debt crisis - 2

	1990	2005
Congo	19.0	2.3
Gambia	11.9	6.3
Côte d'Ivoire	11.7	2.8
Madagascar	7.2	1.5
Malawi	7.1	4.6
Ghana	6.2	2.7
Zambia	6.1	3.3
Guinea	6.0	4.9
Cameroon	4.6	4.7
Tanzania	4.2	1.1
Niger	4.0	1.1
Equatorial Guinea	3.9	0.1
Mozambique	3.2	1.4
Uganda	3.4	2.0

▲ *Levels of debt in some sub-Saharan African countries. The figures show the percentage of GDP taken up by debt repayments in 1990 and in 2005 (following large write-offs of debt by Western countries)*

The impacts of debt cancellation

The cancellation of $1.5 billion of its debt under the HIPC initiative has had major impacts in Uganda:

- Spending on public services has risen by 20% overall – with 40% extra being spent on education and 70% on health care, including the abolition of fees for basic health care.

- Free primary schooling has been introduced – with girls, in particular, benefiting most. Five million extra children have begun to attend school. Enrolment rates for primary schooling increased from 62.3% in 2000 to 92% of girls and 94% of boys in 2006. Before debt relief, there were 20% fewer girls than boys in primary school, but now numbers are almost even.

- 2.2 million people (nearly 10% of the population) have gained access to clean water. Fetching water is usually the responsibility of women and girls, and is often a reason for girls not going to school.

1985	0.420
1990	0.434
1995	0.433
2000	0.480
2005	0.505

◀ *Although there is still a long way to go, big improvements have occurred in Uganda's HDI rating since 1985*

It is a splendid start, and one hopes that they will go on to cancel all debt for most of the countries – I gather it is about 62 countries – which are heavily indebted. But remember that the West had a hand in promoting some of those leaders, because it suited them at the time.
Archbishop Desmond Tutu of South Africa

▲ ▼ *Two reactions to the Gleneagles G8 debt agreement in 2005*

Tomorrow, 280 million Africans will wake up for the first time in their lives without owing you or me a penny from the burden of debt that has crippled them for so long. But the end will not be achieved until we have the complete package ... of debt cancellation, doubling of aid, and trade justice.
Singer and activist Bob Geldof (organiser of the 1985 Live Aid and 2005 Live 8 concerts)

	Before debt cancellation	During and after debt cancellation
Population using an improved water source (%)	44 (in 1990)	60 (in 2004)
Population undernourished (% of total population)	24 (1990-92)	19 (2002-04)
Public expenditure on education (% of GDP)	1.5 (1991)	5.2 (2002-05)
Public expenditure on education (% of total government expenditure)	11.5 (1991)	18.3 (2002-05)
Adult literacy rate (% aged 15 and older)	56.1 (1985-95 average)	66.8 (1995-2005 average)
Young adult literacy rate (% aged 15-24)	69.8 (1985-95 average)	76.6 (1995-2005 average)
Total debt service expenditure (% of income from exports, plus net income from abroad)	81.4 (1990)	9.2 (2005)

◄ *Progress in Uganda following the beginnings of debt cancellation in the late 1990s. The years for each set of data are shown in brackets.*

Population without electricity (millions)	24.6
Population without safe water (%)	40
Estimated earned income, female (PPP$)	1199
Estimated earned income, male (PPP$)	1708
Adult literacy rate, female (% aged 15 and older)	57.7
Adult literacy rate, male (% aged 15 and older)	76.8

▲ *But there's still some way to go ... all data are for 2005.*

The significance of educating girls

Few girls receive secondary education because of its cost. Yet education is the path to independence and greater prosperity. In Ugandan primary schools, the ratio of girls to boys is 1:1, but by age 11 it is 1:2 – and by 16 it may become 1:6 or even 1:10. Girls may marry as young as 13 or 14 in rural communities, and have their first child soon after – hence Uganda's high fertility rate of 6.8.

Women are the poorest Ugandans – by custom they rarely own land and are most likely to work as landless labourers as and when jobs are available. They have little control over how much they earn and even less over any career path. They return to work soon after giving birth. Maternal mortality rates are also high, and unhealthy mothers have babies which are more likely to die in their first five years.

Yet investment in girls' education can have huge social impacts. Educated women often defer marriage until they are older. Therefore, the fertility rate tends to be lower among professional women. Educated women are also more likely to select their career, work in that career before and during marriage, and select their own marriage partner. Infant mortality rates among educated women in Uganda are almost as low as in many developed countries.

Over to you

1 In pairs, draw up a table comparing development indicators for Uganda and the UK.
2 Draw a flow chart to show three parallel strands: **a** how Uganda got into debt, **b** how the money became available to borrow, **c** how debt mounted up to high levels.
3 What are the benefits for **a** Uganda, **b** the G8 nations in cancelling all debts?
4 In pairs, draw up a list of **a** the risks and **b** the advantages of cancelling even more debts – by extending the list of countries eligible for debt cancellation to 62. Would you vote for the cancellation of such debts or not?

On your own

5 Research up-to-date data for **a** GDP and **b** social and economic indicators for Uganda, using the UN Development Report (http://hdr.undp.org/en/statistics/), the World Bank website (http://www.worldbank.org/) or the CIA World Factbook (https://www.cia.gov/library/publications/the-world-factbook/). How far is Uganda improving, compared to indicators in this book?
6 Using the information in this unit, write a 500-word argument to convince a sceptic that debts are worth cancelling.

Exam question: Examine the role played by debt in maintaining a global development gap. **(15)**

Who is to blame for global debt – the ones who borrowed or the ones who lent the money?

What do you think?

In this unit you'll learn about how people's attitudes towards global development and other countries are linked to the development gap.

Just how big is Greenland?

Study Maps A and B, which show two different ways in which the world has been portrayed. Map A was devised by Gerardus Mercator in 1569. It was, and remains, an amazing achievement – drawn without satellite photos or digital surveys, just explorers' accounts and measurements from sailors who were guided by the stars.

Mercator had the same problem that all mapmakers face – how to produce a flat map from a sphere? To overcome this problem, he devised a map projection which is accurately scaled at the Equator, but which stretches and flattens the world out towards the poles, where the lines of longitude converge. The nearer the poles you get, the more stretching is required – and the map becomes more inaccurate as a result.

For over 400 years, Mercator's projection has been the dominant map projection in use, and it has really affected people's thinking. Using Mercator's projection:

- Greenland looks the same size as Africa. In fact, Africa is 14 times larger than Greenland.
- Alaska appears to be seven times larger than it really is.
- Europe and North America seem much larger than they really are.

- Europe has been drawn centrally, because early explorers and cartographers – like Mercator – were European. Geographers refer to maps like this as **Euro-centric**.

Mercator's projection is still generally used in American classrooms and by American companies. Many British TV News organisations use Mercator – and so does Google Maps!

Equalising areas

Now consider Map B – the Peters projection, also known as the 'equal area projection'. Devised in 1973 by Dr Arno Peters to address Mercator's imperfections, Peters consciously set out to equalise areas, and therefore – by implication – their importance. It maps areas close to the poles by reducing the width between the lines of latitude. Although it distorts the **shapes** of places, it does show the correct **areas**. So, compared to Mercator's projection:

- Africa, Latin America and southern Asia dominate in size – just as they do in real life.
- The USA, Canada and Greenland seem relatively smaller.
- Europe dominates less, although it is still at the centre.

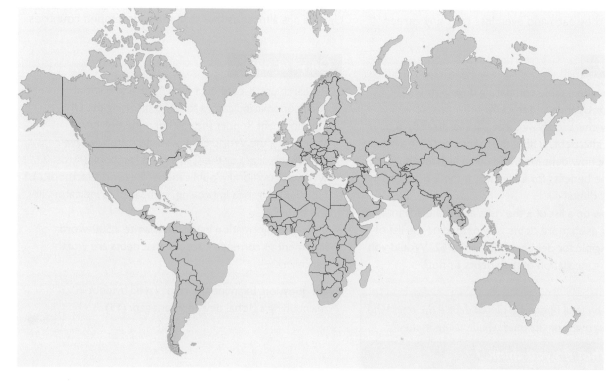

◄ Map A: Mercator's map projection of the world. First devised in 1569, it is still widely used – especially in the USA – despite its distortions.

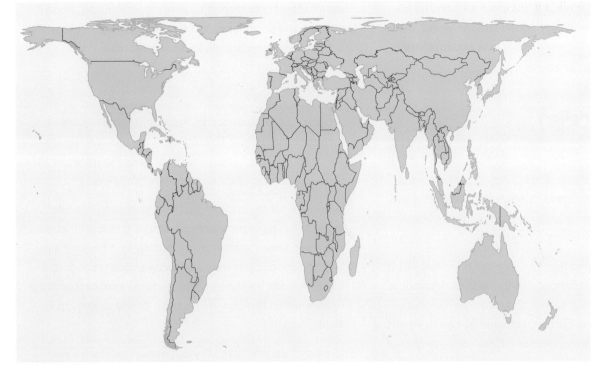

◀ *Map B: Peters' map projection of the world. Often criticised for distorting the actual shape of countries, this projection distorts no more than Mercator's – just differently.*

Peters felt that Mercator's projection exaggerated the size of white-dominated regions, and reflected a racist attitude towards Africa, Latin America and southern Asia. The Peters projection is used in many British schools, and also by several aid organisations, for example within the UN. Its use is greatest amongst organisations which work in, and are aware of, the needs of developing countries.

Does it matter?

It is true that relatively few people consciously use one map projection over another. However, the argument runs that, if people are exposed to maps emphasising particular places, they will be influenced by them. Britain and Europe (the major colonising powers since 1500) are shown, using Mercator's projection, to be larger in relation to their former colonies than in real life. Mercator's projection has resulted in people – even unconsciously – seeing the world as dominated by Europe and North America.

This is further reinforced by the concepts of 'top' and 'bottom'. Who decided that 'the north' should be at the top of the map? Again, it was originally a European decision that stuck! All maps are unavoidably political – there is no neutral!

The world sphere as seen from space. This perspective, with Antarctica at its centre, shows how arbitrary it was to make the north 'top' and the south 'bottom' on world maps. ▶

ODT Inc., an American company which deals with management training and cultural perspectives, claims that it does not really matter which map projection people prefer, so long as they recognise that there are different ways of seeing the world:

> There are many different valid points of view. People communicate better with others when they recognize that there are many perspectives from which to view the world. When you believe that your own view is the only valid one, you cut off communication with others who may not share your assumptions.
>
> **The views of ODT Inc. about maps and perspectives on the world**

How did Western thinking come to dominate?

It is not only through maps that European – and later American – thinking came to dominate the world. Two theories explain how they also became **economically** dominant – firstly through a geographical model, known as 'core and periphery', and secondly through a classical economic theory, known as 'economic man'.

Core and periphery

Wallerstein, an American sociologist and world-systems analyst, developed a theory in 1974 about the capitalist world system and the development gap. He divided the world into two types of economic area: core and periphery.

Core areas

Wallerstein claimed that core regions drive the world economy. These developed as the world's first industrial areas during the eighteenth and nineteenth centuries – initially the UK and Europe, followed by North America and Japan. During the Industrial Revolution, they used capital invested from a wealthy farming economy. In spite of the global shift to China and India, this 'core' still owns, produces and consumes over 80% of global goods and services.

Peripheral areas

Peripheral areas lie at the other extreme. Lacking capital from a prosperous farming sector, they rely on core regions for the **3 Es** – to **explore**, **exploit** and **export** their raw materials. Therefore, unequal trade patterns develop between core and periphery – core regions import, process, add value to, and profit from the processing of raw materials from peripheral regions into manufactured goods. Populations in peripheral regions are drafted in as cheap (or forced) labour in mines or on plantations, making those raw materials cheap.

Recent global shifts mean that manufacturing has now increased in peripheral areas, because of cheaper labour. However, core countries remain dominant because they own the production lines and dictate what is produced and by whom. Investment and decision-making therefore remain within the core. Hence the development gap is maintained.

Core and periphery theory can occur at all scales and apply as much within countries as between them. The two graphs opposite illustrate how much the South East and London make up the UK's core economic region. Compare them with the North East – a peripheral region by comparison – in which industrial decline has reduced its relative importance. Global cores, such as Western Europe, contain smaller cores, such as the M4 corridor. But they also contain peripheral areas, such as Cornwall. Similarly, global peripheries, like Latin America and Africa, contain small cores, such as Rio de Janeiro or Johannesburg.

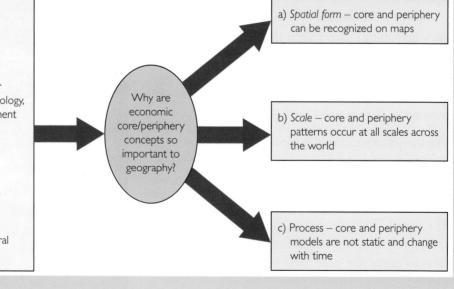

Core/periphery

- Core economic processes are relatively high-technology, high-pay activities, high capital investment

- Periphery processes are at the other end of the scale: relatively low-technology, low-pay activities, low capital investment

- Both core and periphery processes can take place in one industry

- One set of processes cannot exist without the other. The formation of capital and wealth with the core processes is made possible by the labour and resources in the peripheral processes

Why are economic core/periphery concepts so important to geography?

a) *Spatial form* – core and periphery can be recognized on maps

b) *Scale* – core and periphery patterns occur at all scales across the world

c) *Process* – core and periphery models are not static and change with time

▲ *The concept of core and periphery*

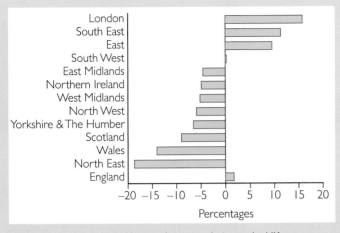

▲ Average weekly household expenditure, in relation to the UK average, 2003/04 – 2005/06, by UK region

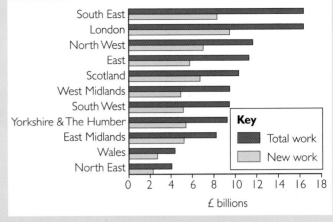

▲ The value of the construction industry in 2006, by UK region

Background

The theory of 'economic man'

At the heart of capitalist economic theory is *economic man*, a classical model of human behaviour. It assumes that people always act in ways that benefit themselves. One of the first modern classical economists, Adam Smith, believed that society benefited if people were free to pursue their own interests, and that the free market worked more efficiently and made the best decisions without Government interference. Smith called this lack of Government interference the 'invisible hand'. Follow Smith's thinking and a development gap is bound to emerge, because some countries win and others lose if everyone is out to look after themselves.

Neo-liberalism

Smith's ideas were revived by **neo-liberalism** in the late twentieth century, which promotes free market theory. One of the UK's most influential think tanks is the Adam Smith Institute, which promotes Smith's ideas in a modern context and advises on how Government influence can be reduced. At the extreme, some believe that even health and education should be private – and not Government – responsibilities. Until recently, the World Bank and IMF have offered loans only on the condition that taxation and Government spending are reduced.

However, many contest Smith's theories. In many developing countries, his thinking is at odds with those who believe that poverty can be reduced only with Government help, e.g. through health and education which are publicly funded. An educated and healthy workforce is seen as an economic asset, as well as being the right thing to do. Hence the HIPC debt initiative.

Others believe that economists have wider motives than simply suiting themselves or their profits. Examples include **altruism**, e.g. a belief in taxing the wealthy to pay for benefits for the sick, elderly or poor for humane reasons.

Over to you

1 Research examples of global data (e.g. GNI) presented on both Mercator's and Peters' projections.
2 In pairs, use these to list advantages and disadvantages of each projection in terms of accuracy, usefulness and appearance.
3 Research and present one other projection to your class.
4 Outline what you see as the advantages and disadvantages of Smith's theory of 'economic man'.

On your own

5 Research one example of a core region. Use data sources such as the CIA World Factbook to describe ways in which it is a core region.
6 In 500 words, explain how core and peripheral regions can change over time.

Should companies and countries be free to follow Smith's ideas?

What do you think?

South Africa - the widening gap

In this unit you'll learn how the development gap can be found within one country, and result from political systems dating back to colonialism.

What lies beneath

South Africa can seem like Paradise on Earth. As Africa's wealthiest country by far, it has among the world's largest reserves of gold and diamonds. Its scenery is striking – from the Drakensburg Mountains (see right) to glorious beaches. For tourists, it has game reserves in abundance.

Beneath that, however, it has some of Africa's most deeply rooted social problems. In January 2003, there were riots in Alexandra, a black township outside Johannesburg. Petrol bombs were thrown at police, who enforced the eviction of families from their homes for non-payment of rent. The protests were aimed at the African National Congress (ANC) government. The crowds were angry that – after nine years of ANC rule and the first votes for black people in South Africa – wide **disparities** still existed between the (mainly white) wealthy population and the (mainly black) poor.

Data from Statistics South Africa, a Government department, show that income inequalities still mainly occur along racial lines, and that they are increasing. South Africa is increasingly suffering from the impacts of these inequalities and an official unemployment rate of 30% (which could be much higher in reality). Rates of violent crime are very high, with over 18 000 murders each year, and other serious crimes – such as car-jacking at gunpoint – also increasing.

Why are there disparities?

Before 1994 and the election of the first black majority ANC government, South Africa's white-dominated economy was heavily State-run and subsidised. To attract overseas investment for the expansion of mining, the new ANC government cut Government spending by privatising State assets and reducing the size of the Civil Service. These measures had several impacts, which are discussed below and in more detail on page 196.

The impacts of the economic changes

Although South Africa's economy has grown rapidly since 1994, its growth has not created many jobs. Most economic growth has been in **capital-intensive**, rather than **labour-intensive**, sectors. This is a particular problem because a high proportion of the South African labour force is unskilled. Companies have also had to cut back on their labour costs in order to remain competitive in global markets.

▲ *The Drakensburg Mountains – some of the superb scenery in South Africa*

> We voted ANC. They said we're going to get the better life! This time I think I could even vote for a white man! He can help me.
> **Protests in 2003 from Sophie Morweng, a 52-year-old domestic worker**

▼ *Riots in Reiger Park settlement outside Johannesburg in 2008 over a lack of jobs and housing, and objections to Zimbabwean refugees*

- **Capital-intensive** means high-cost industries, such as mining, where machines do most of the work and few jobs are created.
- **Labour-intensive** means low-cost industries, such as construction and tourism, where people do most of the work and jobs are created.

South Africa under apartheid

South Africa's inequalities go back to **apartheid**, which legally affected every aspect of life there between 1948 and 1994. The theories underpinning the apartheid laws date back to colonialism under the Dutch Afrikaners and the British. Both set up colonies in what is now South Africa.

> ● **Apartheid** is an Afrikaans word meaning 'segregation', which is used to describe a political and legal system used in South Africa to separate the different ethnic groups there between 1948 and 1994.

The British formed Cape Colony (based on Cape Town) and Natal (based on Durban), while the Afrikaners settled Transvaal and the Orange Free State, which turned out to contain huge deposits of gold and diamonds (see right). They both took over native lands, created separate reserves or homelands for the indigenous peoples, and brought many Indians (later known as 'coloureds' in apartheid South Africa) to build a rail network there. Gradually, South Africa developed a multi-ethnic population.

The Afrikaners believed that they were racially superior to indigenous Africans, and that racial purity was essential. They were supported in the late nineteenth century by pseudo-scientific writing on race in Europe, which argued that white Europeans were racially superior to all other races. A few British supported this view, including Cecil Rhodes, Prime Minister of Cape Colony, who wrote: 'I contend that we are the first race in the world, and that the more of the world we inhabit the better it is for the human race.'

Despite initial major tensions between the British and the Afrikaners, which resulted in the two Boer Wars (1880-81 and 1899-1902), the two groups worked together to extract South Africa's mineral wealth. The Afrikaans language (a Dutch dialect) survived, and the majority of whites continued to believe in racial superiority.

The white minority continued to rule South Africa and deny rights to the black majority.

● **Segregation** was introduced into mines in 1911,

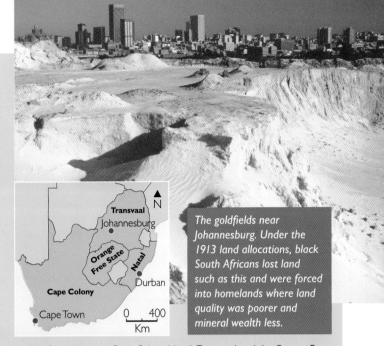

The goldfields near Johannesburg. Under the 1913 land allocations, black South Africans lost land such as this and were forced into homelands where land quality was poorer and mineral wealth less.

▲ The four colonies – Cape Colony, Natal, Transvaal and the Orange Free State – which combined in 1910 to form The Union of South Africa

which banned black South Africans from skilled jobs there.

● In 1913, black South Africans were forced to live in African 'reserves', where land quality and mineral wealth were poorer than in lands owned by whites.

● Whites lived separately from black South Africans – enjoying a high standard of living and keeping to their own clubs and churches.

● In 1948, the Afrikaner National Party took control of the country and began to enact apartheid laws, such as banning marriage between whites and non-whites and declaring some jobs to be 'white only'.

● In 1950, the Population Registration Act racially classified all South Africans as white, black or coloured, which was then used to control what people could do and where they could go. Passbooks were introduced that restricted the movement of the black population.

The African National Congress (ANC) led protests against apartheid from the 1950s onwards, under leaders such as Nelson Mandela – who was imprisoned in 1963. During the 1980s, continued protests – combined with international political pressure and declining foreign investment – led to the end of apartheid. Nelson Mandela was released from jail in 1990 **after 27 years** and the ruling National Party agreed to full and free elections in 1994, which resulted in a resounding win for Nelson Mandela's ANC Party.

The impacts of changes since 1994

The impacts of economic changes since 1994 have made inequalities in South Africa even worse, so that it is now considered one of the ten most unequal countries in the world – as measured by the **Gini Co-efficient** (a global index used to measure the equality of wealth distribution). The Index's scores vary globally from 0.247 (Denmark – the most equal distribution of income) to 0.7 (Namibia – the most unequal).

Inequalities in South Africa have become worse because:

- the liberalisation of the economy (the privatising of State assets and cuts in the Civil Service) has given the wealthy white minority more opportunities to become richer.
- the poor (almost all of whom are black) are getting poorer, because – during the economic transition – millions of jobs were lost and unemployment doubled to 30%, which has affected the black population far more than the white.
- 50% of South Africa's population is under 18, and the economy cannot absorb hundreds of thousands of school leavers every year. Unemployment is especially high among young men.
- exposure to the global market has affected food prices, which were previously subsidised. Annual inflation has varied between 14 and 20% since 2000. This has shown its greatest effects on the poor.
- there are widening differences *within* ethnic groups. A small black middle class has emerged, while poorer whites have lost State benefits.
- cuts in health spending have worsened the impacts of HIV/AIDS, which affects 5.3 million South Africans (out of 43.8 million), with disproportionate numbers among the poor.

Nonetheless, there have also been major successes. The ANC government has provided housing and water for the poor. Between 1994 and 2003, 10 million people – nearly 25% of the population – gained access to clean drinking water. By 2008, the whole population had access to clean water. Women, in particular, no longer have to spend hours each day fetching water.

▲ A largely white housing estate in Johannesburg, with high walls, an electric fence and secure gate. Gated communities like this are recent developments due to soaring crime rates, but the white and black populations always lived separately under the apartheid laws.

▼ New housing, like this estate in Rustenburg north of Johannesburg, has improved the quality of life for many black South Africans. The new houses have water and electricity supplies.

Is apartheid dead?

Although apartheid 'died' in 1994, South Africa's most serious problem remains its degree of inequality. Some of this persists as a hangover from the days of apartheid, while some of it is a result of the changes brought about by exposure to the global economy. Inequality shows itself in three ways:

A There is greater **regional inequality** between the richest and poorest regions (see the map). Those areas benefiting from global growth – such as the goldfields around Johannesburg, or producers of wine and cash crops for export from the Western Cape – have done well. Other areas with a traditional subsistence economy remain poor, with few or no job opportunities, and are reliant on wages sent home by urban-dwelling family members.

B *Ethnic inequalities* still arise because of skin colour. The white population is still significantly better off than the black (see right). But, although white incomes are still markedly higher than those for black South Africans, poverty has also increased for some whites (there is now a small but significant white underclass). Although the data in the table are from the 2001 Census, evidence is that the gap is actually increasing.

C One of South Africa's greatest concerns is *crime and personal safety*, as indicated in the table below right of **Peace Index** scores. This Index is compiled from 24 indicators, which combine external factors such as a nation's military expenditure and its relations with neighbouring countries, to internal such as safety and gun crime. Scores are based on data, e.g. numbers of murders, or legalised access to guns.

Over to you

1 In pairs, produce a short PowerPoint presentation which outlines South Africa's assets as a country. Use atlas maps to research mineral wealth and the CIA World Factbook website for economic and social data. Add the reasons why these assets are distributed so unequally.

2 In pairs, identify five priorities which you think would help to improve life for **a** black South Africans, **b** all South Africans. Justify your choices.

On your own

3 In 500 words, explain why South Africa is becoming a more unequal society.

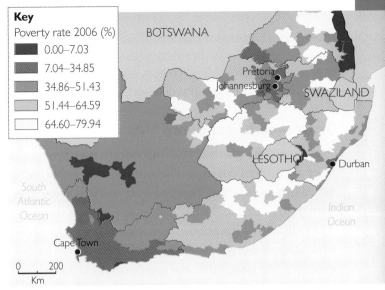

▲ Regional inequalities in South Africa. The map shows the poverty rate (the percentage of households living below the poverty line), by region, in 2006.

Education – % aged 20+ …	Black	White
with no schooling	22.3	1.4
with only primary school education	25.4	2.0
with some high school education	30.4	25.9
Who completed high school and/or higher education	22.0	70.7
Housing – % households …		
with phone and/or mobile phone	31.1	95.4
with a flush or chemical toilet	41.9	98.7
whose refuse is removed at least once a week	45.3	90.8
who use electricity for cooking	39.3	96.6
who use electricity for lighting	62.0	99.2
who own a TV	44.2	92.6
who own a computer	1.8	46.0
Employment and income		
Unemployment % among adults aged 15-65	28.1	4.1
Median annual income; adults aged 15-65 (in rand)	12 073	65 405

Indicator	South Africa	UK
Overall rank out of 140 countries	116	49
Individual scores (1 = most peaceful, 5 = least)		
Level of distrust in other citizens	3	3
Number of internal security officers and police per 100 000 people	2	2
Number of murders per 100 000 people	5	1
Number of prison inmates per 100 000 people	2.8	1.7
Ease of access to weapons, e.g. shotguns	4	2
Level of violent crime	5	1
Political instability	2	1
Respect for human rights	3	2

▲ A selection of Peace Index scores comparing South Africa with the UK in 2008. In the table, individual scores range from 1 to 5.

In this unit you'll learn how rapid development in one Indian city has exposed cultural differences between people.

Travelwatch, Bangalore

Rush hour in Bangalore, India. To Western eyes, it resembles chaos. Three lanes are marked out for traffic but the drivers know better – you can actually squeeze in five. There's a hold-up in the outer lane on the main boulevard leading to the city centre, and traffic is down to two lanes. The reason? The water bowser that keeps the hedge green in the central reservation is being followed by a cow, enjoying the freshly watered greenery. She is sacred so she takes priority. In Bangalore, India's fastest growing city, traditional values sit alongside some of the most explosive urban growth and post-modern architecture in the world.

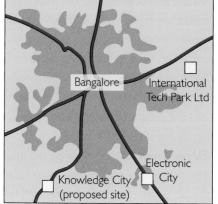

◄ *Traffic congestion in Bangalore*

Bangalore is India's hub in the 'new economy'. Unlike most Indian cities, which owe their growth to ports and industry, Bangalore is the centre of new technology, banking, finance, and the knowledge economy. 40% of India's 1.3 million workers in the IT industry are based in Bangalore. The Indian Government liberalised India's economy in the early 1990s, allowing large numbers of overseas companies to set up operations there. British Airways set up its accounting operations in India in 1996, setting a trend for what has become known as **out-sourcing**. Wages in India are only 10% of those in London, and savings for Western companies like BA are considerable. Now, a whole range of out-sourced operations are based in Bangalore and other large Indian cities, including:

- technical development – e.g. companies such as Deutsche Bank, one of London's biggest investment banks, which has set up teams of software developers in India (IBM, alone, had 73 000 employees in India in 2008)
- support – people who provide technical support, e.g. for BT Broadband
- call centres – e.g. for rail companies in the UK selling tickets or handling refunds

● **Out-sourcing** is the employment of people overseas to do jobs previously done by people in the home country.

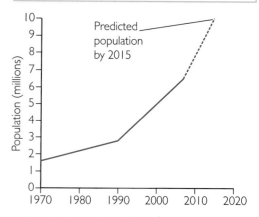

▲ *Bangalore's population. This is the population within the city, and does not include those within Greater Bangalore, up to 30 miles away.*

Bangalore is well placed to take on expanding 'new economy' roles such as these. Its university provides a highly educated workforce, which – since the 1960s – has supplied technical expertise for India's defence and space research industries.

In the 1990s, Bangalore set up designated areas, such as Electronic City, to become 'hubs' of high-tech firms. They were attracted by its tax breaks (i.e. low tax rates), as well as its cheap labour rates. Now Bangalore has grown so much that – what started as off-shore operations for large Western companies – has since spawned home-grown, Indian-owned companies which provide a range of technical and support services.

▲ *The Infosys building at Electronic City in Bangalore*

Further growth is certain. A new 'Knowledge City' is planned in Bangalore. In 2008, Oracle estimated that India would need 8 million extra workers in out-sourced industries! Now the ripple effect is starting, so that other Indian cities – such as Chennai – are also attracting large companies providing similar services.

The impacts of growth

Bangalore now has the highest average incomes in India, and jobs there are plentiful. For example, the IT company Infosys announced in 2007 that it would be hiring several thousand workers there. The city is displaying a number of indicators of increasing affluence:

- There are six new shopping malls.
- Luxury car showrooms selling upmarket brands are booming.
- 500 new bars and cafes opened in 2007 (young, well-paid workers go out more often to restaurants and bars).
- Bar owners cannot get enough staff (staff members' salaries can be doubled if they are good enough to retain).
- Taxi firms are also booming, because the big IT companies pay to transport their visitors around the city and their workers to and from the office.

But is the wealth trickling down? India's caste system still persists (see next page), and the disparities between rich and poor are stark. At night, jobless, low-caste Dalits scratch a living by removing human excrement from pits in the poorer neighbourhoods of Bangalore (see right).

Bangalore's 'night soil' collectors

Bangalore may be India's high-tech capital, but the practise of humans using their bare hands to clean out toilet pits still continues there. 10 000-15 000 Dalits are employed to clean out toilet pits in the poorer areas of the city.

The failure of the city authorities to provide a proper sewage system has forced residents to build these pits. When the pits get clogged, the call goes out for these wretched workers, who are desperate for work. After swigs of alcohol to ward off the stench, one gets into the pit and empties it with buckets or cans and the sewage is then taken away by the others. At the end of the night, they head home with earnings of 300-500 rupees (about US$6-10).

Adapted from a BBC report from Bangalore in September 2002

Booming Bangalore - 2

India's caste system

The caste system is a religious and social class system in India, where classes are defined by birth and family. Found mostly among Hindus, it also exists among some Muslims and Christians.

Although India's Constitution makes caste discrimination illegal, barriers persist. Under the caste system:

- the Dalits have the lowest status. Once known as 'Untouchables', they tend to work in unhealthy, unpleasant or polluting jobs – suffering from social prejudice, segregation, and extreme poverty. Prejudice is greatest in rural areas, where they are isolated, humiliated, and discriminated against. They are not allowed to worship in the same temples as others, and must obtain water from different sources. They cannot stray from their own part of the village, and Dalit children sometimes have to sit at the back of classrooms.
- further discrimination exists within the Dalits, so that there are 'outcasts among outcastes'.
- movement is possible, especially in the middle castes. For example, those born into a lower caste could rise by becoming vegetarian and teetotal – normally associated with higher castes.

'Untouchability' was outlawed in 1950 and has declined; urbanisation has helped to break down caste barriers. President Narayanan of India (1997-2002) and the present Chief Justice each belong to Scheduled castes (see below).

As well as the advantaged Forward castes – forming 25% of India's population – the Government now classifies people by:

- Scheduled castes, or former 'Untouchables' (16% of India's population); the word 'Dalit' is now preferred
- Scheduled tribes, which consist of tribal groups (7%)
- Other backward classes (52%)

▲ Dalit women carrying breeze blocks on a building site in Bangalore

It does this to discriminate **positively** in education and jobs for the most disadvantaged. Some people protest about this, because they believe that negative treatment of Forward castes is socially divisive. They believe that school and employment choice should be based on economic status, as there are now many Dalits who are wealthier and more educated than Forward castes, but who still benefit from positive discrimination.

However, the caste system runs deep and is widespread. Inter-caste marriages are now more common, although caste is still a factor in some marriage choices. Some Indian matrimonial websites and news columns contain caste-based categories, and adverts can state caste as a factor in choice.

Future challenges

Bangalore's growth poses several questions:

- Should more housing be built to keep pace with the increasing population and to bring prices within the reach of more people? Rents are currently beyond the means of most workers on average pay; they can easily spend half their salary to rent one room outside the city. Construction workers often live in squalid roadside tents.

- How can Bangalore's public transport system be improved to keep pace with the city's growth? Five million vehicles currently clog its roads. Bullock carts block the road to Electronic City, which is crammed with trucks, cars and two-wheelers – despite funding for a new expressway.

- How can enough energy be provided? Every IT company in Bangalore has to have a private generator; power failures occur daily.

- Should Bangalore International Airport be enlarged to meet the volume of international traffic? Its terminal is cramped, with just one luggage carousel.

Bangalore is planning for further, but different, growth. The city government plans to decentralise, by building new towns on greenfield sites around the city. It is keen to develop labour-intensive sectors like car manufacturing in Bangalore, and to disperse IT jobs to other cities in India.

▲ *New housing in Bangalore – but for whom? The shortage of new housing has driven prices up and only the well paid can afford them.*

Over to you

1 In pairs, research virtualbangalore.com and YouTube (type in 'Bangalore traffic') on **a** types of companies in Bangalore, **b** what Bangalore is like to live in, **c** problems faced by the city.

2 Adding material from this unit, list the social, economic and environmental benefits and problems brought by Bangalore's growth.

3 What are the potential benefits and problems of an economy with growth driven by foreign companies?

4 In groups, decide what priorities Bangalore should make about **a** its future economic growth (what kind?) **b** how it should house its people, **c** whether it should expand or set up new towns on the edge. Present your findings to the class.

On your own

5 Explain why the caste system is more likely to break down in urban than rural areas.

6 Prepare a class debate on **a** whether positive discrimination should be encouraged to favour disadvantaged groups, **b** whether newspapers should allow marriage adverts in which a preferred caste is stated.

Out-sourcing – good or bad?

What do you think?

In this unit you'll learn how important trade is in enabling economic development to take place.

Trade or aid?

For many decades now, the process of development or managing poverty has been linked to giving aid. With tens of millions of pounds raised after the 2004 tsunami, or for Live Aid, individuals are generous when helping others. However, governments are less so. By 2008, only five countries had actually met a commitment made in 1981 to donate 0.7% of their GDP to international aid (Norway, Sweden, Denmark, Luxembourg, and the Netherlands), with the UK on target to meet it by 2013.

However, in the long term, trade is far more effective than aid in enabling countries to develop. Trade creates long-term employment – which, in turn, provides wages that people spend, thereby creating further demand. This process is known as the 'virtuous cycle of development' (also see page 154). Economists refer to it as the 'multiplier effect'.

Trading in commodities

After Uganda's independence from Britain in 1962, its economy was strong. With fertile soils and a favourable climate, it exported **commodities** (sugar, tea, coffee, cotton and bananas) to the developed world. But the rise of Idi Amin in the 1970s saw these exports collapse, leaving a subsistence economy in which farmers mainly grew crops to feed their own families. During Amin's eight-year rule to 1979, cotton production collapsed from 80 000 to 5000 tonnes a year, and tea from 21 000 tonnes to just 3000. Because of Uganda's political instability, long-term investment in the country also fell.

After Amin's fall from power, Uganda's economy recovered slowly and exports increased. Economic growth averaged 6% throughout the 1990s. However, critics argue that this growth has not filtered down to the poor. To counteract this, the current President of Uganda has emphasised **processing** agricultural products within Uganda. If successful, this would shift Uganda from a primary export, aid-based economy, to one with growing secondary and tertiary sectors. This would help because:

- manufacturing adds value
- fewer major global price fluctuations occur for manufactured goods, when compared to the prices of raw commodities
- manufacturing creates jobs, which is important in a country where 35% of the population live below the poverty line – mostly due to under-employment or unemployment

---

● **Commodity** means raw goods – either farm produce (e.g. coffee, wheat) or minerals (e.g. iron ore, uncut diamonds).

A tea plantation in Uganda. Uganda's regular rainfall pattern and fertile volcanic soils make this rich productive farmland.

Uganda and the IMF/World Bank

When Uganda's current President came to power in 1987, he took over a devastated economy in serious debt. The IMF agreed to assist Uganda by lending money, but on condition that Uganda underwent a programme of **structural adjustment** (see page 186). This involved **trade liberalisation**, by reducing import tariffs and export taxes. 82% of Ugandans earn a living through farming, and trade liberalisation had the greatest impact on them. Many decided to switch from subsistence to cash crops, such as coffee and tea, because at that time global commodity prices for these crops were high. However, a crash in the price of coffee in the late 1990s reversed this.

Structural adjustment had several major impacts on Uganda:

- Trade liberalisation did not lead to more wealth for individual farmers. Most Ugandan farms are small, and the country's poor transport infrastructure made it unrealistic for them to export their own produce. As a result, middlemen could exploit the situation by buying from individual farmers and then selling on the combined produce from several farms at an, often large, profit.
- The owners of the newly privatised industries (where the Government sold off State assets to private companies), cut their costs by reducing the size of their workforces — thus increasing Uganda's unemployment problem.
- Cuts in Government spending badly hit education and health care in Uganda. The Government introduced charges for schooling and health care that were often beyond the means of the poorest Ugandans — just as the massive social impacts of HIV/AIDS were beginning to hit home.

> ● **Structural adjustment** is a combined package of cuts in Government spending, trade liberalisation, and **privatisation** which is imposed on governments seeking debt relief from the IMF.
> ● **Trade liberalisation** – also known as free trade – means removing barriers such as duties or customs. The theory is that the fewer barriers there are to the flow of goods, the greater trade will be.

The IMF holds Uganda up as a success story, because:

- its economic growth averaged 5.6% between 2000 and 2005, which even the recent prolonged drought there has only reduced slightly – in spite of energy shortages (affecting industrial production) and high global oil prices
- the percentage of those living in poverty fell from 56% in 1992 to 31% in 2006. Nonetheless, poverty levels remain high in rural areas, and in Northern and Eastern Uganda.

However, in spite of the supposed success there:

- raw per capita income in Uganda remains low (US$340 per annum, or US$1454 in PPP$)
- most growth has been achieved by the wealthy, and the poor have failed to catch up
- Ugandan academics argue that there has also been an erosion of traditional values – structural adjustment has focused on individualism and monetary gain, both of which run contrary to the traditional African focus on community and family

▼ *Traditional Ugandan housing consists of homesteads built around a family compound, which helps to maintain a strong sense of family and community*

Free trade or fair trade?

Joining the WTO – to benefit or not?

Uganda joined the WTO (see page 156) in 1995. In theory, the removal of trade barriers, such as import taxes and other tariffs, should enable Uganda to export commodities and earn foreign currency. However, the WTO has not been successful in removing EU tariffs, especially the Common Agricultural Policy (CAP). This subsidises EU farmers (especially in France), so that they can sell their produce more cheaply. This discriminates against developing nations by undermining their natural price advantage.

Now, the WTO is toughening its position over subsidies and protectionism in the developed world:

- In November 2005, it put pressure on the EU to cut its annual 1.5 billion euro subsidy to European sugar farmers. In theory, this means that sugar farmers in Uganda should be able to export to the EU on a fairer basis, with EU consumers benefiting from lower prices.
- NGOs such as Oxfam (through their Make Trade Fair campaign) are increasingly campaigning through the WTO to force rich countries to drop barriers to imports from developing countries.

The new face of coffee – fair trade?

Uganda's biggest export crop by far is coffee, which was worth US$350 million in 2007. The bulk of this is grown by smallholder farmers, an increasing number of whom now grow coffee for the fair trade market. One example is the Gumutindo Coffee Co-operative in the Mount Elgon region of Eastern Uganda. Here 3000 farmers – 91% of whom depend on coffee for their main income – produce arabica coffee beans, which then undergo primary processing on the farms themselves. This involves processing the ripe red coffee cherries, by:

- removing the skin and pulp, and drying the bean
- secondary processing (milling) at a nearby warehouse
- packing for export, leaving the final roasting to be done at its destination

EU agrees cut in sugar subsidies

A reduction in subsidies for Europe's sugar farmers has been agreed by EU agriculture ministers, officials in Brussels have confirmed. They have agreed to cut the prices offered to European sugar farmers by 36%, bringing the EU's sugar rules into line with global frameworks. The changes were demanded by the World Trade Organisation, but EU farmers and sugar firms have warned of job losses. The EU had been paying Europe's sugar producers three times the world price.

Adapted from a BBC online article in November 2005

> The money from the fair trade premium last year helped me to pay for my daughter's school fees, which are very expensive. I tell my children and neighbours to spend time producing good-quality coffee. Since the other farmers have seen us receive the fair trade premium, they have tried to emulate what we are doing and the quality is getting better.
>
> Mr Difasi Namisi, a Gumutindo farmer

Economic partnership agreements

Can free trade with the EU help Uganda to develop? In 2001, new trade relations began – known as Economic Partnership Agreements (EPA). They are a series of trade rules between the EU and Africa, the Caribbean and Pacific countries to establish a free trade area. The aim is to promote sustainable development in developing countries by bringing them gradually into the global economy. They are complex, because they vary between countries. However, they do conform to one idea – that of **reciprocity** – meaning that Uganda must remove its import duties and subsidies to gain access to EU markets.

Privatisation and free trade should, in theory, increase Uganda's trade. However, Uganda faces problems before it can trade equally with the EU.

- Its economic capacity is not always sufficient to meet demand.
- Its infrastructure (e.g. banking and transport) is weak; Uganda is also landlocked and has little port capacity.

Therefore, while its agricultural exports find markets in the EU, exposure to open global competition could put Uganda's infant manufacturing industries out of business. Aware of this, many EPAs now have interim periods during which tariffs are removed slowly. However, developing countries still largely export raw commodities without building an industrial base. The real winners are EU manufacturers. Economist Ha-Joon Chang calls this process 'kicking away the ladder' – by which developed countries promote free trade for their own interests. He claims that Asia's 'tiger' economies grew with strong State intervention to promote enterprise and job creation. Perhaps Uganda should only be exposed to global markets if the Government supports its infant industries and technology at the same time.

▲ One of Uganda's manufacturing industries - an electric arc furnace at Jinja Steel Rolling Mills. Should it be given some protection as Uganda eases into the global market?

How far is free trade fair trade?

What do you think?

Over to you

1 In pairs, show the advantages and disadvantages for Uganda of encouraging **a** manufacturing, **b** trade liberalisation (free trade), **c** fair trade, **d** EPAs and **e** Government support for industry.
2 Explain which of these should **a** improve life for most of the population, **b** preserve traditional family lifestyles, **c** improve Uganda's trade, and **d** be most beneficial for Uganda to develop overall.

3 What might the advantages and disadvantages be for Uganda of encouraging TNCs to invest there?

Exam question: Examine the barriers that exist against the expansion of trade in some developing countries. **(15)**

In this synoptic unit you'll learn whether aid or investment is the best way of funding development.

Which way now?

In 2005, the G8 summit at Gleneagles made a major contribution to closing the global development gap, by writing off loans that were crippling the world's poorest countries. However, at national or local levels, the threats posed by disparities in wealth (e.g. the increasing violence in South Africa) remain unresolved. This unit investigates six projects which represent different ways in which the development gap might be closed. Each has impacts at different scales.

Types of development

Many projects in the developing world are financed by Western countries, and create wealth through employment in construction, manufacturing or developing primary products. The argument goes that, by creating jobs, wealth 'trickles down' to the poorest – by gaining jobs they spend more, and thus create further demand. Economists refer to this as the **multiplier effect**.

Many decisions about where development projects will be targeted are made by governments or large organisations – a process known as **top-down development**. This is generally the case where strategic decisions are needed, e.g. improving welfare (such as health care or schooling), or the provision of energy. The decision comes from above and is then managed from there.

▼ *UNICEF's child immunisation programme – an example of top-down development. UNICEF works globally on major child health issues like this.*

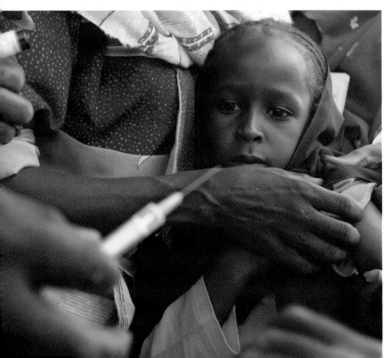

By contrast, **bottom-up development** occurs at community level – people's needs are identified and local projects are designed to meet them. Much of this work is done by **NGOs (Non-Governmental Organisations)**, through charities such as Oxfam, which work with communities to provide either for their long-term needs or for short-term emergencies such as natural disasters. NGOs try to maintain impartiality, although some are founded on particular religious principles, e.g. Christian Aid.

▲ *Equatorial College School, Uganda. Funded by University College School in North London, this example of bottom-up development provides secondary education in a remote rural area. All decisions are made locally, and the school is in regular contact with its London partner.*

Finance for each type of development comes in two forms – aid and investment:

- **Aid** refers to gifts or repayable loans made by one country or organisation to another. These assist in either developing a country or responding to a disaster, such as the 2004 Asian tsunami. Aid can be **bilateral** – from the government of one country directly to another – or **multilateral** – from alliances of several countries or organisations to another. Sometimes aid is given with conditions, e.g. the money must be spent on the donor's own products, so the receiving country has little say or control. This is known as **tied aid**. Western countries agreed to end tied aid in 2001, after the Pergau Dam project (see Resource 4).

- **Investment** refers to repayable loans used to develop a country, but with an expectation of a share in the profits. This usually comes from individuals or companies (e.g. when TNCs invest in a factory).

- Some projects combine aid and investment, e.g. Ghana's Akosombo Dam (see Resources 1 and 2). As this example shows, there is sometimes a fine line between what is 'aid' and what is 'investment'.

Evaluating projects

Resources 1-6 are examples of investment and/or aid projects:

Resource 1   The Akosombo Dam – has elements of both aid and investment

Resource 2   Ghana's aluminium smelter – has elements of both aid and investment

Resource 3   Water provision in Ghana – an investment project

Resource 4   The Pergau Dam, Malaysia – an aid project

Resource 5   Grass roots development in Barlonyo, Uganda – an aid project

Resource 6   The Kingship Project, Moldova – an aid project

In addition, Resource 7 is about land redistribution in Zimbabwe, which was implemented by Robert Mugabe for political rather than development reasons.

Resource 1

Aid and investment – the Akosombo Dam

Built between 1961 and 1966, Ghana's Akosombo Dam flooded the land behind Akosombo to create Lake Volta, the world's largest artificial lake. Designed to provide electricity through an HEP plant for smelting raw bauxite into aluminium, its funding came jointly from World Bank loans (US$40 million), Ghanaian government investment (US$69 million) and aid grants from the USA and UK.

Advantages	Disadvantages
Electricity production	
• The dam gives Ghana the capability to generate electricity for industrial purposes.	• Most Ghanaians still have to live without electricity, because they cannot afford it.
• The HEP plant supplied cheap electricity to the aluminium smelter at Tema for 30 years, thus helping to expand Ghana's industry.	• Although the dam gives Ghana the capability to generate electricity, the smelter built to link up with the dam was mothballed between 1998 and 2006.
• Aluminium improved Ghana's export trade with the rest of the world.	• The HEP plant is owned by Alcan (an American company – see Resource 2), so profits from export sales of electricity go back to the USA, and not to Ghana.
• Electricity is now exported to neighbouring countries, thereby earning overseas currency.	
Communication	
The lake provides an internal waterway, and has increased the use of water transport inland. This is especially useful for linking up with the remote northern region of Ghana.	Lake Volta flooded 4% of Ghana's land area and forced 80 000 people to relocate.
Economic activity and the environment	
• Lake Volta has led to an increase in tourism. Tourists can go on cruises on the lake.	• The reduced river flow as a result of the dam brings less food for the freshwater shrimp population. Therefore, local people now have access to less protein in their diets.
• Lake Volta provides a source of irrigation water for farming.	• Less silt now reaches the fields downstream, which has led to falling crop yields and reduced income for some farmers.
	• Greater poverty caused by the falling crop yields has led to rural-urban migration away from the area and the removal of trees to sell as fuelwood – causing serious deforestation.
	• There has been an increase in water-borne diseases, e.g. bilharzia, as well as malaria and sleeping sickness as a result of the lake.

◀ A Western tourist cruising on Lake Volta, viewing local water transport and traders making use of the lake

Aid or investment? – 2

Aid and investment – Ghana's aluminium smelter

For years, Ghana has tried to develop an integrated aluminium industry – from raw material to finished product – in order to develop its industrial economy. Raw bauxite and cheap electricity are the key inputs, and Ghana has both.

In the 1960s, the US government gave grants to US company Kaiser Aluminum to form VALCO (the Volta Aluminium Company) and build a smelter at Tema, which in theory was intended to process Ghanaian bauxite. Ghana has vast deposits (780 million tonnes) of high-grade bauxite, and is Africa's third largest annual producer. The bauxite is mined by the Ghana Bauxite Company – 80% of which is owned by US company Alcan (a competitor of Kaiser), and only 20% by Ghana. Alcan also controls the two HEP plants at Akosombo and Kpong on the Volta River.

So Ghana had everything – except an integrated aluminium industry. The raw bauxite and electricity were owned by Alcan, and the aluminium smelter by Kaiser. Ghana's raw bauxite was exported to Alcan smelters in Scotland and Canada, while semi-processed alumina for the VALCO smelter at Tema was imported from Jamaica and the USA until 1998 (when Kaiser mothballed the smelter and decided to leave it idle). All value-added from Ghana's aluminium industry was therefore going overseas. Investment capital for both companies was underwritten by the US government.

However, in 2005, the Ghanaian government purchased 90% of the shares in the smelter at Tema from Kaiser. Its intention was to restart processing at the smelter. Aluminium production restarted in 2006 and, in 2007, the Ghanaian government purchased the remaining 10% to give it full ownership and enable it to exploit its own bauxite strip mines (separate from those it owned jointly with Alcan). Just as things were looking up, the plant had to be closed in 2008 because low water levels in the dam (caused by drought) prevented electricity production.

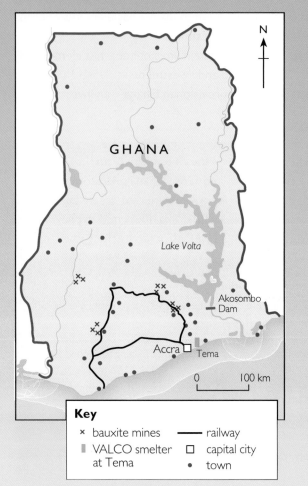

Key

× bauxite mines	— railway
▐ VALCO smelter at Tema	☐ capital city
	● town

▼ The Akosombo Dam provides hydro-electric power to run the smelter at Tema

Resource 3

Investment – water provision in Ghana

Water provision is a challenge for Ghana. In many areas, the supply network has fallen into disrepair because of a lack of investment. In 2008, only 50% of the population had access to safe, treated water; the remainder depended on rivers and untreated wells. Access to safe water varies – 62-70% of urban dwellers have access, but only 35-40% of those in rural areas. Even in urban areas, only 40% of the population has a water tap that actually flows, and affluent Ghanaians with water meters have to buy from private sellers because supply is so irregular. 78% of the urban poor have no piped water.

▲ For many, sources like this one are the only way to obtain water, even though it may be dirty

The impacts of this are considerable on:
- the time taken to collect water – a job normally done by women
- the cost – without a piped supply, buying three buckets of water a day (below the WHO's recommended daily amount for one person) costs 20% of an average family's daily income
- health – 70% of disease in Ghana (e.g. cholera) is caused by poor water quality and sanitation

Ghana's government estimates that it will cost US$800 million to bring clean water to all Ghanaians by 2015. This will require loans from the World Bank, which now only provides money for this if countries' water systems are privatised so that Government expenditure is not increased. So, in 2005, Rand Water Vitens, a joint South African and Dutch company, was contracted to provide water for Ghana's urban centres.

In theory, privatisation increases the profit motive and efficiency, and investment from private sources should provide safe water for all – simply by expanding the market as far as possible. However, there are three problems with this:
- Under privatised water systems, people have to pay for their connection and supply – which many Ghanaians cannot afford.
- Half of the water industry's 4600 employees are threatened with unemployment as the new private company aims to cut costs and maximise profits.
- Current plans are to privatise only Ghana's urban water and sanitation systems. What will happen to rural supplies and the aim of access to clean water for all Ghanaians?

> To privatise water is like handing down death sentences to the urban and rural poor in Ghana, because they cannot afford to pay.
> **The Christian Council of Ghana**

The 2005 G8 conference led to 70% of Ghana's US$6 billion debt being written off, which in 2006 saved the Government US$156 million in interest alone. The write-off came after Ghana agreed to spend the savings on improving education and health care provision. However, this saved money cannot be spent on improving water supplies, in spite of their importance for health improvement, because the World Bank says so!

Resource 4

Aid – the Pergau Dam

The Pergau Dam in Malaysia, near the border with Thailand (see map), illustrates both top-down and tied aid. It was built with £234 million of British aid from the UK Government's Overseas Development Agency (ODA – now DfID), with the stated intention of boosting Malaysian energy production. It remains one of the largest amounts of aid Britain has ever given.

Announced in 1989 by the then Prime Minister, Margaret Thatcher, its construction began in 1994 and was completed by 2000; the dam began operating in 2003. It was always a controversial aid project, and even the ODA opposed it – concluding in 1990 that it was uneconomic. Nonetheless, the Conservative Government agreed to fund it. UK Parliamentary records show that:

- the package was tied to an arms deal – the Malaysian Government agreed to spend £1 billion on British-built fighter jets in return for British funding of the Pergau Dam
- a judicial review in the High Court ruled that this contravened the UK's *Overseas Aid Act* (1966), which forbade the giving of aid in return for arms purchases
- contracts for the dam's construction were awarded jointly – without competitive bidding – to Balfour Beatty (a donor to the Conservative Party) and Cementation International (a company which employed Margaret Thatcher's son, Mark, as an adviser)

The Economist described this project in 1997 as 'the most notorious example of how not to encourage exports'. The Pergau Dam has been widely condemned as a waste of money, which is unlikely to produce any long-term benefits for Malaysia. It cost three times more to build than a gas-fired power station, and also cost the Malaysian government an additional £100 million on top of the British money. Large areas of fertile flood plain were flooded by the dam's lake, and thousands of farmers were displaced (the exact number was never revealed). Finally, its generators can only be used at peak energy hours, because the volume of water in its lake is only enough to allow the dam to operate for a few hours at a time.

As a result of the Pergau Dam project, the giving of tied aid by Western countries was ended at an OECD meeting of Western countries in 2001.

Resource 5

Aid – bottom-up development in Barlonyo, Uganda

The type of aid in Barlonyo typifies bottom-up development, which seeks to develop from the community upwards.

The village of Barlonyo, in the Lira District of Northern Uganda, experienced one of the worst massacres in Uganda's history – when rebels killed over 300 people in 2004. After the attack, the remaining villagers fled to refugee camps where they felt safer, but where they were also forced to depend on food aid. Now, there is peace in the area and the villagers of Barlonyo returned to their land in 2006.

With the support of national and international NGOs, local farmers have linked up to form a democratically run cooperative. All farmers have an equal say in any decisions, including women farmers who previously felt marginalised and are now more empowered. One NGO working in Barlonyo is ActionAid, which describes its approach thus: 'We don't impose solutions, but work with communities over many years to strengthen their own efforts to throw off poverty'.

Almost all farmers in Barlonyo own less than two acres of land. Because, individually, they could not afford to take their harvest to market themselves, they were often exploited by middlemen – who bought up all the separate small harvests cheaply and then sold them on for a profit. The cooperative now enables all the farmers to share the cost of hiring a truck – thus improving their profits by allowing them to sell directly to consumers in the market town of Lira, 26 km away. In 2008, they were able to sell their sesame seed crop for over three times the amount they received in 2007. The extra income has rippled through the local economy and enables farmers to send their children to school – thus promoting long-term development through education and creating a virtuous cycle of development.

NGOs have also helped by contributing items to improve farm output and efficiency, e.g. ox ploughs, hand hoes and high-yield seeds. The Barlonyo farmers have now created a seed-bank to reduce their reliance on TNCs for seed. The cooperative manages the seed-bank sustainably, ensuring that all farmers contribute some of their harvest to guarantee the availability of seed for the next season. They have also built up a food reserve to protect against hunger during times of drought.

Resource 6

Aid – The Kingship Project

The Kingship Project is a joint charity run by students at The King's School in Ottery St Mary, Devon, together with the local Rotary Club and the charity Christian Response to Eastern Europe. Its aim is to build and maintain a community centre for Gura Bi Cului (a large rural town in Moldova, Eastern Europe).

Moldova is Europe's poorest country. Formerly part of the USSR, it has contradictory development indicators (see the fact file) – literacy and health indicators are high, but it has Europe's lowest GNI. Officially, 30% of Moldovans live in poverty but UNICEF claims that the real figure is 60%. 40% of the working population work in agriculture, but they produce only 18% of GDP. Moldova's population is declining because of emigration to EU countries – 33% of its GNI now comes from remittance payments sent home by Moldovans abroad.

In spite of high literacy rates, economic recession has reduced school enrolment to below 75% in some rural areas. Older children look after younger siblings while their parents work. Primary school attendance is only 85%. While some health indicators are good (98% of mothers give birth attended by a health professional), only 52% of rural people have adequate sanitation according to UNICEF.

Initially, the new community centre will have a kitchen area for cooking and food preparation, a heated shower and washing facilities, a study area for schoolchildren, and a community bakery to provide employment and bread at a fair price. The project's long-term aims include medical facilities, an orphanage, and staff to run the centre.

The Kingship Project is run by students from The King's School – they chose the partner charity, Christian Response to Eastern Europe, and work with them and the local Rotary Club to raise money for the centre. Students from the school have visited Gura Bi Cului to help plan the centre. They lead and coordinate work with their partners, organise fundraising events, and also a 'friends' scheme (in which regular donations are made). By 2008, they had raised £20 000, and building work began early in 2009.

Moldova fact file

Area: 33 843 km² (half the size of Scotland)

Population: 4 324 450 (July 2008 est.)

Population growth rate: -0.09% (2008 est.)

Birth rate: 11 births/1000 (2008)

Death rate: 10.8 deaths/1000 (2008)

Net migration rate: -1.13 migrants/1000 (2008)

Infant mortality rate: 13.5 deaths/1000 live births

Life expectancy: 70.5 years

Total fertility rate: 1.26 children born/woman (2008)

Literacy: 99.1%

GDP per capita US$1100 (US$2900 PPP) (2007 est.)

Population below poverty line: 29.5% (2005)

Internet users: 727 700 (2006)

▲ The building purchased by The Kingship Project to be turned into the community centre in Gura Bi Cului

Resource 7 is not about aid or investment. However, it does illustrate how some policies are designed to redistribute wealth, sometimes forcibly, and close the development gap within countries – even if the end results are not exactly as declared.

Resource 7

Populist development – land reform in Zimbabwe

Sometimes policies are adopted for their popularity, rather than their effectiveness – land redistribution is one such policy in Zimbabwe.

The majority of Zimbabweans make their living from agriculture and, therefore, owning land allows them control over production and income. The issue of land ownership is a political one, because land titles did not exist before British colonial rule – so pre-colonial ownership is difficult to prove. When British settlers, led by Cecil Rhodes, arrived in the late nineteenth century (creating Southern Rhodesia), each settler was rewarded with 6000 acres of land for helping to conquer the territory. Commercial farming developed, and black Africans were restricted to owning land in small reserves.

Instead of gaining a negotiated independence from Britain in the 1960s, like most of Britain's other African colonies, Zimbabwe's government was taken over by a right-wing white minority. They worried that black rule would lead to a redistribution of the land now owned by the descendants of the original white settlers. Following guerrilla uprisings (partly led by Robert Mugabe), independent black majority rule was finally established

in 1980, and Britain agreed to fund land purchase from white farmers willing to sell up. However, much of the land acquired in this way ended up in the hands of the leading associates of Robert Mugabe (now elected leader of Zimbabwe), many of whom had no experience of managing farms.

In 1997, Robert Mugabe authorised the compulsory purchase of white farms in order to provide land for the black population, saying that land redistribution would release Zimbabwe from colonial rule. Again his supporters, most of whom had little experience of managing farms, were the main beneficiaries of the seized land.

As a result of the land seizures, food and cash crop production fell dramatically, with a subsequent fall in exports (see the fact file). Formally considered to be Southern Africa's food basket, Zimbabwe began to head towards famine – becoming increasingly reliant on the World Food Programme for food aid.

After the initial seizures of white-owned farms, Mugabe intensified the process of increasingly violent land seizure from the few remaining white farmers, knowing that this policy was politically popular despite increasing economic collapse in the country – with famine and rampant inflation.

Zimbabwe fact file

Population: 11.8 million

Land area: 386 850 km² (50% larger than the UK)

Food production: 8.3% of land is arable (growing corn, livestock, peanuts, sugarcane, wheat, millet)

Export value: US$1.5 billion (2007), US$2 billion (2000)

National HIV prevalence: Estimates vary between 18% and 25%, as aid workers have no official data

Life expectancy at birth: 37 years

Adult literacy: 90%

People living on less than $2/day: 83%

Human Development Index: 151 out of 177

Agricultural production (% GDP): 18.1%

% of the population malnourished: 45%

GNI per capita: US$200 (PPP$)

Synoptic question

a Outline and justify the criteria that you would use to evaluate each of these projects. **(12)**

b Using these criteria, evaluate:
   - the development projects in this unit
   - whether aid, investment, or populist policies are more effective in achieving development. **(18)**

c Based on your evaluation, would you recommend that future development projects be managed in a top-down or bottom-up way? **(10)**

5.9 ## Ways forward 3: the Millennium Development Goals

In this unit you'll learn about the Millennium Development Goals and how far progress has been made in achieving them.

The world greets the new millennium

Midnight, 31 December 1999. The fireworks are going off in Sydney, one of the first in a series of celebrations to welcome in the new millennium. People have paid big money for grandstand seats to watch the fireworks. Meanwhile, Africa also welcomes in the new millennium. The BBC's World Service has asked African children to say what they hope for in the new millennium. Among the contributions are the four on the right. What hope can these children have that their wishes will be met?

The Millennium Development Goals

At a United Nations summit in 2000, eight Millennium Development Goals (MDGs) were agreed to provide a set of development targets for the world to reach by 2015. Every UN member state signed up to these targets, making it the largest-ever multinational attempt to rid the world of extreme poverty. The United Nations Development Programme (UNDP) was given the task of coordinating efforts to reach the targets listed below:

1 Eradicate extreme poverty and hunger
   a Halve the proportion of people living on less than 1 US dollar a day
   b Full employment and decent work for all, including women and young people
   c Halve the proportion of people suffering from hunger
2 Achieve universal primary education, and ensure that all complete it
3 Promote gender equality and empower women; eliminate gender disparity in primary and secondary education by 2015
4 Reduce child mortality for those under five by two thirds
5 Improve maternal health
   a Reduce by three quarters the maternal mortality ratio
   b Achieve, by 2015, universal access to reproductive health
6 Combat HIV and AIDS, malaria and other diseases
   a Halt and begin to reverse the spread of HIV/AIDS, malaria and other major diseases
   b Universal access to treatment for HIV/AIDS for all who need it by 2010
7 Ensure environmental sustainability
   a Integrate principles of sustainable development into country policies and programmes; reverse loss of environmental resources
   b Achieving by 2010 significant reductions in the rate of biodiversity loss
   c Halve the proportion of people without sustainable access to safe drinking water and basic sanitation
   d Improve significantly the lives of at least 100 million slum dwellers by 2020
8 Develop a global partnership for development
   a Develop further an open, rule-based, predictable trading and financial system
   b Address the needs of the least developed countries, of landlocked developing countries and of small island developing states
   c Deal with debt problems of developing countries to make debt sustainable
   d Access to affordable essential drugs in developing countries
   f Make available benefits of new technologies, e.g. ICT

> I want to ask the president why he can't do something for street children in this country who have nothing?
> **Burundi**

> I want to be a teacher when I grow up so that I can help children to read. It is their right to be educated.
> **Kenya**

> I would like to help people, to give them water and electricity because we haven't got any of these. As for children, I would give them books and pens for nothing, all school stationery.
> **Rwanda**

> I dream that there will be peace in Somalia and that everything will be fine.
> **Somalia**

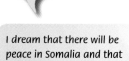

▲ African children's hopes for the new millennium

Progress and the 'Call to Action'

There have already been success stories in achieving the MDGs, including:

- 41 million more children enrolled in primary school
- 2 million more receiving AIDS treatment
- economic growth averaging 6% in sub-Saharan African countries by 2008

However, current improvements are not sufficient to meet the MDG targets by 2015, especially in sub-Saharan Africa. For example, in 2007:

- over 500 000 women died from treatable, preventable complications of pregnancy and childbirth. The odds of a woman dying from these in sub-Saharan Africa are 1 in 16, compared to 1 in 3800 in the developed world.
- 980 million people still live on less than 1 US dollar per day.

To hasten progress, the 'Call to Action' was launched, led by UN Secretary-General Ban Ki-moon and UK Prime Minister Gordon Brown. Its focus was to create a coalition of governments working in developing countries with:

- the private sector, to mobilise technology and job creation
- professionals in medicine, to find cures for diseases that most afflict developing countries (e.g. TB and malaria), and in education
- NGOs, to encourage people in developed countries to continue contributing to good causes, and governments to help financially to help meet the MDGs
- faith groups, who have the power to mobilise millions in eradicating poverty
- cities in the developed world, by consuming in ways that promote development (e.g. fair trade)

Case study 1: progress in Bangladesh?

After China and India, Bangladesh has the world's third largest number of people in poverty. Of its 150 million people (2007), half live in poverty – with 50 million in extreme poverty. With its large, rapidly growing population, Bangladesh faces by far the biggest financial challenge in meeting the MDGs (an average cost of US$14.5 billion **annually** between 2005-15). However, a healthy 5% average growth in GDP allows the Government to raise more than many other countries. International aid is also helping.

UK and UNDP start project to uplift 3 million people from extreme poverty

To improve the livelihoods and living conditions of 3 million urban poor in Bangladesh, especially women and girls, the UNDP and the UK Department for International Development (DfID) have initiated a US$120 million development project with the Government of Bangladesh and city corporations. The project will cover 30 towns until March 2015, including Dhaka (Bangladesh's capital). It will:

- support the development of poverty reduction strategies at town/city level
- link people with local micro-finance bodies
- give access to a range of financial services for community groups for housing and business development
- help urban poor communities to create healthy living environments
- support poor families in acquiring skills to increase their incomes

The Millennium Development Goals - 2

Bangladesh faces huge challenges:

- In education, the number of teachers will have to rise from 350 000 in 2005 to 815 000 by 2015.
- In health, the number of doctors will have to double to 58 000, and nurses and midwives increase by a factor of four to 145 000.
- Its surface water supplies are also badly infected; 20 000 people die annually from diarrhoea alone.

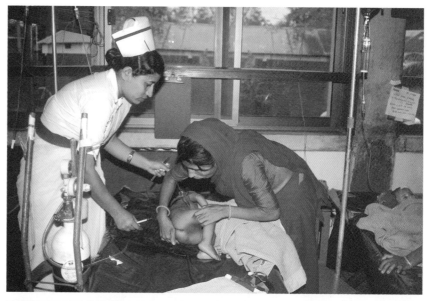

▶ *MDG 4: A mother and nurse caring for a sick child in the Child Healthcare Centre in Dhaka*

▶ *MDGs 3 and 8f: Professor Muhammed Yunas, founder of the Grameen Bank in Bangladesh, visits a Grameen Bank Centre. This bank has pioneered the concept of small, or micro, loans to women - allowing them to start businesses and gain economic power. The bank also finances the Grameen Phone Project, which aims to equip every Bangladeshi village with a telephone.*

> **Is it in the interests of everyone to close the development gap?**
>
> **What do you think ?**

Goal	Progress
1: Eradicate extreme poverty and hunger	The poverty reduction rate is 1.2% a year.
2: Achieve universal primary education	Significant progress has been made. It has improved by 3.4% in primary and 13.2% in secondary – against targets of 3.1% and 10.3% respectively.
3: Promote gender equality and empower women	The gender gap in primary and secondary level education has already been closed.
4: Reduce child mortality	Mortality among under-1s and under-5s has been reduced by 2.8% and 2.3% respectively. If maintained, the under-5s rate will decrease by two-thirds by 2015.
5: Improve maternal health	The maternal mortality rate is still high and remains a challenge.
6: Combat HIV/AIDS, malaria and other diseases	There have been improvements in containing the spread and fatality of malaria and TB. The spread of HIV/AIDS remains low.
7: Ensure environmental sustainability	Maintaining wetlands and biodiversity is still a challenge. There has been considerable progress in ensuring safe drinking water and sanitation in urban areas.
8: Develop a global partnership for development	A significant improvement in debt has been achieved. Telephone access has increased significantly. Unemployment among young adults remains a challenge.

▲ *Progress in Bangladesh towards achieving the MDGs by 2008*

Case study 2: progress in Uganda?

Sub-Saharan African countries, like Uganda, were at the heart of the reasoning behind the MDGs. In the early 1990s, 56% of Uganda's population lived in poverty, it had huge external debts, and the incidence of HIV/AIDS had reached an epidemic level. However, by 2000, a combination of the 'Drop the Debt' campaign, an aggressive educational campaign against AIDS – and support from external donors – had made the country a beacon of hope in sub-Saharan Africa. It was expected to be one of those most able to meet the MDG targets. So how is Uganda performing?

Target	Progress	Comment
1: Eradicate extreme poverty and hunger	Uganda is likely to halve the number of people living on less than US$1 a day by 2015. Currently 31% of its population lives in poverty, so the target of 28% is within reach.	Campaigners criticise the target for being too low; the Government is aiming for 10%.
2: Achieve universal primary education	Primary schooling is now free.	The Government is trying to persuade parents to send their children to school, and not to keep them at work on subsistence farms.
3: Promote gender equality and empower women	This is likely to be met due to legislation, e.g. a minimum of 30% of Uganda's Parliament must consist of women.	
4: Reduce child mortality 5: Improve maternal health	Under-five mortality rates are 137 per 1000 live births, and maternal death rates remain high.	Only 41% of births are attended to by a skilled professional. Maternal mortality is disappointing because it is largely preventable. Its causes include: severe bleeding, eclampsia, unsafe abortions, obstructed labour, malaria, AIDS.
6: Combat HIV/AIDS, malaria and other diseases	50% of reported illness in 2006 was malaria, down from 56% in 1999 but way off the target set in MDG 6. HIV prevalence remains successfully lower than most other African countries.	Lack of insecticide-treated nets is one of the key health issues in the country, and is a simple way of dramatically reducing malaria rates.
7: Ensure environmental sustainability	Current environment challenges include declining soil fertility, deforestation, pasture degradation, decreasing fish stocks, water pollution caused by industrial discharge and domestic waste. This has a huge impact on the livelihoods of poor communities by constraining their ability to increase incomes sustainably.	
8: Develop a global partnership for development	See Unit 5.7	

▲ Progress in Uganda towards achieving the MDGs by 2008

Over to you

1 In what ways are the eight MDGs linked? Draw a web diagram showing the MDGs. On it label the links, e.g. between poverty and education.
2 Research progress in achieving the MDGs in other developing countries – use the UK DfID (dfid.gov.uk/mdg), UNDP (undp.org/mdg) and UN Millennium Goals (unmillenniumproject.org/goals/index) websites. How does progress elsewhere compare with Bangladesh and Uganda?

3 Assume that all the MDGs have been achieved by 2015. In pairs, discuss:
   a What should the next set of Goals be for 2015 to 2030?
   b Will the development gap still be a matter of concern then?

On your own

4 How likely does it seem that the MDGs will be met by 2015? Justify your reasons.

Chapter summary

What key words do I have to know?

There is no set list of words in the Specification that you must know. However, examiners will use some or all of the following words in the examinations, and would expect you to know them and use them in your answers. These words and phrases are explained either in the Glossary on pages 330-333, or within this chapter.

aid
altruism
apartheid
bilateral aid
bottom-up
   development
capital-intensive
commodities
core and periphery
debt service
development
development gap
disparity

economic man
Euro-centric
food security
formal economy
Gini Co-efficient
Gross Domestic
   Product (GDP)
Gross National Income
   (GNI)
Human Development
   Index (HDI)
informal economy
investment

labour-intensive
Millennium
   Development Goals
   (MDGs)
multilateral aid
multiplier effect
neo-liberalism
Non-Governmental
   Organisation (NGO)
OPEC
Out-sourcing
Peace Index
per capita

privatisation
Purchasing Power
   Parity (PPP)
reciprocity
segregation
Structural
   Adjustment
   Packages (SAPs)
tied aid
top-down
   development
trade liberalisation

Films, books and music on this theme

Music to listen to
'Do they know it's Christmas?' (1984) by Band Aid
'Radio Africa' (1985) by Latin Quarter

Books to read
The White Tiger by Aravind Adiga (The Man Booker Prize Winner in 2008)
The Debt Boomerang (1991) by Susan George

Films to see
Water First - Reaching The Millennium Development Goals (2008) directed by Amy Hart. An inspiring story from Malawi which shows that clean water is essential for the achievement of the UN's Millennium Development Goals
'Cyclo' (1995) – a Vietnamese film directed by Tran Anh Hung
Tsotse (2005) – a highly acclaimed, but very violent, analysis of life in South Africa's black townships.

Try these exam questions

1 How far are patterns of global trade responsible for maintaining the development gap? **(15)**

2 Assess the view that economic development is not possible without causing environmental degradation. **(15)**

6 The technological fix?

What do I have to know?

This chapter is about the extent to which technology can manage and solve some of the issues facing the world today. Access to technology is closely related to level of development. Just as development is uneven, so is the geography of technology. Many people rely on technology to solve problems, while others lack access to technology at basic levels. The use of technology has costs as well as benefits. Technology varies between large-scale top-down mega-projects, and small-scale intermediate approaches. The Specification has three parts, shown below, with the examples used in this book.

1 The geography of technology

What you need to learn	Examples used in this book
• Defining technology and how it varies between countries and regions. • Technology is linked to levels of economic development.	• Defining technology. • High-speed rail in France and Africa. • Global distribution of Internet and mobile phone technologies.
• Access to technology varies.	• Access to technology in farming – global comparisons between GM foods, Sri Lanka (agro-technology), and Carfocial, Colombia.
• Inequality of access to technology is due to, e.g. cost, patents, education, or religious denial of access.	• Natural hazards and technology – the Thames Barrier compared with flood prevention in Dhaka, Bangladesh. • Variations in the treatment of HIV/AIDS globally.

2 Technology and development

What you need to learn	Examples used in this book
• The widening technology gap between knowledge-based economies and those lacking technology.	• Technology and development, drug treatments for HIV/AIDS. • Government sponsorship of research in Higher Education; patent laws.
• Technological leapfrogging can overcome barriers to development.	Reducing the technology gap: • mobile phones in Afghanistan • Solar power in India and Pakistan • GM crop use in Latin America and Africa.
• Technological innovation can have unforeseen costs and benefits. • The externalities of technology use (e.g. pollution) are costed in some economies but not in others.	Technology sometimes comes at a price: • GM crop technology • environmental issues; pollution and responses in Western Europe.

3 Technology, environment and the future

What you need to learn	Examples used in this book
• Intermediate technology and mega-projects as models for development.	• Tigray, Ethiopia – high-, intermediate- and low-technology solutions to water supply.
• Technology and overcoming global environmental issues • Is increasing technology use environmentally sustainable? • Technological futures – a wealthy 'technologically fixed' world and a 'technologically poor' world?	• Technological fixes and the potential to solve climate change; renewable energy technologies and carbon capture technology. • Energy futures and technology in Slovakia. • Exploring a range of technological futures, ranging from 'business as usual' to global technological convergence.

The geography of technology

In this unit you'll find out how the geographical spread of modern technology varies around the world.

The day the Internet died for millions

In January 2008, tens of millions of people in the Middle East and Asia were left without access to the Internet, after an undersea cable was accidentally cut. The cable that runs under the Mediterranean Sea from Italy to Egypt is part of the longest undersea cable in the world. It is 24 500 miles long – stretching from Germany, around the Middle East to Australia and Japan.

The result of the break in the cable was a **digital blackout**, which meant that many Internet users in the Middle East and Asia struggled to get any Internet access at all. As many as 70% of Egypt's Internet users were unable to get online, and in India there was a 50% cut in bandwidth – causing major problems for many of the rapidly expanding high-tech companies based there.

The breakdown in Internet communications meant that many Egyptians could not communicate with the outside world. One blogger, Amr Ghaerbeia, said: 'When I woke up this morning there was no Internet at home, and when I visited two or three other places during the day, they had no access either'. Amr was worried that the Egyptian authorities might have blocked the Internet in order to censor access to information (as happens with some websites in Iran).

In the end, it only took a few days to fix the cable and restore the Internet. However, the economic impact was felt by many businesses, including banks and stock-exchanges. British Airways was one of the worst-affected British companies, because their telephone call centres in India were unable to function.

▲ Surfing the Internet – part of our daily routine?

What impact could the sudden shutdown of the Internet in Britain have on individuals and companies?

What do you think ?

▼ International communications via the Internet are routed through submerged cables deep under the world's seas and oceans

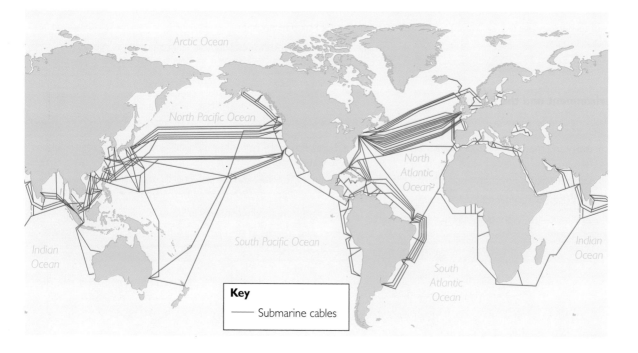

Arctic Ocean

North Pacific Ocean

North Atlantic Ocean

Indian Ocean

South Pacific Ocean

South Atlantic Ocean

Indian Ocean

Key

—— Submarine cables

The spread of technology in the 21st century

Many people think of '**technology**' in the sense of **high-technology (high-tech)**. This often presents itself in the form of **information and communications technology (ICT)**, through the use of computers, telephones and other digital media.

> ● **Technology** means the development of knowledge, techniques and systems which can be used to help solve problems and extend human capabilities. In geographical terms, it can mean ways in which people innovate, change, or modify the natural environment to supply human needs and wants.

Background

Technology levels vary spatially:

- Some areas of the world are **technology rich** – with the infrastructure to develop communications, or process goods and services, efficiently and quickly.
- Other areas are **technology poor** – without the infrastructure found in the world's wealthier countries.
- However, most countries rely on **intermediate technology** – processes and systems which are less costly or technically complex, but which do the job. Examples of this include traditional weaving looms, or re-using resources such as old rubber tyres as soles for shoes.
- Technology is **pervasive** – it can be found in some form or other wherever you find people living on the planet.

The **technological fix** relates to the expectations of people that continuing technological developments will help the world to tackle new problems as they arise. For example, the development of catalytic converters has reduced the quantity of toxic fumes released by car exhausts.

The Digital Access Index

In 2003, the International Telecommunication Union produced the Digital Access Index (DAI), which measured the access to ICT of people in 178 countries. The Index showed that there is a significant variation in ICT access between countries and continents.

The results of the Digital Access Index, 2003 ▶

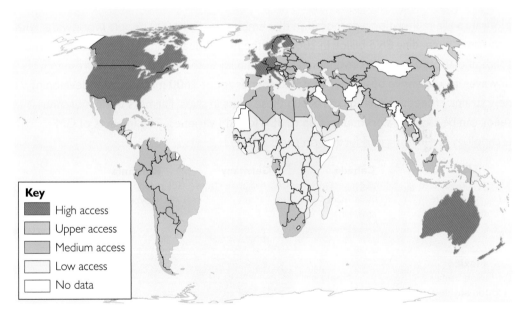

Key

	High access
	Upper access
	Medium access
	Low access
	No data

Over to you

1. Study the map of undersea Internet cable connections.
   a. Describe the locations of the main concentrations of undersea cables.
   b. Which countries generate the most Internet traffic?
   c. Are there any countries that are not connected? Where are they located?
2. How could the January 2008 digital blackout in the Middle East and Asia have affected the banking industry?
3. Using the map of the Digital Access Index, explain how ICT can be said to be pervasive.

On your own

4. Produce a report entitled 'Global digital access', using the data provided at http://www.itu.int/newsroom/press_releases/2003/30.html. In 500 words, describe the global patterns of access to digital technology.

Distribution of technology

In this unit you'll investigate the geographical distribution of technology in three areas: farming, telecommunications and high-speed rail transport.

Evolving farming technology

Perhaps the most important factor in the development of humans has been our continued ability to devise better and better agricultural technology. Technological developments in farming have now been adopted in many countries, and are continually being improved.

- In 1830, it took about 300 hours of work on a typical farm in the USA to produce 100 bushels of wheat – using a walking plough, a harrow, a hand spread of seed, a sickle, and a flail.
- By 1987, the whole process took just 3 hours of work to produce the same amount of wheat. The high-tech farm machinery available today allows farmers in Western countries to cultivate many more acres of land than is possible using simple hand-operated machinery and tools.

Genetic technology in farming

Genetically modified (GM) foods are produced by altering the **DNA** of seeds by the process of genetic engineering. Food that was produced from GM crops first became available in the early 1990s. The most common genetically modified crops are: soybean, corn (maize), canola, and cotton. Two of the most well-known GM food companies are based in the USA: Calgene in California and Monsanto in Missouri.

Monsanto is a multinational corporation that employs over 18 000 people worldwide, with a turnover exceeding $8.5 billion in 2007. It is the market leader in genetically engineered seeds and also the largest conventional seed company in the world. GM crop research is, however, an expensive business. Monsanto spends about $600 million a year developing new strains of seed, and needs to recoup these costs in sales. Farmers using Monsanto seeds can be charged a premium of up to 35% for GM varieties. The impacts of GM technology are investigated further on pages 242-243.

▲ Different technologies used to plough the land today: Zambia (top) and the USA (bottom)

The Cornell Agriculture and Food Technology Park

This is a new research and development park in the USA, linked to Cornell University. Its purpose is to help create new agricultural technologies. A number of high-tech food and agricultural companies have located there. The companies located at the park work closely with researchers at Cornell University – focusing on areas such as plant germ and seed technology, and precision agriculture.

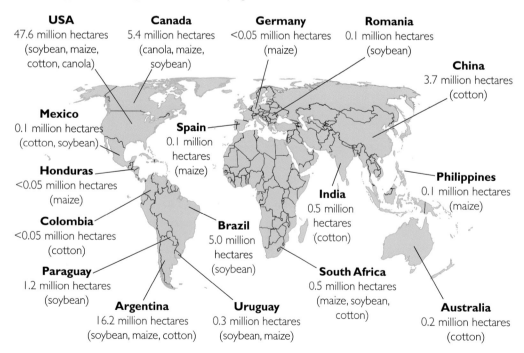

USA
47.6 million hectares (soybean, maize, cotton, canola)

Canada
5.4 million hectares (canola, maize, soybean)

Germany
<0.05 million hectares (maize)

Romania
0.1 million hectares (soybean)

China
3.7 million hectares (cotton)

Mexico
0.1 million hectares (cotton, soybean)

Spain
0.1 million hectares (maize)

Honduras
<0.05 million hectares (maize)

Colombia
<0.05 million hectares (cotton)

Brazil
5.0 million hectares (soybean)

India
0.5 million hectares (cotton)

Philippines
0.1 million hectares (maize)

Paraguay
1.2 million hectares (soybean)

Argentina
16.2 million hectares (soybean, maize, cotton)

Uruguay
0.3 million hectares (soybean, maize)

South Africa
0.5 million hectares (maize, soybean, cotton)

Australia
0.2 million hectares (cotton)

◀ The 17 countries that were growing GM crops in 2004

Continuing developments in farming technology

New technology is not limited to the world's wealthiest economies. Technological developments in twenty-first century farming go beyond improvements to machinery or genetics. The three case studies below illustrate some of the advances being made in farming technology in developing economies.

The India-African Union Summit

At the end of the India-African Union Summit, in April 2008, Indian Prime Minister Manmohan Singh said that India and Africa must meet their food needs through domestic production. He promised to help Africa with technology to increase farm productivity. African leaders said they are ready for investment and technology from India's more mature economy. This will involve the sharing of experiences and information on appropriate food storage and processing technologies, and will encourage the uptake of African- and Indian-developed technologies for farming and agricultural products.

CARFOCIAL, Andean Highlands, Columbia

In Columbia, a farming community is working with scientists to experiment with high-tech innovations to encourage more Latin American farmers to adopt high-tech farming practices. The introduction, in 1990, of a 'local agricultural research committee' in the village of San Bosco – high in the Andes – has led to a mini-boom in the cultivation of maize there. The success of this NGO-funded project relies on the fact that laboratory made technological advances in maize selection and production have been shared with 114 farming families there, who are acting as community based researchers for the International Center for Tropical Agriculture.

What is different about this project is that it has encouraged poor farmers to adopt new technology and incorporate it into the way they farm their land. Usually, when new research is carried out, many farmers are unwilling to try new technologies – or adoption rates are low – because the farmers are worried that the new technologies may damage their land or traditional way of farming. This Columbian approach has involved the farmers in both research and decision-making.

Agro-Technology Park, Gannoruwa, Sri Lanka

The development of this state-of-the-art technology complex is designed to ensure that the latest recommendations in crop cultivation and technology – generated by both public (Government departments and universities) and private institutions – are adopted by Sri Lankan farmers. The Park is leading research into rice development in Sri Lanka, including new improved and hybrid varieties, and also organising their release for general cultivation. The Park also houses a number of technological research gardens, as well as a food technology centre. The centre is initially being funded by Government money, but private companies are being encouraged to rent buildings for their own use.

Key

1 Vegetable garden
2 Root and tuber garden
3 Plant genetic resources
4 High-tech agriculture
5 Fruit garden 1
6 Banana garden
7 Integrated agriculture
8 Bee keeping
9 Seed sow
10 Farm machinery
11 Leafy vegetable garden
12 Dry zone crops
13 Fruit nursery
14 Tissue culture
15 Food technology
16 Organic agriculture
17 Citrus garden
18 Rice garden
19 Kamatha
20 Indigenous aqua culture
21 Local food court
22 Lake and irrigation demonstration
23 Traditional chena
24 Soil science
25 Soil conservation and Holy plants
26 Spice garden
27 Plant quarantine unit
28 Insect zoo
29 Mushroom house
30 Agriculture museum
31 Model home garden
32 Seed, planting material leaflet sales centre
33 Jack fruit garden
34 Fruit garden 2
35 Seed museum
36 Bio diversity park

Getting connected: telecommunications - but for whom?

The Internet

The nominees for the 2008 Millennium Technology Prize included two university researchers from the Bell Laboratories in the USA and an academic from the UK's Southampton University – with an invention that is revolutionising the Internet. Their invention amplifies light as it travels through fibre-optic cables, thus allowing much higher Internet communication speeds over huge distances. This has led to the rapid global development of the Internet, which in turn has had a huge impact on business, education and leisure for billions of people around the world.

The Internet began as a network of computers designed to allow American academics and researchers to share files. Since 1995, it has grown to include hundreds of millions of users around the world. It allows instant communication and information sharing by, for example, businesses contacting partners or clients, or schoolchildren communicating across continents and learning about different cultures. Before the advent of the Internet, a document being couriered from London to Birmingham could expect to take at least three hours, but that same document – when sent by email – can now arrive within seconds of being sent. The benefits of this technological breakthrough in global communication have been enormous and have permeated the lives of people all over the world.

1	Iceland	64.9
2	Sweden	57.3
3	South Korea	55.2
4	USA	55.1
5	Japan	54.5

▲ The top five countries as measured by the number of Internet users per 100 people (in 2007)

Access to the Internet in East Africa

Ugandans currently spend $18 million a year for Internet access – Uganda has the highest access costs in East Africa. It costs $2300 each month to access 512 mega bits a second, whereas in Kenya – which has had assistance from the World Bank – access costs have fallen to $500 per month. The Ugandan Communications Commission (UCC) aims to develop ICT in a 'critical mass', as India has done. Its aim is to install ICT facilities in all educational institutions, Government departments and local and international institutions.

This aim is only possible because of the development of the Eastern Africa Submarine Cable System (EASSy), which began construction in March 2008. The high-bandwidth, fibre-optic cable will run from Mtunzini in South Africa to Port Sudan in Sudan. There will be landing points in six countries, and connections will then spread to landlocked countries like Uganda. Ugandan Internet users will then no longer have to rely on expensive satellite systems to provide their Internet access. The cable system is due for completion at the end of 2010, and should see a dramatic reduction in the cost of accessing the Internet for Ugandans to between $45 and $200 a month. The cable project is being funded by the Development Bank of Southern Africa and the World Bank.

Mobile telephones

Most people in Britain own at least one mobile telephone, and many people own two – or even more. Mobile telephones are part of modern life, and allow people to keep in touch wherever they are. The growth of the mobile telephone industry has been rapid and has revolutionised communication. By the end of 2007, mobile operators had accumulated more than 3.25 billion subscribers worldwide (that's nearly half the world's population!), with more than 1000 subscribers signing up *every minute*.

There has been a huge boom in demand in Africa, and from the emerging middle classes in both India and China. This is due to the fact that, for many telecommunications companies, mobile technology avoids the expense and difficulties of installing landlines. The growth rate of mobile telephone sign-ups continues to accelerate. It took over 20 years to connect the first billion mobile telephone subscribers, only 40 months to connect the second billion – and the three billion milestone was passed in just two years, in July 2007.

A common sight on any British street in the twenty first century! ▶

▼ *The global distribution of mobile telephones in 2008*

Key

■ 0–1000
■ 1001–10 000
■ 10 001–100 000
■ 100 001–1 000 000
□ Above 1 000 000
□ No data

High-speed passenger rail transport

High-speed rail is defined as passenger rail travel at speeds in excess of 200 km/h (125 mph). Most modern high-speed trains do not exceed 350 km/h, and trains that can go beyond this speed encounter problems and challenges that are difficult to overcome.

Rail infrastructure is an essential part of economic development and environmental planning. The Japanese government realised this when undertaking its post-war reconstruction. Japan was the first country to introduce high-speed rail when it introduced Bullet Trains on the Tokyo to Osaka line in 1964. Technological developments in electrical power and track design allowed the Japanese to develop a fast, reliable and safe high-speed rail network on dedicated high-speed rail tracks.

High-speed rail in Europe and Africa

Developments in Europe ...

European high-speed rail travel really took off in September 1981, when the French Train à Grande Vitesse (TGV) started services between Paris and Lyon – on dedicated high-speed track. Today, there are 3000 km of high-speed railway lines stretching across Europe, with nearly 970 trains and more than 100 million passengers a year. There are currently plans to build an additional 6000 km of high-speed lines by 2020 – introducing a new age of high-speed rail travel across Europe. This is expected to lead to the creation of railway 'hubs' – based on the air travel model – and the creation of fleets of high-speed trains.

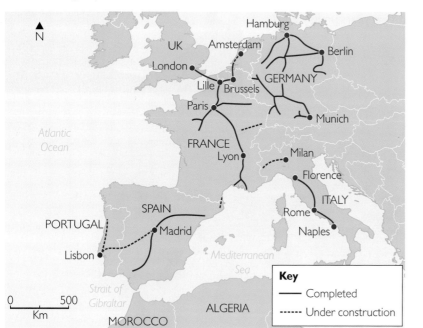

▲ Europe's current high-speed rail network

▲ A high-speed train in Europe – France's TGV

Rail holidays are having a renaissance, thanks to a host of new high-speed links to Europe and a growing determination among travellers to become more environmentally friendly. Demand doubled in 2006, compared with 2005.

The development of new high-speed rail links across Europe has come at just the right time. When the Eurostar terminal at St Pancras opened in 2007, journeys from London to Paris, Lille and Brussels were reduced by up to 20 minutes. The new TGV rail link from Paris to Strasbourg [which opened in 2008] cuts the journey time from 4 hours to 2, and from Paris to Basel and Stuttgart from 5 hours to 3 hours 30 minutes. In tests, the train has reached 553 km/h – breaking the world rail speed record. The normal TGV service averages 320 km/h.

Other major rail developments are in the pipeline. High-speed lines across Spain and Italy, and links from Spain to Portugal and France, are all under construction [see the map]. And if plans for a rail tunnel under the Strait of Gibraltar from Spain to Morocco come to fruition, Europe could be linked to Africa by 2025.

Adapted from an article in *The Observer* by Gemma Bowes, 2007

... and in Africa

High-speed rail travel is due to begin operating in Morocco in 2013 – with the £1.5 billion high-speed rail line linking the northern city of Tangier to the country's commercial capital, Casablanca. The development will cut the journey time between the two cities to 2 hours and 10 minutes, instead of the current 5 hours and 45 minutes – with trains travelling at speeds of up to 300 km/h. The service aims to carry 8 million passengers a year.

In October 2007, Morocco and France signed an agreement to allow French companies to design, build, operate and maintain the high-speed rail link. The technology for this project will be the same as that used in the French TGV network. The new rail network will make Morocco the first African country to have a high-speed rail infrastructure to match Europe's.

Algeria's rail operator, SNTF, also has plans to introduce TGV-style trains – with eight high-speed rail line proposals due to get under way from 2009. In South Africa, the 80-km Gautrain line is expected to increase speeds to 180 km/h (112 mph) in 2010, and could be upgraded later by another 20 km/h (12 mph), so that it is officially classed as high speed.

Work on the Arab world's first high-speed train – stretching from the Mediterranean to the door of the Sahara Desert – could begin next year, the head of Morocco's rail company said. The trains would travel at up to 300 km/h (186 mph), slashing travel times from Tangier in the north – via Marrakesh – to Agadir in the south, and from Casablanca on the Atlantic coast to Oujda on the Algerian border.

Improving transport links from the centre of the kingdom to long-neglected outlying regions is an important part of the Government's attempt to stimulate the economy and reduce unemployment and poverty. Journeys from Casablanca to Marrakesh could be cut from over 3 hours to 1 hour and 20 minutes, and from the capital Rabat to Tangier from 4 hours and 30 minutes to 1 hour and 30 minutes.

Plans for a rail tunnel from Europe to Africa under the Strait of Gibraltar are still on course, so one day it may be possible to take a direct train from Madrid to Marrakesh.

Adapted from an article in *The Khaleej Times*, United Arab Emirates, 2006

Over to you

1 Are you connected to technology? In pairs:
   a List the technological gadgets you use in your everyday lives.
   b Assess the extent to which these are (i) a necessity, (ii) a luxury.
   c Does everyone in the world have equal access to technology? List any barriers that may prevent access.
2 Explain how technological developments have improved farming methods since 1830.
3 a Who are the major investors in agricultural technology?
   b Who gains from technological developments in agriculture?
4 Look at the map of the 17 countries that were growing genetically modified crops in 2004 (page 222).
   a Describe the countries' geographical spread.
   b What pattern is emerging in the geographical distribution of these GM crops?
   c Which continent is largely missing out on GM developments so far? Why do you think this is?
5 Compare the map on page 221 (showing countries' access to ICT in 2003, according to the Digital Access Index) with the map on page 225 (showing the global distribution of mobile telephones). What patterns are evident between ICT access and mobile telephone distribution?
6 Look at the map of the high-speed rail network in Europe, and read the two newspaper extracts.
   a Describe the extent of the high-speed rail network in Europe.
   b Describe the extent of the proposed high-speed rail lines for Africa.

On your own

7 Referring to the information about the digital blackout on page 220 and the laying of the new EASSy cable on page 224, draw up a list of the possible advantages and disadvantages that the new Internet cable system might bring for East African Internet users.
8 Reviewing the information on farming, telecommunications and high-speed rail transport, write a 500-word report for the United Nations assessing the extent to which you agree that some parts of the world remain disconnected from technological developments.

Access to technology – 1

In this unit you'll compare how two capital cities are using different levels of technology to counteract the effects of flooding.

Natural hazards and technology

Natural hazards can present devastating risks to people. 2008 saw several of these natural events strike parts of the world, with terrible consequences:

- Cyclone Nargis struck Burma on 3-4 May, killing as many as 130 000 people in storm surges, and leaving up to 1 million more homeless.
- The earthquake that struck China on 12 May killed an estimated 70 000 people, with another 400 000 missing and injured.

In the twenty first century, there is a belief that technology can help to minimise the threat of nature. So, does technology have the power to conquer the hazards that bring misery to so many people?

The Thames Barrier – protection for London

When work began on the Thames Barrier in late 1974, it was hailed as a technological marvel. The final design was selected from the 41 proposals put forward, because it was compact, attractive, practical, and environmentally sensitive. The Barrier cost £535 million to build, and became operational in 1982 – with an expectation that it would last until 2030. It is the world's largest moveable flood barrier, and spans 520 metres across the River Thames at Woolwich.

In the event of a flood risk to London, the Thames Barrier puts a wall of steel right across the river. This stops the incoming tide that would otherwise sweep up the Thames towards Central London. By July 2007, the barrier had been put into operation 103 times to prevent London from flooding.

Why protect London?

London is a city with over 7 million people (12% of the UK's total population), and has the highest population density in Britain (over 4500 people per square kilometre). It is a major centre for international business and commerce, with the sixth largest city economy in the world, and it generates 20% of the UK's total Gross Domestic Product (GDP). It is also built on a floodplain along the banks of the tidal estuary of the River Thames. As such, the city has always been at the mercy of the river and has flooded many times throughout its history.

The Thames Barrier at Woolwich, London

Background

Technology used by the Thames Barrier

The Thames Barrier is made up of nine concrete islands (or piers), which contain six openings for vessels to pass between them, plus four smaller openings that remain open when there is no flood risk. The piers are built on a base of solid chalk 17 metres below the riverbed. Curved steel gates sit on the riverbed in concrete sills between the piers and, when necessary, are rotated upwards to form vertical walls to stop floodwater from pushing up the River Thames. The four main gates are over 20 metres high, and each weighs about 3700 tonnes. Each one can withstand an overall load of 9000 tonnes. They are 61 metres' wide, allowing larger vessels to pass through the gap between the piers, and there are two smaller openings 31 metres' wide to allow smaller vessels through. The four remaining openings are fitted with falling gates.

In order for the Thames Barrier to protect London, the 50 staff who operate the gates need reliable information to decide when they should be closed. Tide-level information is collected every 15 minutes from five monitoring stations on the East coast of England.

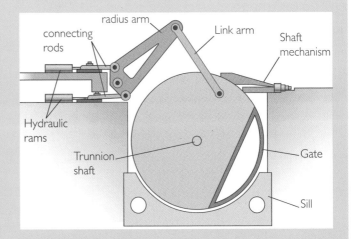

▲ How the Thames Barrier's main gates work (also see opposite)

Information about the height of the river is also collected every minute from 11 stations along the Thames – stretching from Southend to Richmond. This data is then fed into a computer system in the control tower on the south bank. Computer models of possible flood risks are produced, and a decision is then made about whether the Barrier needs to be closed. If a flood risk is detected, computer software is used to activate the equipment on each pier to close the gates.

Is a new barrier needed?

In 2005, scientists for the 'Thames Estuary 2100 Project' – set up by the Environment Agency to assess flood risk and river management over the next century – came to the conclusion that a new flood barrier would be needed to protect London in the future. With the strong possibility that global warming will lead to rising sea levels, the scientists predicted that the level of flood defence offered by the current Thames Barrier will not be enough to withstand future flood surges.

In 2007, plans were put forward by Government ministers for a new £20 billion flood defence scheme for London, which would be built to the east of the current Thames Barrier. This new scheme is intended to offer London protection from a one-in-one-thousand-year flood event. In 2009, a final decision will be made about whether the new scheme will go ahead.

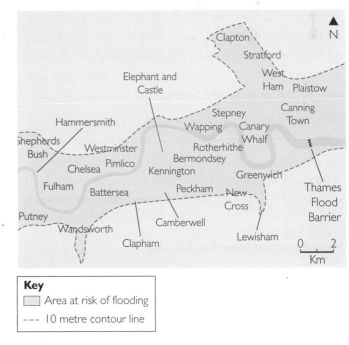

Key

▢ Area at risk of flooding

--- 10 metre contour line

▲ A flood risk map for London

Welcome to Dhaka

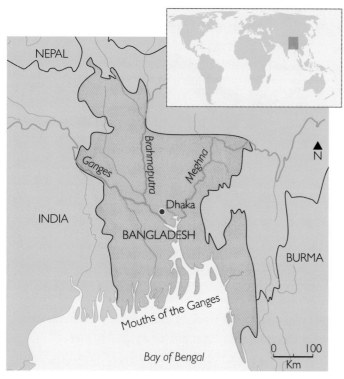

Area	145 km²
Population (city proper)	5 378 023
Population (metropolitan area)	12 560 000
Population density	14 608 people per km²
Height above sea level	4 metres
January temperature range	13°C to 26°C
July temperature range	26°C to 31°C
January average daily rainfall	8 mm
July average daily rainfall	399 mm

▲ Dhaka, facts and figures

▼ Dhaka is the centre of political, cultural and economic life in Bangladesh. Like many of the world's capital cities, it is built on the floodplain of a major river.

Dhaka is the capital city of Bangladesh and the commercial heart of the country's economy, with a current GDP of £26 billion (projected to rise to £65 billion by 2020). However, the average wage in the city is estimated at just £300 a year, with 50% of the population living below the **poverty line**, and around 23% unemployed. It is believed that 50% of the workforce is employed in the informal sector; that 800 000 work in the textile industry, with another 400 000 working as rickshaw drivers. The city is made up of two sections: an older area, which is a maze of narrow lanes and crowded bazaars, and an orderly and well-planned government centre.

Dhaka is located on the floodplain of the Buriganga River, in central Bangladesh, on the lower reaches of the Ganges-Brahmaputra delta. The city is flat and, like much of Bangladesh, lies very close to sea level – making it vulnerable to flooding from monsoon rainfall and from tidal surges.

Flood protection for Dhaka

After the disastrous floods that struck Bangladesh in 1988, badly affecting Dhaka, the Dhaka Integrated Flood Protection Project (DIFPP) was set up as part of the national Flood Action Plan (FAP). This project focused on structural measures, such as building embankments and levees to hold back floodwaters. Initially, the plan was to cover 260 km² of Dhaka, but money could not be raised for the entire project area. Therefore, it was decided to start by protecting the more densely populated western part of the city.

However, in 1998, Dhaka was badly flooded again, because engineering solutions on their own were not enough to protect the city. Water entered the protected part of the city through buried sewage pipes and through breached and incomplete floodwalls. The city's pumping stations were inadequate and could not cope with the excess water. There was also a serious lack of coordination between the agencies responsible for flood protection and drainage of the city. The repair costs were estimated at well over $200 million, and disease spread throughout Dhaka because much of the drinking water was contaminated by dirty floodwater.

After 1998's floods, Phase II of the DIFPP was introduced, with the aims of:

- protecting the eastern part of Dhaka, which had suffered the most flood damage
- starting to implement non-structural solutions – such as better flood forecasting and warning, land-use planning restrictions, flood zoning, and improved emergency responses
- improving coordination between the agencies responsible for flood protection and drainage of the city, which could significantly reduce the flood damage

Yet, despite all the attempts at protection, Dhaka flooded again in 2004 (see below). For nearly three days in July, life in the city came to a standstill. People were forced to stay at home, businesses came to a halt, and there was another shortage of clean drinking water. The rapid growth of the city, due to urbanisation, has meant that the current protection measures are no longer able to cope with the city's vast expanse.

Flood approaches Dhaka

Floodwaters are approaching the eastern fringes of Dhaka, as the water levels of all major rivers surrounding the capital continue to rise. Due to the absence of a flood-control embankment in the east of the city, 75% of Bashabo, Nandipara and Trimohini are under water. Boats have become the only means of transport in those areas. Sources at the Flood Forecasting and Warning Centre (FFWC), said that more parts of Dhaka are likely to be inundated in the next few days.
Adapted from BangladeshNews.com, 5 August 2007

▼ *In 2004, flooding again brought the city of Dhaka to a standstill*

With global warming set to raise sea levels worldwide, only the wealthiest cities will be able to afford technological schemes to protect them from flooding.

What do you think?

Over to you

1 In pairs, discuss and list the reasons why the British Government felt it was necessary to build a flood defence barrier in London.
2 Using the website www.floodlondon.com/floodtb.htm explain how the Thames Barrier operates.
3 Using the website http://www.thedailystar.net/magazine/2004/09/04/cover.htm explain why Dhaka's flood defences continue to fail to protect the city from flooding.
4 Discuss and complete a table showing whether London is more geographically advantaged, or technologically advantaged, than Dhaka to enable flood protection measures to be successful.

On your own

5 The film *Flood* speculates how London could flood if the Thames Barrier were to be breached in the future. Research information on the Internet to decide whether London will need a new barrier to cope with rising sea levels.
6 To what extent do you think that the inhabitants of Dhaka are left to suffer from **environmental determinism**?

Exam question: 'Some are able to access new technology to solve environmental problems while others are left to suffer from environmental determinism'. Referring to examples, assess the validity of this viewpoint. **(15)**

In this unit you'll investigate the extent to which technology is being used to tackle the spread of HIV/AIDS, and whether technological breakthroughs for tackling the disease are accessed equally.

The HIV/AIDS pandemic

▲ Children in Botswana infected with HIV/AIDS

▲ Estimated HIV infection rates for adults in 2007

Key

Adult prevalence (%)

- 15.0–28.0
- 5.0–<15.0
- 1.0–<5.0
- 0.5–<1.0
- 0.1–<0.5
- <0.1
- No data available

- According to UN estimates, there were about 33 million people living with HIV worldwide in 2007 (see the line graph).

- Sub-Saharan Africa is the region of the world that has most felt the impact of the HIV/AIDS pandemic (see the map). In 2007, 67% of those infected with HIV (22.1 million), and 72% of those who died of AIDS (1.5 million), lived there.

- Southern Africa accounted for 32% of all new HIV infections worldwide in 2007.

- With about 5.5 million people living with HIV, South Africa is the country with the largest number of infected people.

- However, HIV prevalence in sub-Saharan Africa has now stabilised – and even begun to show signs that it may be declining (see the line graph).

Key

- Global number of people living with HIV
- Global % HIV prevalence adult (15–49)
- Sub-Saharan Africa-number of people living with HIV
- Sub-Saharan Africa % HIV prevalence adult (15–49)

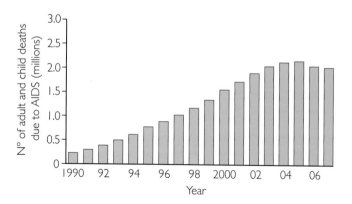

▲ The estimated number of global deaths due to AIDS, 1990-2007

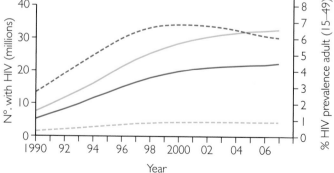

▲ The estimated number of people living with HIV, plus the adult infection rate – globally and in sub-Saharan Africa, 1990-2007

Addressing the HIV/AIDS pandemic

In 2006, there was a UN conference in New York about tackling the global HIV/AIDS pandemic, which included representatives from 180 countries. The conference had one aim – to provide HIV/AIDS drugs to anyone needing them by 2010. Kofi Annan, UN Secretary-General, said that most countries had fallen 'distressingly' short of meeting their targets. The conference ended with a decision to work towards universal access to HIV/AIDS care by 2010 – at a cost of $24 billion.

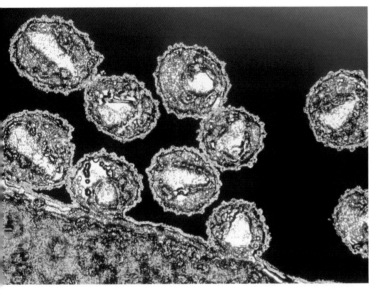

▲ The HIV virus slowly breaks down the body's immune system

Key findings by UNAIDS in 2008

- Every day, over 6800 people become infected with HIV and over 5700 die of AIDS – mostly because of inadequate access to HIV prevention and treatment services.
- The HIV pandemic remains the most serious challenge to public health by an infectious disease.
- The global percentage of people infected with HIV has stabilized since 2000 (see the line graph), although the global number of people living with HIV is still increasing because of global population growth, continuing new infections and longer survival times for those infected.
- There has been a reduction in AIDS-associated deaths (see the bar chart), which is partly attributable to the recent increase in access to treatment, and also a global reduction in the number of annual new HIV infections (from 3 million in 2001 to 2.7 million in 2007).
- Sub-Saharan Africa remains the most seriously affected region, with AIDS remaining the leading cause of death there.

Adapted from the 2008 Report on the global AIDS epidemic by UNAIDS

Background

HIV/AIDS

HIV: The Human Immunodeficiency Virus (HIV) is a retrovirus that attacks the immune system, thus preventing the body from being able to fight infection. It is known as a 'slow virus', because it can take a long time to produce adverse effects in the human body.

AIDS: Acquired Immune Deficiency Syndrome is diagnosed when a person has developed one of several opportunistic diseases, such as Kaposi's sarcoma (a form of skin cancer), pneumonia, and dementia.

HIV is difficult to transmit, because the virus cannot live for long outside the human body. However, it can be transmitted through bodily fluids – through unprotected sexual intercourse, injecting from infected needles, or through infected blood transfusions. Mothers can also pass HIV to their unborn children, and it can be passed through contaminated breast milk.

The first recognised cases of AIDS were reported in the early 1980s in New York and California in the USA. A small number of men were found to have developed rare opportunistic infections and cancers that were resistant to any treatment – and it became clear that they were all suffering from a common syndrome. At the same time, an AIDS cluster came to light in Haiti. Fear of the new disease led to a large number of Haitian immigrants living in the USA losing their jobs and being evicted from their homes – as Haitians were added to homosexuals, haemophiliacs and heroin users to make the 'Four-H Club' of groups considered to be at high risk of AIDS.

Using technology to combat HIV/AIDS

The HIV/AIDS pandemic has seen a surge in technological developments to combat the disease. Although there is currently no preventative vaccine – or cure – **antiretroviral drug** development continues to be the leading way of countering the effects of HIV by slowing down the progress of the infection. The information on these two pages shows how recent technological developments are being used to tackle the impacts of both HIV and AIDS.

Treatment for HIV can result in patients taking a cocktail of drugs in order to maintain a 'normal' life ▶

New HIV drug trials in the UK, January 2008

A new drug, called Raltegravir, is to be made available to an estimated 73 000 HIV patients in the UK. The drug works by blocking an enzyme that is essential for HIV to be able to replicate itself. It means that doctors will now have a further treatment option for patients who have built up a resistance to existing drugs. A clinical trial of Raltegravir found it to be effective in patients who had been taking regular antiretroviral HIV drugs for about 10 years. Drug treatment in the UK has been helping people to live longer, but resistance to commonly used drugs has been a growing problem, which Raltegravir looks set to tackle.

Nanoparticle technology, February 2008

Johnson and Johnson are working with Tibotec to develop their antiretroviral HIV drug, Rilpivrine (currently only available in daily oral tablet form) into a slow-release drug that can be injected once a month (or possibly less) – thus giving HIV patients freedom from the daily chore of taking a cocktail of tablets. The drug is being developed using **nanoparticle** technology (a nanometre is one millionth of a centimetre), which allows the drug to be injected into muscle tissue to be released slowly over a period of up to six months.

Using ICT to track HIV treatment in resource-poor countries, October 2006

Tackling HIV in those countries which do not have access to the latest developments in medical science, does not mean that available technology cannot be used to combat the disease. Medical researchers in Australia have been developing information-management technologies which link information between different aid agencies working in some of the hardest-hit African nations. Using cost-effective information systems, plus the involvement of local people in the development and management of the project, it is hoped that lives can be saved through direct care and preventive care programmes.

Oral vaccine technology to tackle HIV, June 2007

The Department of Trade and Industry awarded a grant of £1.1 million to Cobra **Biomanufacturing** PLC (working with researchers from Cambridge University and Royal Holloway University) to develop an oral HIV vaccine. The technology stems from research that was used to produce an oral vaccine for both anthrax and plague, and, should it prove successful, it could pave the way for the first HIV-vaccine trials.

Eliminating HIV from semen, September 2007

With HIV infection no longer a death sentence – due to antiretroviral drugs – many patients are able to live relatively normal lives and make plans for the future. This has led to an increase in the number of couples made up of HIV-positive men and HIV-negative women who are seeking medical help in order to have a healthy uninfected child. Scientists in Japan have developed a process that 'washes' semen – eliminating the HIV virus from it. They have been successful in extracting HIV-free sperm from the washed semen and using this to artificially inseminate the female partner. The process has yet to go beyond the testing phase, but it offers the hope of starting a family to many couples with an HIV-positive male partner.

HIV-free breast milk, May 2007

A cheap and simple method of preventing breast milk from transmitting HIV from mothers to their children has been devised in Lusaka, Zambia. The process involves the 'flash-heating' of infected milk, which inactivates the virus – giving hope that breastfeeding could become much safer in poor countries with high HIV-infection rates. National banks which collect, store and disperse human milk already **pasteurise** it, but they commonly use a method which relies on thermometers and timers that can be hard to get hold of in resource-poor communities. The new method involves simply heating a glass jar of expressed milk in a pan of water over a flame or single burner, so it can be used easily by mothers at home.

Over to you

1 The UN states that $24 billion could ensure universal access to HIV/AIDS care by 2010. Why do you think that, despite the USA pledging $50 billion in 2008, this may still not ensure universal care?

2 Look at the map on page 232 showing worldwide adult HIV infection rates in 2007.
   a What patterns can you see at both global and regional scales?
   b How does the global spread of HIV compare with the map showing digital access on page 221?

3 Do you think that there is a link between the availability of technology and the impact of HIV? How can this link be explained?

4 Explain the reasons that may have contributed to the concentration of the HIV epidemic in sub-Saharan Africa.

5 Are the six technological developments presented on these two pages available to all? Copy and complete the table below:

Technological development	How it works	Available to whom?	Barriers that may prevent access

On your own

6 Visit the UNAIDS website at: http://www.unaids.org/en/ and investigate the global impact of the HIV pandemic.

7 Produce a 500-word report to discuss the following statement: 'Despite all the technological developments that are taking place to reduce the impact of HIV/AIDS, the world pandemic has still not been brought under control.'

The technological fix?

In this unit you'll learn that there is a link between economic development and technological innovation.

Technology promotes development

In May 2008, *The Growth Report: Strategies for Sustained Growth and Inclusive Development* was published by the independent Commission on Growth and Development (made up of key policy makers and economists). The report highlighted the importance of technology, innovation and higher education in achieving economic growth. It stated that: 'Fast, sustained growth is not a miracle – it is possible for developing countries'.

The Growth Report identified the key factors for growth as:
- engagement with the global economy
- specialising exports
- the transfer of key technologies

The report commented that, in all 13 cases of sustained high growth which it refers to (see the table on the right), those high-growth economies rapidly absorbed know-how, technology, and – more generally – knowledge from the rest of the world. It is accepted that much of the technology did not originate in those countries, but their economies had to assimilate it very quickly in order to develop. A key finding was that economies can learn faster than they can invent from scratch and, therefore, knowledge gained from the global economy is the basis of economic catch-up and sustained growth.

Country	Period of high growth	GDP per capita income in constant 2000 US$ (taking inflation into account) ...	
		... at the beginning	... and in 2005
Botswana	1960–2005	210	3800
Brazil	1950–1980	960	4000
China	1961–2005	105	1400
Hong Kong (China)	1960–1997	3100	29 900
Indonesia	1966–1997	200	900
Japan	1950–1983	3500	39 600
South Korea	1960–2001	1100	13 200
Malaysia	1967–1997	790	4400
Malta	1963–1994	1100	9600
Oman	1960–1999	950	9000
Singapore	1967–2002	2200	25 400
Taiwan	1965–2002	1500	16 400
Thailand	1960–1997	330	2400

▲ The 13 countries which have achieved fast and sustained growth since 1945 (both India and Vietnam may soon be joining this group). This table demonstrates that growth is possible – because all 13 economies have achieved it – but it also demonstrates that it is not easy to achieve, because **only** 13 economies have ever done it. Indeed, some people view these cases as 'economic miracles'.

Innovation and higher education promote development

The Growth Report also stated that, as economies begin to develop, there must be increased investment in both innovation and higher education to ensure that the growth can be sustained, and to see a transition from middle-income to high-income status. The two tables on the right show the top ten countries as measured by the number of patent grants issued (i.e. new technical innovations) and the number of students who enrolled in higher education in 2006.

Country	Number of patents registered
USA	157 496
Japan	125 880
Germany	41 585
France	36 404
South Korea	34 956
UK	33 756
Italy	19 652
Russia	17 592
Netherlands	17 052
Spain	15 809

▲ The top 10 countries for registered patents, 2006

Country	University enrolments per 100 000 adults
Canada	5997
South Korea	5609
Australia	5552
USA	5339
New Zealand	4508
Finland	4190
Norway	4164
Spain	4017
Ireland	3618
France	3600

▲ The top 10 countries with the highest university enrolments, 2006

Patent grants

A **patent** is the exclusive right of ownership and possession of intellectual property, as granted by a country to an inventor for a fixed period of time (usually a year). A patent can be awarded for the design of a new machine, an improvement to an existing machine, or the development of a process. Patents can be awarded, for example, in the areas of science, engineering, business, and software.

The Growth Report highlighted that many countries may not have the ability to invent as quickly as they can adopt and learn new technologies. This idea is supported by the fact that, of the 13 high-growth economies highlighted in the first table, only South Korea and Japan feature in the top-ten list of countries by patents registered in 2006.

Participation rates in higher education

University education is integral to long-term economic development, by providing a country with the highly skilled labour force required if that country is to produce sustained growth. The table opposite indicates that only South Korea from the 13 high-growth countries breaks into the list of top-ten countries by university enrolments in 2006.

The importance of higher education to economic success is now increasingly recognised around the world. China is currently seeing a surge in university applications, and will overtake the USA as the country awarding the most university degrees within 10 years. One of the final recommendations made in *The Growth Report* was that industrialised countries, such as those in Europe and North America, have a responsibility to finance the expansion of Africa's university education to make up for Africa's 'brain drain'.

▲ *James Dyson invented the bag-less vacuum cleaner but, unlike a songwriter who owns his or her songs, an inventor must pay to renew his or her patents each year*

Is it feasible that most of the 13 countries which have achieved sustained high growth since 1945 will eventually be able to break into the lists of top ten countries for patent grants or university enrolments?

What do you think ?

Over to you

1 In pairs, use the Commission on Growth and Development website (http://www.growthcommission. org/) to research the factors that promote economic development. Report your findings back to the class.
2 Explain how technology, innovation and university education can help to promote development and growth.

On your own

3 a On a blank political world map, plot the 13 countries which make up the success stories of sustained high growth since 1945.
   b Does any pattern emerge?
   c On the same map, now plot the top-ten countries by both patent grants awarded and university enrolments.
   d What pattern do you notice for these countries, and how does it compare with the pattern of the 13 countries which have achieved sustained high growth?

Technological leapfrogging

In this unit you'll find out what technological leapfrogging is and how it can be used as a tool for development.

Mobile phones provide a future for Afghanistan

Meet Abdul Wakil (on the right), who lives in the village of Daw Koo in Afghanistan, about 40 km north of the capital, Kabul. He first used a mobile telephone in 2008 and says that it has now transformed the way in which he runs his small dry goods store. 'We used to go all the way to the city to order products, but now it is only a phone call away and the costs are much less' he says.

Abdul is one of a growing number of mobile telephone subscribers in Afghanistan. 72% of Afghans are now covered by a mobile telephone signal, whereas only 1% has access to a fixed telephone line. The more-or-less constant conflict in Afghanistan since the Russian invasion of 1979 has left the country littered with landmines and unexploded bombs, so laying and maintaining landline telephone cables is an extremely dangerous business which few companies wish to attempt. The terrain in Afghanistan – of deserts and mountains – is also very challenging, aside from any security issues from the ongoing conflict. Therefore, any attempt to put new telephone cables in the ground would take decades and be prohibitively expensive.

The country's largest telephone provider, Roshan, has therefore 'leapfrogged' the need for a fixed-line telephone infrastructure and moved into the safer and cheaper wireless mobile sector. The Chief Executive of the company, Karim Khoja, says: 'There is no fixed-line infrastructure here – because of the landmines – and there probably never will be, so we have seen a leapfrogging of technology'. In 2008, Roshan had 2 million mobile telephone subscribers in Afghanistan – making up 43% of the total market.

▲ Mobile telephone technology has transformed the way that Abdul Wakil runs his business in a rural village in Afghanistan

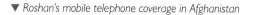

▼ Roshan's mobile telephone coverage in Afghanistan

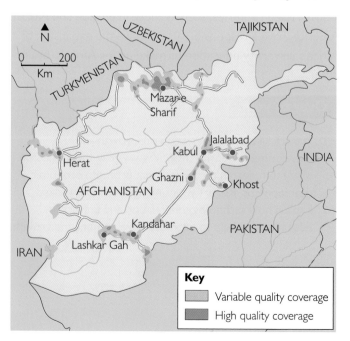

Key
- ▢ Variable quality coverage
- ▢ High quality coverage

Mobile telephones: the leapfrogging process

The rapid growth of mobile telephone usage in Afghanistan is not an isolated case. Many developing countries are now establishing mobile networks. In 2006, 68% of the world's mobile telephone subscriptions were in developing countries. Mobile telephone networks allow developing countries to avoid the need for expensive cabled networks of landlines – often across vast geographical areas. Some of the benefits of this technological leapfrogging can be seen below:

- In China, there has been a surge in mobile telephone usage, because the laying of the country's infrastructure of cable landlines has been too slow to keep up with its huge economic growth and the demand for better communications.
- In Kenya, a low-cost mobile telephone service, called M-Pesa, allows Kenyans to send and receive money via text messages (see the photo). The service is popular, because it removes the need to travel long distances to deliver or receive cash.
- In parts of rural Rwanda, healthcare workers use mobile telephones with special software, which allow them to enter data into the telephone about drug stocks and AIDS patients and then transmit this information, via text messages, to health officials in Rwanda's capital, Kigali.
- In Uganda, The Village Phone Initiative (see page 285 of the AS textbook) has opened up telecommunications in the country.

What is technological leapfrogging?

Leapfrogging is the process by which some countries may advance their development by jumping over inferior, less-efficient, or more-expensive technologies, in order to move directly to the best and most-appropriate solutions.

An example of leapfrogging is where countries develop their energy provision by avoiding a reliance on fossil fuels and jumping directly to solar power. Examples of this can be seen on pages 240-241.

In the case of telephone communications, leapfrogging is the process by which a country moves from having no telephone service to having a modern mobile telephone network – by avoiding ever installing a system of cable landlines or outdated mobile technology. Therefore, the country can bypass the developmental process involved in getting to that point by skipping: the telegraph, manually switched telephony, direct-dial telephones, brick-sized mobile telephones, and analogue mobile networks – just moving straight to 3G – leapfrogging the enormous human and capital investments that it took to get there.

Technological leapfrogging may be the process by which the development gap between countries is reduced – because it is the richer, more technologically advanced nations which have paid the costs to develop the new technology. Competition in these developed countries then ensures that the new technology becomes cheaper, so poorer countries can then obtain the benefits.

Solar energy in India and Pakistan

Solar Home Systems for villages in Pakistan

In December 2003, the Alternative Energy Development Board (AEDB) in Pakistan launched the first of its Solar Home Systems (SHS) in the village of Alipur Farash, on the outskirts of Islamabad. The AEDB – funded through private investors – fitted 100 homes with solar panels, in order to provide electricity as part of a pilot that has led to more than 1000 houses in 11 other rural villages being supplied with solar-powered electricity over the following three years (see the table). The AEDB developed this project to provide villagers across Pakistan with the comforts of lighting, cooking and water disinfection through solar-energy technologies. The project demonstrates that solar energy can be effectively used for small-power requirements in remote areas.

Each individual household was installed with its own SHS, and has been made responsible for operating and maintaining it. Each one was provided with: an 80-watt solar panel, charge controller, battery, 4 CFL lamps, 2 LED lights, a 12-volt DC fan and a TV socket. The solar panel charges the battery during daylight hours and the stored energy is then used to provide light for homes and streets, and to operate fans and TVs. The user is only required to switch the lighting system on or off, as is done in normal home lighting systems. The households were also supplied with solar-concentrating cooking facilities and a solar water disinfector.

Pakistan's goal is to have 10% of its national electricity generation come from alternative sources by 2010. The AEDB believes that providing electricity to villages and communities outside the reach of grid-based suppliers will lead to a faster scale of development in the country.

▲ The Solar Home System in a village in Pakistan

▲ Solar-concentrating cooking in Pakistan. Using direct solar heat, the solar concentrating cooker can cook food in the same time as gas cookers. It has been developed by local manufacturers.

Village name	Province	Number of houses
Narian	Punjab	53
Korian		57
Lakhi bhair	Sindh	135
Basti Bugha		100
Pinpario		100
Bharomal		115
Takht	Balochistan	
Killi Mama Macherzai		100
Allah Baksh Bazar Dandar		121
Jhanak	Northwest Frontier	120
Shnow Garri		100

▲ The 11 Pakistani villages electrified using solar power by the Alternative Energy Development Board, 2004-2006

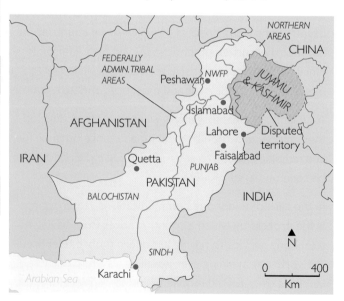

India's Barefoot solar engineers

The Barefoot College, also known as the Social Work Research Centre (SWRC) – an NGO based in Rajasthan – covers nearly 27 000 m² and is powered completely by solar energy. The solar energy is used to run fans, pump water and operate 20 computers. However, the real innovation about the college is that it trains people to build, operate and maintain their own solar-power systems. Most importantly, it trains women – some of the poorest in India. Over 500 people are trained each year to assemble, install and maintain solar technology. The skills of the Barefoot graduates serve over 100 villages and more than 100 000 people.

The Barefoot solar engineers work across eight Indian states to develop solar-energy systems in areas where the electricity supply is either non-existent or unreliable. The majority of the engineers are illiterate or semi-literate, but they have skills that are transforming the lives of the rural poor in India. The project has empowered women – giving them employment, freeing them from the constant need to search for fuel-wood, and also reducing the health and environmental hazards of burning wood fires. This project has demonstrated how solar energy can provide a solution not just for cooking and lighting but also for education, agriculture, health, and income generation.

▲ Barefoot solar engineers in India, usually women, are adept at fixing and maintaining solar-energy systems

Over to you

1 List the possible advantages for Abdul Wakil (page 238) of having a mobile telephone connection. Would the same advantages apply with a fixed landline?
2 What human and physical barriers make the creation of a fixed landline infrastructure in Afghanistan difficult?
3 In what ways would the provision of a reliable electricity supply in rural areas of Pakistan and India improve the quality of life there?
4 Explain the process of technological leapfrogging.

On your own

5 To what extent is leapfrogging helping to reduce the technology gap that exists between countries.
6 Mobile telephones and solar power are good examples of how technological leapfrogging has taken place. Use the Internet to research another form of technological leapfrogging. Produce a 300-word report detailing the benefits that it has brought to people in developing countries.

Exam question: Examine the importance of technological leapfrogging for developing countries. **(15)**

Impacts of technology

In this unit you'll investigate the impacts of technological innovation, in this case genetically modified (GM) foods.

A picnic with a point?

It was a hot August weekend, in 2007, in the town of Verdun-sur-Garonne in south-west France. The perfect weather for a picnic. But this particular picnic was happening in the middle of a field of GM crops. More specifically, it was in a field of maize (corn) from a variety called MON810 (developed by the American company Monsanto) – genetically modified to be immune from insect attack. The 'picnic' consisted of hundreds of protestors, who were campaigning to see an end to the growing of GM maize. The protesters claimed that they had a right to destroy crops which they believed threatened to cause ecological disaster.

Things very quickly turned ugly when the farmer who owned the field arrived – along with many members of the French farmers union, the FNSEA. The police were called to the scene to prevent the angry farmers from crossing a bridge to the field where the protestors were 'picnicking'. Tempers hit boiling point when the farmers tried to force their way across the bridge. The police finally had to resort to tear gas to push the farmers back and keep the two groups apart.

This clash of ideologies was a confrontation between two different approaches to agriculture. The FNSEA supports a scientific and highly productive approach to agriculture, while the GM protestors support a traditional, small-scale, GM-free approach.

GM foods: what is all the fuss about?

Genetically modified foods were first introduced in the USA in 1994. By 1999, 60% of all food on British supermarket shelves contained some form of GM ingredient. However, this had fallen to just two products by the end of 2007. How did such a change occur and why did public opinion turn against GM foods so rapidly? Particularly since, so far, no health hazards can be attributed directly to GM foods.

> Monsanto is an agricultural company. We apply innovation and technology to help farmers around the world to be successful, produce healthier foods, better animal feeds and more fibre – while also reducing agriculture's impact on our environment.
> *Adapted from a statement on the Monsanto website*

▲ *Protestors destroy GM maize in southern France*

> The current generation of genetically modified crops unnecessarily risks the health of the population and the environment. Present knowledge is not sufficient to safely and predictably modify the plant genome, and the risks of serious side effects far outweigh the benefits. We urge you to stop feeding the products of this infant science to our population and to ban the release of these crops into the environment, where they can never be recalled.
> *Adapted from a statement by the Organic Consumers Association, read out at the Joint International GM Opposition Day, 2006*

What are the benefits?

GM foods were developed in the laboratory as a technological response to tackle the most pressing issue facing humanity – that of feeding an ever-growing population. It is argued that genetic modifications have many benefits, for example disease-resistant seeds, higher-yielding crops, or herbicide-tolerant varieties, which enable a farmer to use weed-killer rather than have to hoe a field by hand. Scientists in the biotech industry argue that GM development could see the removal of allergens from peanuts, develop fat-free chips, and alter the composition of fruit and vegetables so that they can be used to tackle cancer or heart disease. Some scientists and economists believe that it is important not to delay or stop the use of GM crops and the benefits that they can bring.

... and the costs?

Both Greenpeace and the Organic Consumers Association have actively campaigned against GM crops, because they believe that the long-term effects of these crops have not been fully tested. They point out that it is impossible to know what effects these crops may have on the health of people eating them over a long period, and they also believe that their long-term environmental effects could be devastating. The fear is that, if GM crops escape into the wild and mix with non-GM crops, there is no knowing how this might affect plant and animal life. There have been some arguments that GM crops have not been fully scientifically tested before going into production (also see page 285).

Golden Rice – a saviour for millions?

Golden Rice was developed in 1999 at the Institute of Plant Sciences in Switzerland to contain added Vitamin A through the process of genetic modification. This was done by adding DNA from the daffodil plant and a form of soil bacteria to the DNA of the rice plant. The purpose of developing this rice was to tackle the issue of Vitamin A deficiency in children across the world. UNICEF estimates that the annual number of child deaths caused by Vitamin A deficiency is 1.5 million. It is also estimated that a further 500 000 are blinded each year because of a lack of Vitamin A.

Many children who are Vitamin A deficient live in parts of the world where rice is a major part of the diet, so the development of Golden Rice is seen as a simple and cheap way of providing extra Vitamin A. It is thought to be a more-reliable method of providing Vitamin A than through vitamin supplements or trying to promote the eating of green vegetables.

Field trials of Golden Rice began in Taiwan and the Philippines in 2004. They showed that the rice grew well in local conditions and contained a significant amount of Vitamin A.

... or just another Frankenfood?

Friends of the Earth point out that there are still concerns about Golden Rice, including that:

- it might take as long as 5 years before trial crops can be devised that will suit local climates in many countries
- Golden Rice does not contain enough Vitamin A – a child would have to eat 5 kg of rice per day to meet their recommended daily allowance (RDA)

▲ Rice fields like these in the Philippines could look the same in future, but the people living off them could be much healthier

- children should be encouraged to eat a balanced diet – the RDA of Vitamin A can be met with just two tablespoons of yellow sweet potatoes, half a cup of dark green leafy vegetables, or two-thirds of a medium-sized mango
- GM crops usually require herbicides to develop fully, which can often have a negative effect on other crops, poison rice paddy waters and kill fish

Over to you

1 Technology is not always 'neutral'. Give reasons why GM foods cannot be considered to be a 'neutral' technology.
2 Complete a table to show the positive and negative impacts of GM foods on the economy, society and the environment.

On your own

3 To what extent could you argue that the impacts of GM foods are 'unforeseeable'?
4 Consider the impacts of GM foods and research the debate using www.organicconsumers.org and www.monsanto.com. Write 300 words explaining whether you believe that GM foods will bring benefits to some of the poorest people in the world.

The effects of technology

In this unit you'll investigate how the effects of technology use are accounted for in some countries but not in others.

A world of cars

In 2007, there were nearly 600 million cars on the world's roads. That is nearly one for every ten people – and the prediction is that this figure will double by 2030. Nearly all of those cars are powered by the internal combustion engine, which releases carbon dioxide into the atmosphere. This car-caused CO_2 is a major contributor to global warming. The early pioneers of the car in the 1880s could never have foreseen the **externalities** (environmental costs) of their technological achievement.

Polluter Pays Principle

The 'Polluter Pays Principle' (PPP) is the idea that those who cause pollution should bear the economic cost of the damage they are doing to the environment. The PPP can be implemented through two different approaches:

● Command and control – where new technologies, enforced by law, are introduced to limit pollution By law in the USA, all cars built since 2004 must be fitted with a catalytic converter to reduce their emissions.

● Market-based – where governments introduce pollution taxes, carbon trading permits and product labelling. One such scheme is carbon offsetting which can be read about in the AS textbook on pages 52-55.

Many governments have sought to impose financial penalties on polluters. In the case of motor vehicles, the British government has introduced Vehicle Excise Duty (VED) bands, based on the amount of CO_2 that a vehicle emits (see the table). The aims of this are to encourage people to drive cars which produce less carbon dioxide, and also to invest the revenue gained from the tax in improving public transport.

The environment as a sink

For some burgeoning economies, such as China and India, the wealth of their citizens is just beginning to increase. In countries that have experienced high levels of growth, such as the 13 highlighted in *The Growth Report* on page 236, the aspiration to own a car is high. Rapid industrial growth, coupled with rising incomes, will lead to more cars on the roads in these countries. With the need for economic development seen as more important than environmental concerns, the CO_2 emissions of these extra cars is likely to continue unchecked and unabated – treating the atmosphere as a 'sink' to soak up their emissions.

▲ *25% of all carbon dioxide emissions in the UK come from car exhaust fumes*

> ● **Externalities** are external effects of a human process or activity, which can often be unforeseen or unintended.

VED band	CO_2 emissions (g/km)	2009/10 rate	2010/11 rate
A	Up to 100	0	0
B	101 – 110	£20	£20
C	111 – 120	£30	£35
D	121 – 130	£90	£95
E	131 – 140	£110	£115
F	141 – 150	£120	£125
G	151 – 160	£150	£155
H	161 – 170	£175	£180
I	171 – 180	£205	£210
J	181 – 200	£260	£270
K	201 – 225	£300	£310
L	226 – 255	£415	£430
M	over 255	£440	£455

▲ *UK Vehicle Excise Duty (VED) bands*

▼ *World CO_2 emissions by country, 1990–2030 (projected)*

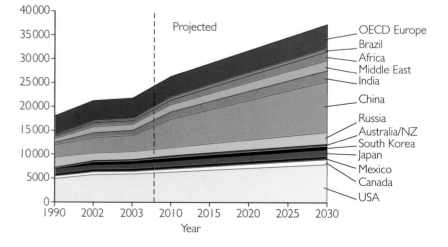

How Western Europe tackles cars' CO_2 emissions

London
In 2003, the London Congestion Charge was introduced (see page 53 of the AS textbook for details) and in 2008 the Low Emission Zone (LEZ). The LEZ charges high-polluting, usually older, lorries with diesel engines weighing over 12 tonnes £200 a day to enter the Greater London area. From July 2010, this will include buses and coaches weighing over 3.5 tonnes.

United Kingdom
Vehicle Excise Duty (VED) is charged yearly per vehicle, based on the amount of CO_2 released per kilometre.

The Netherlands
A new registration tax is charged when a car is sold to its first buyer. The tax is divided into seven categories, based on fuel consumption, CO_2 emissions and an efficiency indicator. Cars with low emissions are given a registration tax discount of €1000, while cars emitting high CO_2 can easily pay in excess of €5000.

Stockholm
After a trial period running throughout 2006, the Stockholm congestion tax has been in permanent operation since August 2007. The purpose of the system is to reduce congestion and improve the air quality in Stockholm. The charging zone covers the whole of Stockholm city centre and drivers must pay up to £5 to enter the city.

France
In January 2005, the Environment Ministry introduced a green road tax on new car purchases. Vehicles emitting over 180 g/km of CO_2 are charged up to €3500, while smaller, cleaner vehicles with emissions lower than 140 g/km receive rebates of up to €700.

Berlin, Cologne, Hanover
The introduction of 'Environmental Zones' means drivers must display a coloured sticker to enter these three city centres. The stickers cost a one-off charge of €10 and are compulsory for foreign drivers as well as locals. Fines of up to €40 can be charged for anyone not displaying a sticker. The Government plans to introduce the scheme to all German cities.

Portugal
10% of a car's registration tax is differentiated according to CO_2 emissions in four classes – with hybrid vehicles earning their owners a 40% reduction. The rates range from no tax (for vehicles with CO_2 emissions of less than 120 g/km) to nearly €5150 (for vehicles with emissions over 210 g/km).

Germany
In January 2005, Germany introduced an electronic heavy goods vehicle toll system, covering its entire motorway network. Lorries must carry a satellite-tracked on-board transmitter, and drivers are then charged about €0.15 per kilometre. Plans to introduce, in 2009, a car tax linked to CO_2 emissions are currently on hold, due to opposition in the German parliament.

Madrid
Plans are in place to ensure that only cars that emit low levels of CO_2 will be able to enter the city centre from 2010. Residents might be exempted from this measure. It could cut the amount of CO_2 released by cars in the city by a third.

Switzerland
In 2001, the Swiss Heavy Vehicle Fee was launched for all vehicles over 3.5 tonnes. The Government provides an on-board unit, which is fitted to all lorries, and they are then charged using a formula of their weight and the distance travelled. The aim of the scheme is to protect the delicate alpine environment and shift the burden of the cost of damage back to the freight-haulage companies.

Milan
The 'Eco Pass' system began in 2008 and charges drivers up to €10 to enter the inner area of the city. 8 km² is covered and is policed by cameras and gates at 43 fixed points. The scheme operates on weekdays, between 0730 and 1930. The aim of the charge is achieve a 30% cut in pollution and a 10% reduction in traffic.

Rome
The 4.6 km² historic city centre of Rome is now protected by a gate system. Drivers must fit an on-board unit to their car, which communicates with the gates – allowing them to open as the driver approaches. A smart card fitted to the on-board unit is used to pay the fee. Access is free to residents.

Montenegro
Summer 2008 saw the launch of a tax on all foreign cars, buses and trucks entering Montenegro, with the aim of reducing carbon emissions in the country. Cars and minibuses pay €10, while buses and lorries pay between €30 and €150. Montenegrins pay an annual car tax of €5 to use their cars, and the money raised is used to improve environmental protection.

The Indian car sales boom

The world's cheapest car

In January 2008, Tata, India's leading motor manufacturer, announced the launch of the 'people's car' – the Tata Nano. The Nano sells for 100 000 rupees (about £1250), and is the cheapest new car in the world. Tata initially produced 250 000 cars and expects demand to soar to one million vehicles a year. The car is marketed as a safer way of travelling for many Indians, who have – until now – had to transport their families balanced on the back of their motorbikes. Tata designed the car to target India's booming middle class, which is expected to grow from 50 million in 2008 to 583 million by 2025, because of the continued growth of India's economy.

What makes the Tata Nano so cheap?

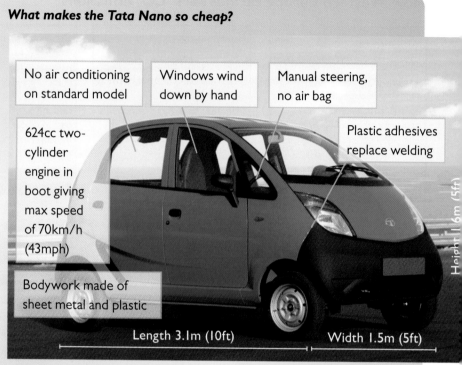

No air conditioning on standard model

Windows wind down by hand

Manual steering, no air bag

Plastic adhesives replace welding

624cc two-cylinder engine in boot giving max speed of 70km/h (43mph)

Bodywork made of sheet metal and plastic

Height 1.6m (5ft)

Length 3.1m (10ft)   Width 1.5m (5ft)

▲ The Tata Nano was the world's cheapest new car when it went on sale in India in 2008 – and this is why

Just over 1 million cars were sold in India in 2007 – which is projected to rise to 3.8 million by 2013. India has an ambitious programme to become the main global car producer by 2016. Much of this growth will be aimed at producing cars for the domestic market. As India's Gross Domestic Product (GDP) is expected to rise by 8% a year for the next decade, the number of drivers is set to increase from 7 per thousand to 11 per thousand. This will mean a doubling of Indian car owners since 2000 – when just 0.5% of Indians had a car. By 2017, it is projected that there will be more car owners in India than in Japan and, with sustained growth, Indian car ownership could top 140 per thousand by 2030.

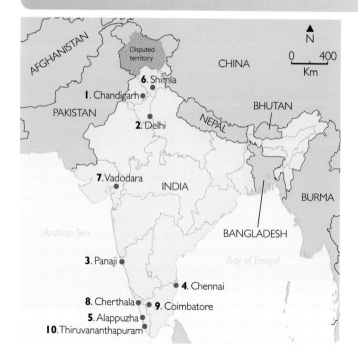

Rank	Town/city	Cars per 1000 people in 2004
1	Chandigarh	82
2	Delhi	54
3	Panaji	48
4	Chennai	43
5	Alappuzha	43
6	Shimla	40
7	Vadodara	40
8	Cherthala	37
9	Coimbatore	35
10	Thiruvananthapuram	32

◀ The ten Indian towns/cities with the highest levels of car ownership in 2004

Externalities of the Indian car boom

In 2004, India released 266 million tons of carbon into the atmosphere – making it the world's fourth largest CO_2 emitting country. Since 1990, India's CO_2 emissions have almost doubled, and are currently increasing by nearly 6% a year. However, with the world's second-largest population (1.15 billion people in July 2008), India's **per capita** emission rate in 2004 (of 0.34 metric tons of carbon) was well below the global average of 1.23.

Scientists are concerned, however, that India's CO_2 emissions will accelerate further as a result of the rapidly increasing car ownership there. The introduction of the Tata Nano has led environmental critics to raise concerns that the car does not meet current European emissions standards and will lead to a massive increase in the amount of CO_2 released into the atmosphere. As European countries introduce measures to reduce CO_2 emissions from vehicles, it appears that growth in India could negate any global benefit from the European schemes.

▲ *Traffic congestion in New Delhi in 2008*

▼ *India's per capita emissions of CO_2, 1950-2005*

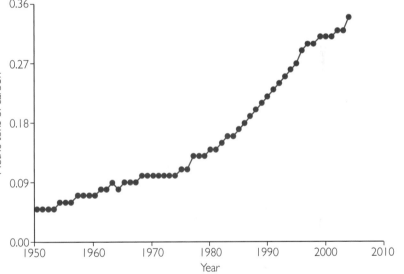

Should those who pollute the environment bear the financial cost of the damage that they are causing?

What do you think ?

Over to you

1 How much impact will the increase in Vehicle Excise Duty in the UK have on car ownership, or reducing CO_2 emissions?

2 Look at the map of Europe (page 245) and the schemes introduced to reduce CO_2 emissions.
   a To what extent, if any, will these European schemes have an effect on global CO_2 emissions?
   b Do these schemes really start to tackle the issue of global warming?
   c Some people argue that such schemes are used to raise money rather than tackle the environmental issues. To what extent do you think this is true?

3 If increased car ownership in China and India will lead to higher CO_2 emissions, is there any point in European governments trying to reduce their own CO_2 emissions?

On your own

4 Research an example of a city in Europe that is tackling CO_2 emissions (see the examples on page 245). Devise a presentation to give to the rest of the class, including the issues the city faces, its environmental standards, how it is trying to tackle these, and with what success.

5 Using a researched example of a city from Asia (such as Beijing, Kolkata, Jakarta, or Kuala Lumpur), identify the problems that unchecked car growth is having for the city.

6.9 Contrasts in technology

In this unit you'll investigate how a range of technologies are being introduced to tackle water shortages in the Tigray region of Ethiopia.

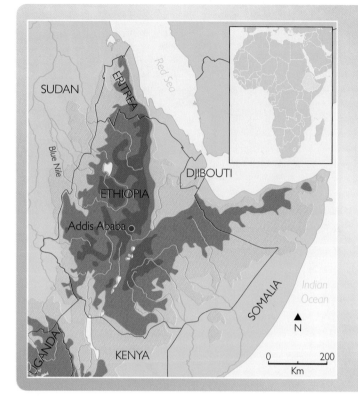

Ethiopia fact file	
Capital	Addis Ababa
GDP (US$)	19.43 billion (2007 estimate)
Population	78 254 090
Life expectancy (years)	49
Birth rate (per 1000)	36.8
Death rate (per 1000)	14.5
Infant mortality rate (per 1000)	90.2
Literacy rate (%)	42.7

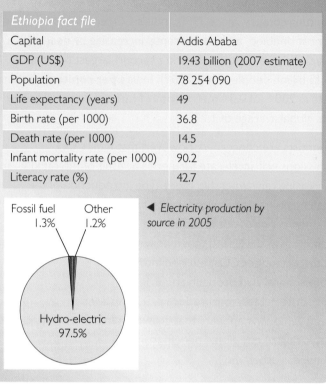

◀ Electricity production by source in 2005

Fossil fuel 1.3%
Other 1.2%
Hydro-electric 97.5%

Electricity and water – a dangerous mix?

In 2008, drought struck again in Ethiopia. One consequence for Addis Ababa was power rationing. Electricity was available for only five days a week at best – and in some parts of the capital was restricted to only six hours a day. While many rural Ethiopians were struggling to find drinking water, those living in the capital were concerned about the lack of bottled mineral water – a relatively new phenomenon in Addis Ababa.

The lack of bottled water was not due directly to a lack of available water arising from the drought, but to the lack of power supply. One of the consequences of the drought was that Ethiopia's hydro-electric dams were not able to operate at full capacity, which caused a reduction in available electricity supply. Bottled water companies require a lot of electricity and were forced to cut back on their production as a result. Highland Spring cut back from 700 000 bottles a week to 620 000, because its machines had to shut down twice a week. Similarly, Abyssinia, another major brand of bottled water, cut back its production by half, from 40 000 to 20 000 bottles a week, because of power cuts of up to 15 hours a day.

Ethiopia appeals for urgent aid

Ethiopia has launched an urgent appeal to international donors for more than $300 million (£154 million) of emergency aid. Because of the drought that struck most of the country in the early part of this year, a total of 4.6 million people are now thought to need food aid. In some parts of the country, health centres and feeding clinics are already being overwhelmed by large numbers of severely malnourished children.

Adapted from an article by Elizabeth Blunt, BBC News, 2008

Ethiopia: a burgeoning economy?

Ethiopia is considered to be one of the poorest countries in the world. Yet, since 2003, it has seen a growth in its Gross Domestic Product that has been equal to – and at times greater than – most other developing economies. According to the International Monetary Fund (IMF), Ethiopia's GDP growth for 2007 was an impressive 11.4%. It was much higher than the previous five-year average of 3.4% from 1997 to 2002, and the sub-Saharan average of 6.6%. The big question for Ethiopia now is whether the severe drought of 2008 and its associated problems can be overcome, and whether technology has the answers.

According to the World Bank, Ethiopia is poor because it does not use its enormous water potential properly. The World Bank has developed a Country Water Resources Assistance Strategy to help Ethiopia to develop a water storage capacity to provide year-round access to water. This has led to an array of projects from the high-tech Tekeze Dam to the low-tech 'harvest the rain' strategy.

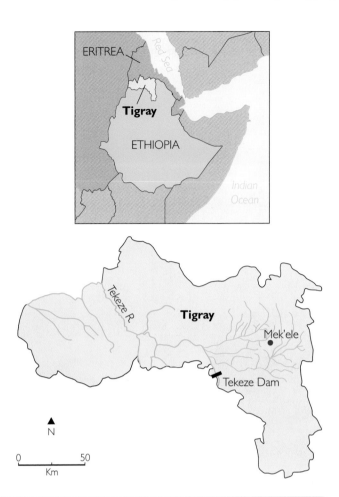

▶ *Tigray is the northernmost region of Ethiopia and borders Eritrea to the north. It has a population of just over 4 million and the provincial capital is Mek'ele. It is a successful agricultural region of Ethiopia, but regularly suffers from drought.*

Low-tech solution: harvesting the rain

Tadesse Desta is a farmer in the Tigray region of Ethiopia. Every year, in June, he hopes the rains will come early, so that he can plant his crops. Only a few weeks later, he is worrying that the harvest will fail through lack of water. In 2006, the Tigray Regional Government suggested to him the option of 'harvesting the rains'. Following advice, he dug a small pit and borrowed money to pay for a blue plastic sheet to line it. The Regional Government's advisors told him that the pit would hold 60 m³ of water and would help him to save his crops if the rains failed.

This initiative is part of a Government-backed 'rainwater harvesting' programme, which has set targets for local officials and left the countryside punctuated with blue pits of water. However, the project has attracted criticism, because the pits have become breeding grounds for mosquitoes and have left many farmers with debts that they cannot afford to repay. It does, however, help to improve the water supply situation for Ethiopia's rural poor.

▲ *Digging a water catchment pit to store run-off rainwater in Tigray*

Ethiopia's dam options

High-tech solution: the Tekeze Dam

▲ *The Tekeze Dam, in northern Ethiopia, will provide 300 megawatts of electricity when it is finished*

Ethiopia has abundant resources of water but is often affected by drought. Northern Ethiopia has a successful agricultural economy, specialising in coffee production, which contributes about 10% of Ethiopia's GDP and employs over 15 million people. Large-scale irrigation is needed to ensure its continued success.

The Tekeze hydro-electric project aims to construct the highest dam of its type in Africa. It is being built on the Tekeze River, which flows through the Tigray region of northern Ethiopia, Eritrea and Sudan – where it eventually joins the Nile (see the map on page 249). The Tekeze is over 600 km long and has created one of the deepest gorges in the world – reaching a depth of over 2000 metres. The dam will hold back 9 billion m³ of water that will be used to irrigate 60 000 hectares of land, as well as generate electricity.

The construction of the Tekeze Dam began in 2002. When completed, it will be the highest dam in Africa – at 185 metres tall (10 metres higher than the famous Three Gorges Dam in China). The cost of the project is estimated to be $224 million, and, when finished, it will be the tenth hydro-electric power plant in Ethiopia.

The dam is being built by the Chinese Water Resources and Hydropower Engineering Company (CWHEC), which was awarded the contract by the Ethiopian Electric Power Corporation (EEPCo). CWHEC was chosen to build the dam (beating bidders from Germany, Italy and Japan) because of its experience in working on the Three Gorges Dam. There are 500 Chinese expatriate workers at the site, working alongside 2000 Ethiopians.

Electricity pylons stretching 105 km have been built to connect the dam to the regional capital, Mek'ele (population 96 000). Once complete, the project will supply electricity to Ethiopia's national grid, as well as being sold to neighbouring countries to earn much needed foreign currency.

Intermediate technology: small dams

The Relief Society for Tigray believes that the priority for water development in Ethiopia should be for many thousands, even tens of thousands, of small- and medium-sized dams, rather than large-scale projects like the Tekeze Dam.

A small dam (around 15 metres high and 300 metres long) was constructed near the village of Adis Nifas, not far from Mek'ele, to meet more targeted needs than those of the huge Tekeze Dam. This small dam was built by the local people, with support from the Relief Society for Tigray, which provided machinery and money while the villagers supplied the labour. The dam is made from local materials. The reservoir is close to the village's fields and manages to retain water for most of the year, even during the dry season.

Each family in the village has been given a quarter of a hectare of irrigated land, as well as fruit tree seedlings and elephant grass to plant into the earth walls that divide up the fields. These help to stop the walls eroding away. The irrigated areas are lush with crops and have provided food security for the people of Adis Nifas.

▶ *Opening the irrigation valves on a small local dam*

6.10 The global fix

In this unit you'll investigate whether there are technological fixes to overcome global environmental issues, such as global warming.

▲ The launch of the Virgin Earth Challenge prize, 9 February 2007

Is technology the answer?

In February 2007, Sir Richard Branson and Al Gore (former US Vice-President) launched the Virgin Earth Challenge prize, which offers a $25 million reward for the best idea to remove at least 1 billion tons of carbon dioxide from the Earth's atmosphere each year. The prize has encouraged people across the world to focus their minds on tackling one of the greatest challenges facing humanity.

You have already studied strategies for dealing with climate change as part of your AS course, including possible solutions such as carbon offsetting (through the European Trading Scheme). As part of this chapter, you have also looked at the Polluter Pays Principle (PPP) on pages 244-245. Yet, technological development could be pivotal in tackling the issue of global warming.

Boserup's view

Esther Boserup (1910-1999) was a Danish economist who worked for the United Nations and who is most famous for her views about population and food supplies. She believed that, as population grows, the demand for extra food and resources will automatically encourage scientists to develop new solutions to the lack of supply. However, in 1981, she followed up her original thesis with 'Population and technological change', in which she argued that rapid increases in population are also accompanied by rapid technological change. Writing before global warming became a global environmental concern, Boserup's theory that technology can be used to overcome global issues serves as a good basis for underpinning the view that technology can provide a global fix.

Can technology fix global warming?

The stated aim of the Virgin Earth Challenge prize is 'to encourage a viable technology which will result in the net removal of anthropogenic, atmospheric greenhouse gases each year for at least ten years without countervailing harmful effects'.

Since the announcement of the Virgin Earth Challenge prize, a number of scientists have begun to suggest possible solutions to global warming (see the three examples opposite). Some of the proposed solutions seem far-fetched, but the reality is that ideas are moving forward quickly from the drawing board to prototype – backed by an assortment of universities, private companies and government organisations.

Technological solutions to global warming

Example 1: Giant sunshade for the Earth

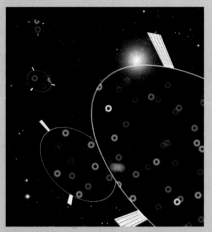

▲ *Professor Roger Angel, an astronomer at the University of Arizona, presented his idea for a 'solar sunshade' to the US National Academy of Sciences in April 2006 – and won funding from NASA research in July 2006. The 'sunshade' would be made up of 16 trillion metre-wide flat discs, launched into space via 20 million rockets to sit between the Earth and the Sun and reflect solar energy away from the Earth. The cost has been estimated at $4 trillion over 30 years (equivalent to double the UK's annual GDP).*

Example 2: Iron fertilisation of the oceans

A Californian company, called Planktos, began testing this idea in early 2007. Their ship, the *Weatherbird II*, dumped 50 tons of iron dust into the Pacific Ocean in order to fertilise the sea and encourage the growth of marine plants (phytoplankton).

The theory is that by pouring iron into the sea, large blooms of plankton could be encouraged to grow at a rapid rate – consuming and removing vast amounts of excess carbon dioxide from the atmosphere and taking it deep down into the ocean as organic matter when the plankton eventually dies. If done on a large enough scale, it is believed that this process could remove enough CO_2 from the atmosphere to reduce the impact of global warming. The estimated financial cost for such an endeavour is set at $100 billion, but it is the unknown cost to the oceans that has sparked the most debate amongst scientists.

Example 3: An artificial volcano

During the 1991 volcanic eruption of Mount Pinatubo in the Philippines, huge amounts of sulfur were released into the atmosphere. This enhanced the reflection of solar radiation into space and cooled the global climate by about 0.5 °C over the following year. Nobel laureate, Paul Crutzen, from the Max Planck Institute for Chemistry in Germany and the Scripps Institution of Oceanography, University of California, has used this event as the inspiration for his idea.

The idea is relatively simple – to artificially release sulfur particles into the atmosphere in order to reflect radiation from the Sun. This could be done using jet engines, balloons or large cannon. According to the US National Center for Atmospheric Research (NCAR), this method would require tens of thousands of tons of sulfur to be released into the atmosphere each month – at an estimated annual cost of $100 million. It is believed that the process could take as little as 5 years to reduce global temperatures to pre-1900 levels. However, once operative, the system could never be shut down without causing immediate and sustained global temperature rises.

▲ *The eruption of Mount Pinatubo in 1991 is the inspiration behind the idea to release sulfur into the atmosphere to reflect radiation from the sun and reduce the effects of global warming as a result.*

Carbon capture

Fixing the climate with artificial trees

Wallace S. Broecker is credited as one of the first scientists to alert the world to the idea of global warming – in 1975. His latest book, *Fixing Climate* (published in May 2008, and written with Robert Kunzig), focuses not only on the science of climate change but – more importantly – on the technological fix for global warming.

As you saw on the last page, there are already examples of technological solutions to global warming in development. However, Broecker has worked with his colleague Dr Klaus Lackner, a physicist at Columbia University, to develop the idea of a carbon capture artificial tree which can remove CO_2 from the atmosphere. Broecker believes that this solution is the technological answer to global warming.

Carbon capture technology

Early plans

The idea for artificial trees is based on the simple premise of finding a man-made way of replicating a natural process. In 2003, Dr Lackner envisaged a synthetic tree that would do the job of a real tree and remove CO_2 from the atmosphere. He estimated that a single artificial tree could remove 90 000 tonnes of CO_2 a year – the equivalent of 15 000 cars' worth of emissions. The carbon would be retained and would then have to be disposed of.

The technological background

There are two aspects to this technology – collecting the carbon and then storing it. It is possible to collect some CO_2 as it is emitted, e.g. from power stations. However, Dr Lackner realised that this is not possible for all CO_2 emissions, such as those from vehicles. A car cannot capture and store its CO_2, because it would require a huge storage tank. Therefore, he believes that the alternative option is to capture CO_2 from the atmosphere.

The carbon capture artificial tree – acting like a filter – would capture CO_2 as the wind blew. A coating of limewater would be used on slats to collect the CO_2. The limewater would then form limestone that would need removing regularly. The key issue with the system was what should be done with the carbon once it was collected.

▼ *The 2008 vision of what CO_2 scrubbers would look like in the landscape (see opposite)*

2008: The 'CO₂ scrubber' prototype is announced

In 2008, Global Research Technologies, in conjunction with Dr Lackner, began to build a prototype CO_2 scrubber – based on the designs for the carbon capture artificial tree. The prototype is being built in Arizona at a cost of $100 000, and is expected to take two years to build. A further technological breakthrough has been the development of the 'ion exchange membrane' to extract CO_2 from the air, instead of the limewater (see below). The CO_2 will then be pumped from the scrubber directly into greenhouses in order to intensify plant growth.

> I'd rather have a technology that allows us to use fossil fuels without destroying the planet, because people are going to use them anyway.
>
> **Dr Klaus Lackner, Columbia University**

Air and CO₂

Air containing dissolved CO_2 is pumped in

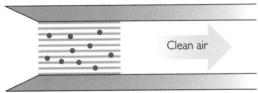

Clean air

The air passes over an ion exchange resin on a membrane. The CO_2 sticks to the resin and clean air is pumped out

Humid air

Humid air and CO₂

The ion exchange is washed with humid air, which cleans off the CO_2. The CO_2 can now be extracted and either buried or used in other ways, such as in greenhouses

> *Can a technological fix really be found for global warming?*
>
> **What do you think?**

Over to you

1 In groups, produce a table to assess the costs and benefits of each of the technological solutions to global warming put forward in this unit.

2 In pairs, reach a rank order of which solution is most likely to be developed as a technological fix for global warming.

3 In groups, debate the view put forward by Klaus Lackner that it is better to find solutions to take CO_2 from the atmosphere than to try and stop people using fossil fuels.

On your own

4 Research **(a)** the problems **(b)** the possible solutions of taking CO_2 out of the atmosphere and storing it. Start by using the BBC science website: http://www.bbc.co.uk/sn

5 In 600 words, assess whether technology to remove CO_2 from the atmosphere is the technological fix to the problem of global warming?

In this synoptic unit you'll investigate the future opportunities for energy security and independence in Slovakia, which draw together this chapter.

Energy security in Eastern Europe

The Union of Soviet Socialist Republics (USSR) came to an end in 1991. With its demise, many Eastern European countries – which had previously been linked to the USSR – gained independence and freedom from Soviet political control. However, many were still economically tied to Russia via their infrastructures. Some of these countries, such as Ukraine and Slovakia, contained nuclear power stations built during the Cold War era to supply energy to Russia – with Russia supplying oil, natural gas and other resources to them in return.

▲ The Mochovce nuclear power station in Slovakia

Resource 1

Oil dependency on Russia

The table shows the top ten countries relying on Russia to supply part of their oil requirements in 2007. Five of the countries were formally under the control of the Soviet Union: Hungary, Slovakia, Poland, the Czech Republic and (East) Germany.

The *Druzhba* (meaning 'friendship') pipeline supplies countries in Eastern Europe with oil from Russia. The 2500-mile pipeline was originally built during the Cold War to supply oil to Eastern European allies of the former Soviet Union. Today it supplies 1.2 million barrels a day to European countries which include Germany, Poland and Slovakia. In January 2007, Russia temporarily shut down the pipeline for two days, because of a disagreement with Belarus. For Slovakians, this highlighted their dependency on imported oil, and raised fears that the pipeline could be shut down at any time without warning for reasons beyond their control.

2007 rank	Country	Russian oil as a percentage of all oil used
1	Hungary	83.5
2	Slovakia	82.2
3	Finland	79.1
4	Poland	77.2
5	Czech Republic	49.3
6	Belgium	31.8
7	Sweden	29.4
8	Germany	26.2
9	Netherlands	25.3
10	Italy	18.1

▲ Top 10 users of Russian oil

Putin stands firm on oil blockade

Oil refineries in Germany, Poland, Hungary, Slovakia and the Czech Republic, all of which rely on the *Druzhba* oil pipeline, have reported that their crude oil processing has been affected. They have been forced to start drawing on operational stocks.

Adapted from *The Times*, 10 January 2007

Resource 2

Slovak Republic fact file

The Slovak Republic (usually referred to as Slovakia) became an independent state on 1 January 1993, when it split away from the Czech Republic – breaking up the former Czechoslovakia. Slovakia became a member of NATO in March 2004 and joined the European Union in May 2004.

Capital	Bratislava
GDP (US$)	74.99 billion (2007 est.)
Population	5 455 407 (2008 est.)
Life expectancy (years)	75
Birth rate (per 1000)	10.6
Death rate (per 1000)	9.5
Infant mortality rate (per 1000)	7
Literacy rate (%)	99.6

Resource 3

Slovakian energy production

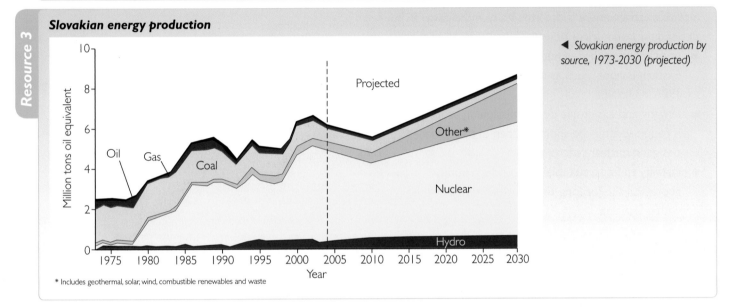

◀ Slovakian energy production by source, 1973-2030 (projected)

* Includes geothermal, solar, wind, combustible renewables and waste

Resource 4

Slovakian energy policy review, 2005

In 2005, the Slovakian Government carried out a detailed energy policy review. Some of the key recommendations were:

To ensure energy security

- Consider ways to diversify oil, gas and nuclear fuel supply.
- Develop a wide range of supplies for crude oil.
- Adopt a wider range of energy supply sources.

To tackle energy pollution

- Continue to reduce the level of emissions of local pollution and enhance the monitoring system of local pollution.
- Develop an energy research and development strategy to target those technologies that can help the country to achieve its specific energy goals – in particular, improving energy efficiency and reducing CO_2 and pollutant emissions.

To reduce dependence on nuclear power

- Prepare the shutdown and decommissioning of the two nuclear power stations at Bohunice, in line with previous commitments.

To increase the use of renewable energy

- Ensure that renewable energy is used as part of a wide range of energy sources, and that it is developed to be cost effective.
- Consider temporary tax, regulatory and financial incentives, in particular for market and project studies and renewable energy investment projects.
- Consider the introduction of legislation to ensure that energy companies must use renewable energy sources as part of their supply.

Adapted from the Slovak Republic energy policy review, 2005

Technology and the future - 2

Is nuclear power the answer?

The former Czechoslovakia pursued a large-scale nuclear energy programme throughout the 1970s and 1980s, using Soviet-designed nuclear reactors. The first nuclear power station was located at Bohunice (about 70 km from Bratislava), and was commissioned in 1972. A number of accidents saw one of the reactors close by 1977 and it has been undergoing decommissioning since 1995. As part of the 2005 energy policy review, the Slovakian government is determined to move away from dependency on nuclear power. This is because of the:

- poor safety record of nuclear power stations in Eastern Europe, such as the Chernobyl disaster of 1986 in Ukraine
- lengthy and expensive decommissioning process that is required after the short life span of the Soviet-designed nuclear reactors
- inability to find a suitable deep underground, geological site for long-term nuclear waste disposal within Slovakia.

2006 rank	Country	Nuclear electricity as a percentage of total electricity	Continent
1	Lithuania	78	Europe
2	France	77	Europe
3	Belgium	58	Europe
4	Slovakia	53	Europe
5	Ukraine	46	Europe
6	Sweden	44	Europe
7	Bulgaria	42	Europe
8	Hungary	39	Europe
9	Slovenia	39	Europe
10	South Korea	39	Asia

▲ Top 10 countries with most reliance on nuclear power

	Capacity (MW)	Status	Commercial date	Planned shutdown	Planned end of decommissioning
Bohunice A1	(144)	Decommissioning	1972	1979	2037
Bohunice V-1	440	Operational	1980	2006	2025
Bohunice V-1	440	Operational	1981	2008	2025
Bohunice V-2	440	Operational	1985	2025	2043
Bohunice V-2	440	Operational	1985	2025	2043
Mochovce, Unit 1	440	Operational	1998	2038	…
Mochovce, Unit 2	440	Operational	2000	2040	…
Mochovce, Unit 3	(440)	Construction halted	…	…	…
Mochovce, Unit 4	(440)	Construction halted	…	…	…

▲ The status of nuclear power reactors in Slovakia, April 2005

Solar and biomass wait for State support

European Commission rules bind Slovakia to increasing the share of the energy it produces from renewable sources to 14%. Currently, the actual figure stands at around 6-7%. Solar energy and energy from biomass – which the Slovakian economy minister has identified as the country's most promising renewable resource – could contribute to reaching this goal.

Adapted from The Slovak Spectator, 24 March 2008

Renewable energy projects in Slovakia

The current contribution of renewable energy to the energy supply in Slovakia is marginal at 3.4% (except for large hydro-electric power generation). Other renewable energy sources, such as geothermal, wind, solar, and biomass, also have valuable potential that is yet to be fully developed in Slovakia.

▶ *Renewable energy potential in Slovakia, 2003*

	Technical usable potential	Current utilisation		Potential not used	
	GWh	GWh	%	GWh	%
Geothermal	6300	340	5.4	5960	94.6
Wind	605	0	0	605	100
Solar	5200	7	0.1	5193	99.9
Small water	1034	202	19.5	832	80.5
Biomass	16794	3192	19.0	13602	81
Forest biomass	2828	494	17.4	2334	82.6
of which fast grown energy forests	343	103	30	240	70
Forest industry	4881	2638	54	2243	46
Agriculture (excl. biofuels)	6586	60	0.9	6526	99.1
Energy utilisation of waste	3535	1251	35.4	2284	64.6
Sludges from sewage tanks	230	13	5.6	217	94.4
Municipal waste	1775	368	20.7	1407	79.3
Other waste	1530	870	56.9	660	43.1
Biofuels	2500	330	13.2	2170	86.8
Total (excl. big water)	**31288**	**5322**	**17**	**25966**	**83**
Big water	6607	5093	77.1	1514	22.9
Total (incl. big water)	**37895**	**10415**	**27.5**	**27480**	**72.5**

Wind farm project heads to eastern Slovakia

The wind farm will comprise 23 wind turbines, with an output of between 46 and 63 megawatts (MW), depending on the type of turbines, and will generate between 96 and 114 gigawatt hours of electricity annually.

Ventureal will begin construction in March 2009, with a goal of completing the farm by September 2009 and putting the turbines into operation in November 2009. It estimates the lifespan of the wind farm to be 25 years.

Ventureal, which was established in the Czech Republic in 2001 and later joined Austrian power concern Bawag, is involved in planning, building and operating facilities that generate energy from renewable sources – with a focus on wind energy. It operates eleven wind power facilities in Austria and the Czech Republic.

Adapted from The Slovak Spectator, 7 April 2008

Notes:
1 Gigawatt (GW) = 1 billion watts of energy. A gigawatt-hour (GWh) is the amount of energy equivalent to a steady power of 1 gigawatt running for 1 hour.
Big water = large-scale hydro-electric power schemes
Small water = small-scale, often localised hydro-electric power schemes

Synoptic question

a Explain the factors that led to Slovakia's energy supply being dependent on imported oil and technology. **(12)**

b Assess the sustainability of Slovakia's current energy sources. **(16)**

c To what extent do renewable energy resources provide a technological fix that will provide energy security and independence to Slovakia. **(12)**

Chapter summary

What key words do I have to know?

There is no set list of words in the Specification that you must know. However, examiners will use some or all of the following words in the examinations, and would expect you to know them and use them in your answers. These words and phrases are explained either in the glossary on pages 330-333, or within this chapter.

antiretroviral drugs
biomanufacturing
carbon capture technology
Digital Access Index (DAI)
digital blackout
DNA
environmental determinism
externalities

genetic modification
high-technology
information and communications technology (ICT)
intermediate technology
nanoparticle
pandemic
pasteurisation
patent

pervasive (in respect of technology)
Polluter Pays Principle
poverty line
technological leapfrogging
technology
technology-poor
technology-rich

Plus
Examples of technology – e.g. solar technology, mobile phone technology, biomedical technology, genetic modification, iron fertilisation, CO_2 scrubber technology

Films, books and music on this theme

Music to listen to

'The tide is turning (after Live Aid)' (1987) by Roger Waters

iTunes has several downloadable podcasts worth listening to on contemporary issues dealt with in this chapter, such as HIV/AIDS and GM technology. Key in phrases such as 'Vital Voices' for HIV/AIDS.

Books to read

Fixing Climate (2008) by Wallace Broecker and Robert Kunzig

There is no such thing as a natural disaster (2006) by Chester Hartman and Gregory D Squires (about the New Orleans flood and the extent to which this was a human as well as a natural disaster – and the reliance upon technology as well as the failure of social and political systems to cope.

Films to see

'Flood' (2007) – a film about the role of technology in saving London!

A.I. (2001) – Artificial Intelligence

Try these exam questions

1 The development of technology is a possible response to future resource shortages. Assess the possible costs and benefits of this approach. **(15)**

2 Assess the view that economic development is not possible without appropriate technologies. **(15)**

Introducing Unit 4

This unit is very different from others in the AS/A2 course.

- It is a **research** module. It is as much about developing *your own abilities* to research and present findings, as it is about geographical content.
- It offers **Options** for study, so that you and your teachers can choose according to your interests.

How much choice you have will depend on your school or college – which in turn will depend on what resources are available, and how far your teachers' interests match the Options.

What are the choices?

Six research Options are available in Unit 4. You have to select and study **ONE** of the following:

- Option 1: Tectonic activity and hazards
- Option 2: Cold environments – landscapes and change
- Option 3: Life on the margins – the food supply problem
- Option 4: The world of cultural diversity
- Option 5: Pollution and human health at risk
- Option 6: Consuming the rural landscape – leisure and tourism

For some of you, the chance to explore in more depth a theme that you have studied before will be attractive, e.g. 'Tectonic activity and hazards'. Alternatively, 'Pollution and human health' may be something that you have never been taught, and you might like to study that as a new topic. The Option choices range from themes with a strong physical geography focus, to those that are focused on environmental, social and cultural geographies.

Glacial landscapes – how much influence have these had on present landscapes? One of the questions explored in 'Cold environments – landscapes and change'. ▼

What will lessons be like?

Because this is a research unit, lessons may well be different in style – with more research done by you than content taught by teachers. Time might well be spent in the computer room or library, as well as in class. The Options are designed to provide opportunities to study from a range of geographical sources – books, journals, video, Internet reports, or fieldwork. Your teachers do not have to be expert in helping you with the content, but they will offer you assistance with researching and presenting the information in different ways.

Many Option topics will be unfamiliar to you. Therefore, consider all the choices before selecting, and perhaps read about two or three of the choices here before making a decision.

▲ *Volcanoes differ in their shape and form. One of the issues explored in 'Tectonic activity and hazards'.*

The future – GM or organic? One of the issues explored in 'Life on the margins – the food supply problem'. ▼

Unit 4: Geographical research

How is Unit 4 assessed?

The exam is different from others in the course. It is worth 40% of the A2 course. It is a 90-minute examination, where you will have to write a single essay. In the exam, you will be given one question on each of the six Options. You should select and answer the question about your own Option – there is no choice. You will be asked to write **one long essay**, in which you can demonstrate the results of your research.

Examiners are looking for depth of study and your ability to structure material and present your findings in the form of an essay. Your essay will be marked out of 70, and then converted to a Uniform Mark System (UMS) mark out of 80.

To prepare for this exam, you will receive advance notice from Edexcel **four weeks** before the exam takes place, stating which part of your Option will be assessed. For example, the Option 'Pollution and human health at risk' consists of four Themes:

1 Defining the risks to human health
2 The complex causes of health risk
3 Pollution and health risk
4 Managing the health risk

Two of these Themes will be selected by examiners to be assessed in the exam – and this is what you will be notified about. Pages 319-329 explain further.

(**Remember:** you will **NOT** be able to take any materials into the exam room.)

How to use this part of the book

In this part of the book, Chapters 7-12, you will be able to read something about each of the six Options. In each case, the purpose is to give you a **'flavour'** or **'taster'** of the Option – it does **NOT** provide enough to cover the requirements of the whole Unit!

Each 'taster' chapter has eight pages, which are divided into:

● an introduction – to tell you what to expect
● six pages of resources with guidance and some references to help you explore further
● a summary – with guided activities to help you research and essay questions of the kind you might meet in the exam

▲ In 1984, several thousand people were killed, and many others were blinded, when toxic gases escaped from a chemical plant at Bhopal in India. One of the issues explored in 'Pollution and human health at risk'.

▲ Great food or symbol of globalisation? One of the issues explored in 'The world of cultural diversity'.

Tourism – beneficial or problematic? One of the issues explored in 'Consuming the rural landscape – leisure and tourism'. ▶

7 Tectonic activity and hazards

Introducing this Option

The movement of the Earth's tectonic plates can be hazardous for human activity. Volcanic eruptions, earthquakes and tsunami often grab the news headlines when many lives are lost. The short-, medium- and long-term impacts of these tectonic events vary in relation to the intensity and frequency of the event, and the nature of the location affected. Levels of economic development, methods of prediction and preparation, and population densities, often determine the severity of hazardous events. How people cope and recover depends on the scale and nature of the event.

- Some events are dramatic. The 2004 Boxing Day tsunami, which was caused by an earthquake with a magnitude of 9 to 9.3 under the Indian Ocean (near the coast of Sumatra) killed over 150 000 people. Although the earthquake was localized, the effects of the surge of water were felt over a vast area and reconstruction will take years.
- Some events are small scale and go unnoticed by the rest of the world. Minor earth tremors and volcanic eruptions occur throughout the year in Iceland. They usually only cause short-term disturbance and help to generate valuable tourist revenues as people deliberately visit the country to watch the geysers, while Icelanders enjoy cheap central heating and year-round salad crops and fruit grown in glass houses heated by the geothermal activity.

What is this Option about?

This Option will probably appeal to you if you like studying physical geography, geology, biology, environmental studies, and ecology, or have an interest in managing the impacts of natural hazards.

The Specification lists four Themes in this Option. If you choose it, you will learn about:

1 Tectonic hazards and causes: What are tectonic hazards and what causes them?

In which you will learn about:
- tectonic hazards and disasters and what makes them hazardous
- the frequency, magnitude, duration and areal extent of hazards
- the causes of tectonic hazards
- tectonic activity associated with different types of plate boundaries

2 Tectonic hazard physical impacts: What impact does tectonic activity have on landscapes and why does this impact vary?

In which you will learn about:
- the varying impacts of extrusive and intrusive igneous activity
- the formation and shape of different volcanoes and their eruptions
- the effects that earthquakes can have on landscapes

3 Tectonic hazard human impacts: What impacts do tectonic hazards have on people and how do these impacts vary?

In which you will learn about:
- why people live in tectonically active areas and their levels of economic development
- the range of hazards associated with tectonic activity
- specific impacts of tectonic activity in countries at different levels of development
- trends in frequency and impact over time

4 Response to tectonic hazards: How do people cope with tectonic hazards and what are the issues for the future?

In which you will learn about:
- the varying approaches to coping with tectonic hazards in countries at different stages of development
- specific strategies involved in adjustment and the range of approaches and strategies used in locations at different stages of development
- the effectiveness of different approaches and methods of coping and how approaches have changed over time

How to use this chapter

This chapter contains six pages of resources.
- Page 264 introduces the topic with a case study of Montserrat.
- Pages 265-266 investigate the issue further using resources. You will be guided but you will need to think about each resource carefully.
- Page 267 provides background information about tectonic hazards and risks.
- Pages 268-269 investigate the issue in other areas of the world, again using resources.
- Page 270 provides activities, useful websites for further research, and examples of the kinds of questions that you will meet in the exam (each worth 70 marks).

Montserrat – a taste of paradise?

Years of violent tectonic activity have undermined Montserrat's claim to be an island paradise. As part of an island arc of volcanoes in the eastern Caribbean, Montserrat always faced the risk of destruction, and, following two years of small eruptions between 1995 and 1997, the Chances Peak volcano fully erupted in June 1997. Massive **pyroclastic flows** – molten lava, ash and mud at temperatures in excess of 500 °C – cascaded down the sides of the Soufrière Hills, engulfing everything.

Advanced warnings about this devastation had been coming for two years, with an 'exclusion zone' being established across the southern part of the island. In August 1995, huge pressures built up within the volcanic cone, as molten rock heated the groundwater – turning it into steam and causing what is known as a **phreatic eruption**. Plymouth, Montserrat's capital, was smothered in dense clouds of ash and all shipping and flights were diverted away from the island. Two years of 'gentle' eruptions followed and most people either left Montserrat completely or moved to the north of the island, out of harm's way. When the major pyroclastic event occurred on 25 June 1997 (see the photo), around 6 million m³ of material was dumped onto the island in just a few minutes.

Although Plymouth was destroyed, advance preparations and warnings by the Montserrat Government saved many lives. Just 19 people, who stubbornly continued to farm their land in the evacuated zone, were killed. Eruptions continued for months and the volcanic dome eventually caved in, sending fast-flowing walls of mud down the mountainside covering much of the south of the island.

Aftermath

By 2003, Montserrat's population had fallen from 12 000 to 4000, due to emigration and evacuation – with most remaining people living in the north, where the risks were lower. This dislocation of the population caused problems for Montserrat's fragile economy – tourism was suspended, farms and commercial assets were destroyed, and investment dried up. As a **British Overseas Territory**, under British sovereignty, Montserrat received millions of pounds for aid and reconstruction projects – but not enough to bring back the former island paradise. The Government's four-year Sustainable Development Plan was launched in 2003, with the aim of restoring confidence in the island and rebuilding the **infrastructure** to support future growth.

▼ Montserrat is part of the Lesser Antilles island arc in the eastern Caribbean, which has formed above a subduction zone where two tectonic plates converge

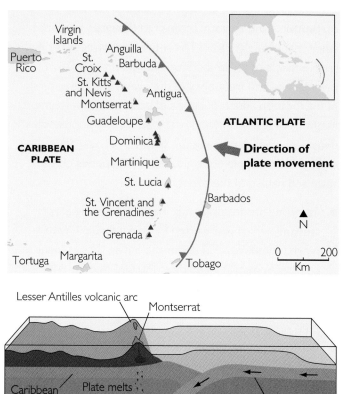

So ... how safe is Montserrat now?

Consider these resources:

- Resources 1-2 are about Montserrat's revival.
- Resource 3 is about funding the recovery.
- Resources 4-5 consider how the risks of living on Montserrat are being managed.

Resource 1

Turning adversity into an asset?

In 1997, Montserrat's tourism industry died out but, in 2005, visitors began to return. Today, Montserrat's 13-square-mile 'safe area' – beyond the threatening grip of the volcano – is on the rebound. With British support, the $18.5 million Gerald's Airport opened in February 2005. Plans for a new capital to replace Plymouth, and a nine-hole golf course, are in the works – and construction continues to rebuild the tourism infrastructure.

Resident volcanologists lead hour-long tours of the Montserrat Volcano Observatory twice a week, and a larger interpretive centre with photos, videos and models will open soon. Escorted tours of the abandoned streets of Plymouth, billed as a present-day Pompeii (see the photo), are also available (when the volcano is considered safe enough).

The remnants of buildings not burnt down by the hot ash can be seen on a boat trip which sails round the coastline to Plymouth. This is now the only way that you can get to

▲ Plymouth – Montserrat's old capital – now abandoned and covered in ash, with the volcano looming in the background

the remains of the former capital. The buildings sit in the middle of the acrid brown ash fall which surrounds and clings to everything in sight. Frozen details still remain: upstairs windows are still intact, road signs point the way to nowhere.

Montserrat's past is painfully locked in time, but life goes on. The Montserrat Tourist Board puts an optimistic spin on the calamity by describing the island as '39 square miles – and still growing', on account of new land being formed by pyroclastic flows spilling into the sea.

Adapted from the New York Times, October 2005

Resource 2

Perception and reality

There are common misconceptions that Montserrat is uninhabited, that tourism has not returned, and that there is nothing left of the island since the eruption of the Soufrière Hills volcano. Nothing could be further from the truth – new housing developments, new banks and supermarkets, roads, a cultural centre and luxury villas, and ash manufacturing are all expanding ... along with tourism ...

An advert produced by the Montserrat Tourist Board ▶

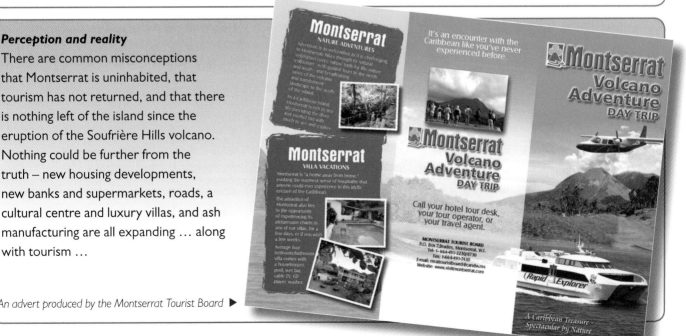

Managing the future

Laying the foundations for recovery

Musicians who had previously recorded music at George Martin's AIR Studios in Montserrat, staged a concert in 1997 to raise £500 000 for short-term aid and future reconstruction. The British Government has contributed £160 million in assistance to Montserrat since the start of the crisis in 1995, with a further £55 million pledged from 2008 onwards (amounting to nearly £50 000 for each of the island's remaining residents). Basic items, such as electricity and water, have had to be restored, plus:

- transport improvements – with a new harbour at Little Bay, a new airport, and replacement road links
- a new cliff-top housing estate in the north, called Lookout, where 25% of the country's population will be housed
- a strategic fuel depot, with enough petrol, diesel and cooking gas to last for two months, should the island's supplies be cut off by disaster

Adapted from the UK Overseas Territories Association News, 2008

Uncertain future

The recovery of Montserrat is now threatened, because insurance companies who are not willing to insure new developments are pulling out of the island. Without insurance in place, banks will not lend money for mortgage loans, which blocks new private construction. The insurance companies say they cannot afford to provide coverage in such a hazardous location.

However, some new private developments have been started in Montserrat. The Woodlands Hotel, a joint venture with Landbase International Ltd., was the first substantial private project to be built since the volcano erupted. The Montserrat New Town Trust has also been formed, which plans to build a new town in northern Montserrat.

Adapted from Rising from the Ashes, David Hanington, 2005

> Is it worth funding the future for just 4000 people on Montserrat?

What do you think ?

Living with danger – being prepared

The Montserrat Volcano Observatory (MVO) was set up in 1996 to monitor the volcano's activity. The MVO scientists notify the civilian authorities about every move that the volcano makes. They are at the cutting edge of disaster preparedness.

Techniques used include:

- a series of remote-controlled **seismic** stations, powered by solar panels and car batteries, dotted around the island to 'broadcast' seismic activity back to the observatory – minute by minute.
- daily reports and advice which are passed on to the local radio station, ZJB.
- ongoing ground deformation studies – plotting the profile of the mountain and the growth of the dome itself with global positioning systems (GPS), using satellites which can detect any changes in the volcano's shape within a few centimetres.
- gas and environmental monitors ('Cospec' – correlation spectrometer) to fly beneath the smoky 'plume' of the volcano, measuring sulfur dioxide emissions.

Extracts from Paradise Postponed *by Richard Herd (Scientific American) and the MVO website, 2008*

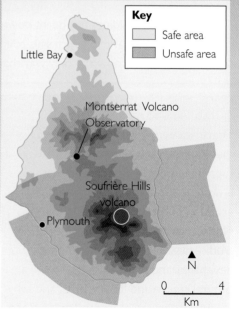

Key
- ☐ Safe area
- ☐ Unsafe area

Little Bay

Montserrat Volcano Observatory

Soufrière Hills volcano

Plymouth

N

0 — 4 Km

▲ A risk map for Montserrat in September 2007 (including offshore areas at risk from pyroclastic flows) – based on data issued by the MVO. The MVO regularly updates its risk assessments as a result of its scientific monitoring.

Background

Living with risks

A potential disaster in Montserrat was avoided because scientists and the Government there recognised the risks of a catastrophic eruption. Two-thirds of the island was declared an exclusion zone before the 1997 eruption, and lessons were learned which could help to protect people living in similar vulnerable areas.

However, the people of Montserrat still face an uncertain future, and the island's economy has still not recovered. For many of the evacuated residents, most of whom now live in the UK and the USA, avoidance appears to be the best form of defence. Those who decided to remain on the island, or who came back, still face health issues despite the volcano's present dormancy, because quartz in the volcanic ash causes the lung disease silicosis. As the risk map opposite shows, large areas of the island remain vulnerable because the risk of an eruption is always there. However, it is unlikely that any eruption will turn into a human disaster because of the observations and warnings provided by the MVO. Information about the behaviour of the volcano has enabled advances in prediction techniques and – for an island the size of Montserrat – warnings of just a few hours will save most lives.

Hazards, risks, and disasters

Tectonic activity is defined as hazardous when it has the potential to cause the loss of life or property. People living in naturally volatile areas are therefore exposed to increased risks and are considered vulnerable. When vulnerability and hazard coincide, there is a high risk of disaster – as the Venn diagram shows.

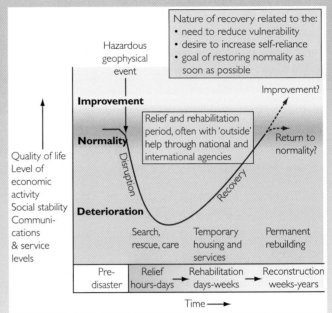

▲ The impacts of hazards and disasters on people. Sudden impact hazards (like earthquakes) and creeping hazards (like the Montserrat volcano) may have similar long-term effects but significantly different immediate consequences

Moving into danger

Despite advanced technology, the ability to accurately predict the timing and scale of any tectonic incident remains elusive. As global population grows, more and more people are now living in danger zones – urbanisation and economic development are forcing more people to live close to volcanoes or in earthquake-prone areas. It is not that the events themselves are becoming more frequent, just that more people are now vulnerable and are suffering the consequences. Approximately 20% of the world's population now live within volcanic zones.

In developing countries, like Indonesia, population growth, competition for land and increasing migration are moving people into previously uninhabited volcanic regions. The threat from southern Italy's active Mount Vesuvius casts an ominous shadow over 4 million people living in the Neapolitan Riviera. A Montserrat-style evacuation of this area is simply not viable. The costs of evacuation anywhere are enormous, and generally carry further risks of inadequate housing, food supplies, overcrowding, health and civil order problems.

▲ Vulnerability to natural hazards – turning a hazard into a disaster

Earthquakes in China and Japan

The next two pages look at the impacts of earthquakes in areas with differing population density.

- Resources 7-9 are about the Chinese earthquake in 2008.
- Resources 10-11 look at the effects of earthquakes in Japan.

Earthquake in Sichuan Province, China, 12 May 2008

Towns have been cut off by smashed roads and landslides in the mountainous earthquake zone. Thousands may lie buried by debris in Wenchuan County, where 110 000 residents have failed to make contact with the outside world. Few buildings in this remote region have been built to withstand earthquakes. One in ten in Beichuan County, with a population of 161 000, may have been killed or injured.

The magnitude 7.9 earthquake struck at 2.28 pm. Tremors rocked buildings in Beijing, sent ripples of fear through Bangkok, Thailand, and caused near panic in Hanoi, Vietnam. Some skyscrapers in China's financial hub, Shanghai, swayed violently and had to be evacuated.

In Juyuan, just south of the epicenter near Dujiangyan, a school has collapsed and buried up to 1000 staff and students. Houses and other buildings, including schools, hospitals, power stations and chemical plants, have collapsed across the region. Many dams have been weakened and are threatening to burst, adding flooding to the existing problems faced by rescuers. Deaths have also been reported in three neighbouring provinces and in the mega-city of Chongqing.

▲ *The Sichuan earthquake*

Adapted from reports in The Times *and* The Independent *on 13 May 2008. The final death toll from the Sichuan earthquake is uncertain but is thought to be around 70 000.*

The earthquake hit Beichuan City and set off a series of landslides, which dammed river valleys – causing floods

Olympic pressure helps to foster a caring response to the earthquake

After previous natural disasters, China's rulers tended to cover up and disguise the extent of the problems caused, anxious that high casualties could be perceived as signs of failure. However, their response to the Sichuan earthquake is proving to be very different. There is little sign of a cover-up, as there was during the SARS outbreak in 2003 – when secrecy triggered rumour and panic. The leadership, aware that its people have access to increasing amounts of information on the Internet, is becoming less defensive.

Chinese State television has interrupted normal programming to run live updates of the rescue efforts. The evening soap operas have been replaced by interviews with residents and survivors. On the Internet, official news agencies have issued reports to provide the latest death tolls. Details of rescue operations, of missing children, and damaged hospitals have not been concealed.

China's leadership knows that, with the Beijing Olympics drawing near, it cannot afford to blow its international reputation. They do not want China to come in for criticism for slow or secretive handling of this disaster – criticism which has been levelled at Burma in the aftermath of Cyclone Nargis, which affected 1.5 million people in the Irrawaddy Delta at the beginning of May.

Adapted from The Times, *May 2008*

Surprise earthquake shows Japan's vulnerability

Japan, which lies at the junction of four tectonic plates, experiences 20% of the world's powerful earthquakes and is constantly striving to protect itself against major tremors. By monitoring seismic waves, the country has started a world-first early warning system for earthquakes. This system kicked in on Saturday and provided a short warning.

Saturday's powerful earthquake in Miyagi was triggered by the build-up of pressure where the Pacific Plate meets the Japanese archipelago. The earthquake struck at a previously unknown fault line – raising new alarm that the dreaded 'Big One' could hit anywhere in the country, experts said. The 7.2-magnitude earthquake, which was the most powerful to strike inland Japan in eight years, killed at least 12 people and triggered massive landslides.

'Seismically speaking, major earthquakes can strike anywhere and anytime in Japan' – said a seismologist from the University of Tokyo – 'Therefore, it would be no surprise if another earthquake matching the one on Saturday occurs in any part of the country in the near future. We had not been aware of faults in the area where the earthquake occurred this time'.

Despite the massive strength of the tremor, damage to facilities and numbers of casualties were limited, because it hit a rural area. About 9.7 million people live in the rice-growing region – accounting for 7% of Japan's population. By contrast, a government study in 2006 warned that a 7.3-magnitude earthquake in Tokyo would kill 4700 people and damage 440 000 buildings. The last giant tremor in Tokyo was the Great Kanto Earthquake of 1923, which left 142 807 people dead.

ABC News, 15 June 2008

Why are the after effects of an earthquake often as hazardous as the earthquake itself?

What do you think ?

Urban and rural Japan – contrasting impacts

▲ *The 2008 Miyagi earthquake – which measured 7.2 magnitude – killed 12, caused damage to bridges and rural roads, and triggered landslides*

▼ *The 1995 Kobe earthquake, which measured 7.3 on the Richter scale, killed 6434 people. It caused $213.75 billion in economic damage – considered the highest ever to a single country from a disaster*

Option summary

● On your own

Look at Resources 1-11 on pages 265-269, and then answer the following questions. Each answer should quote named examples of people, places or incidents as evidence, as you will have to do this later in the exam.

1 Discuss the factors which affect the impacts of tectonic events on people and the environment.
2 Referring to examples, show how well the risks posed by tectonic hazards are managed.
3 How far can prediction aid decision-makers faced with tectonic risks?

Useful websites for research

Remember that this unit is a taster of what 'Tectonic activity and hazards' is about. Although it shows how tectonic incidents have different impacts, it will not cover the full requirements of Unit 4! The following websites will help you to extend your research further.

News items about tectonic hazards can be found at:
- BBC Science website – http://news.bbc.co.uk/1/hi/sci/tech/4126809.stm
- British Geological Survey – http://www.earthquakes.bgs.ac.uk
- US Geological Survey – http://earthquake.usgs.gov/regional/neic/

For more news on Montserrat, China and Japan try:
- The Montserrat Volcano Observatory – http://www.mvo.ms/Volcanic_activity_this_week.htm
- Kobe 10 years on – http://www.dpri.kyoto-u.ac.jp/web_e/index_e.html
- Sichuan – http://shanghaiist.com/2008/05/12/earthquake-hits-wenchuan-sichuan.php

Use key terms to search from this study. For example: subduction zone, epicentre, tectonic activity, aftershocks

Other useful websites about tectonic activity include:
- For volcanoes – http://www.bbc.co.uk/scotland/education/int/geog/envhaz/home/flash/volcanoes/index.shtml
- For earthquakes – http://tsunami.geo.ed.ac.uk/local-bin/quakes/mapscript/home.pl
- General – http://www.nationalgeographic.com/forcesofnature/

Other tectonic incidents

Building up a file of tectonic hazards will help you to understand issues about their impacts and management. Useful incidents include:

- In 1963, Icelanders were able to witness the birth of Surtsey, a new volcanic island across the mid-Atlantic constructive plate boundary.
- The Pakistan earthquake in Balakot, which initially killed 19 000 in October 2005, with a further 60 000 dying in the aftermath following a harsh winter in an area of poverty.
- The largest earthquake in the UK for 25 years on 27 February 2008 at Market Rasen.
- 15 000 Colombians were forced to evacuate as the Nevado del Huila Volcano began to erupt in April 2008.

Films books and music on this theme

Music to listen to
'Sonification' (2006) New Earthquake Warning Systems at – http://www.youtube.com/watch?v=FNRQ_LuzMt4

Books to read
Richter 10 – Taming the Earthquakes (1996) by Arthur Clarke & (Late) Mike McQuay
Volcano – Nature's Inferno (1997) National Geographic

Films to see
'Volcano' (1997)
'Dante's Peak' (1997)
'Earthquake' (1974)

Try these exam questions

When you have completed a programme of research as preparation for the examination, try these questions:

1 How far do human responses to tectonic hazards reflect the frequency and magnitude of the events? **(70)**

2 'The short-term consequences of major tectonic activity are inevitable; the longer-term ones are entirely avoidable.' Discuss. **(70)**

3 With reference to examples, explain how tectonic activity can lead to distinctive landforms. **(70)**

To understand how these questions are marked refer to pages 327-328.

Introducing this Option

We know that the Earth's climate is currently changing; scientific research shows that it often changes. In the last 2.6 million years of geological time (known as the Quaternary period), major shifts of global climate have occurred several times. Each time, ice sheets extended into temperate latitudes – bringing glaciers to upland areas, altering the landscape, changing sea level, and shifting plant and animal habitats. This all happened in a period shorter than 0.05% of the Earth's total history!

Cold environments (i.e. those with temperatures permanently <0°C) give rise to distinctive landscapes, which not only display the power and presence of ice but also restrict human activity. Beneath and beyond the ice masses, natural systems reflect changes in global climate, which can be:

- rapid – like the collapsing Hornbreen glacier in Svalbard, part of which crashed into the Arctic Ocean in August 2007, causing a huge wave which injured 17 British tourists aboard a nearby sightseeing ship.
- slow – such as the gradual retreat of glaciers and ice sheets down the valley of the Grossglockner (Austria), or in most of southern Iceland, uncovering glacial deposits.
- preserved – as landscape features from past ice ages, such as the 3 km-deep Sognefjord (Norway), or the more modest lumps of Lake District rock (called erratics) in Stoke-on-Trent.

What is this Option about?

This Option will probably appeal to you if you like studying physical geography, geology, biology, environmental studies, politics, ecology, or have an interest in current social, economic and environmental issues.

The Specification lists four Themes in this Option. If you choose it, you will learn about:

1 **Defining and locating cold environments: What are cold environments and where are they found?**
In which you will learn about: • glacial and periglacial environments   • the concepts of landscape systems, glacial systems, mass balance, frequency/magnitude, equifinality and dynamic equilibrium   • the varying nature of different cold environments   • the past and present distribution of cold environments, especially in the British Isles

2 **Climatic processes and their causes: What climatic processes cause cold environments, and what environmental conditions result?**
In which you will learn about: • the climatic causes of cold environments   • how long-term global climate change alters the distribution of cold environments   • the meteorological processes associated with cold environments   • the spatial and temporal relationships between glacial and periglacial environments

3 **Distinctive landforms and landscapes: How do geomorphological processes produce distinctive landscapes and landforms in cold environments?**
In which you will learn about: • the role of ice and sub-aerial processes in glacial environments   • the role of geomorphological processes in periglacial environments   • the distinctive landforms produced by these processes, including relict landforms

4 **Challenges and opportunities: What challenges and opportunities exist, and what management issues result?**
In which you will learn about: • the terms 'challenges' and 'opportunities', and the links between them   • the challenges and opportunities of cold environments – past and present   • how humans are realising the opportunities by overcoming the challenges of the cold   • the effectiveness of different approaches to using and managing cold environments – considering the attitudes of different groups and the conflicts that can exist between them

How to use this chapter

This chapter contains six pages of resources.

- Page 272 introduces the topic with a case study about the Arctic region.
- Pages 273-274 investigate the issue further using resources. You will be guided but you will need to think about each resource carefully.
- Page 275 provides background information about the distinctiveness of cold environments.
- Pages 276-277 investigate the impact of past processes in Britain, again using resources.
- Page 278 provides activities, useful websites for further research, and examples of the kinds of questions that you will meet in the exam (each worth 70 marks).

The Arctic region extends across eight countries and contains a frozen ocean that has been making headline news for several decades. In 2007, the extent of the Arctic Ocean's ice sheet fell to a record low for September. The average for summers since 1979 had been an extent of 6.7 million km² but, in 2007, a mere 4.13 million km² remained by September. Winter measurements reinforced this downward trend – an extent of 14.7 million km², compared to the average of 15.6 million km². The permanent year-round ice also showed signs of thinning, with the Beaufort Sea's perennial ice layer reduced from a thickness of 3.3 metres to only 0.5 metres, with most being lost from below as warmer water caused melting.

This 'Big Thin' has set the climate change alarm bells ringing. Meanwhile, there are economic opportunities to be taken advantage of, as the Arctic region emerges from the ice and once fairly inaccessible parts of the world become less of a challenge. The Northwest Passage – a shortcut connecting the Atlantic and Pacific Oceans via the Canadian Arctic (see the map below) – was entirely navigable in 2007, and there is talk of the Arctic Ocean being completely ice free in the summer months by 2013 if current melt rates continue. Such potential accessibility is raising international tensions, because the Canadians have claimed 'ownership' of the Northwest Passage route, while the USA and the EU are demanding full international access for trade.

There are also tempting reserves of oil and gas beneath the Arctic Ocean (see the map), and the thinning ice makes extraction of these resources a great deal easier. International agreements govern control of the seabed and allow countries sole mineral exploitation rights within a 370-km zone extending out from their coastlines (with an additional 280 km if their continental shelf extends that far). The countries of the Arctic region are now gearing up to take advantage of the easier access provided by the ice melt, and disputes about the extent of their territory (and therefore their exploitation rights) have grown more intense.

The Lomonosov Ridge, which extends under the North Pole from the area of Greenland (Danish) and Ellesmere Island (Canadian) towards the islands of Siberian Russia, became a matter of dispute when the Russians claimed it in 2001 and then the Danes established research projects into its resources. On 2 August 2007, the Russians used two mini-submarines to plant a Russian flag 4200 metres below the North Pole – claiming it as theirs and causing comments about a new 'Cold War' as tensions in the region began to increase.

▼ *The Arctic region and its potential economic importance for the future*

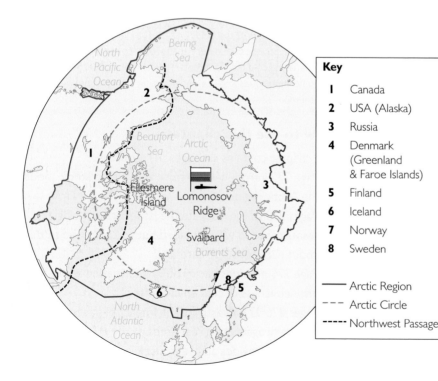

Key

1	Canada
2	USA (Alaska)
3	Russia
4	Denmark (Greenland & Faroe Islands)
5	Finland
6	Iceland
7	Norway
8	Sweden

—— Arctic Region
- - - Arctic Circle
---- Northwest Passage

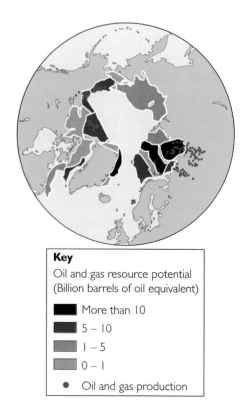

Key

Oil and gas resource potential
(Billion barrels of oil equivalent)

- ■ More than 10
- ▨ 5 – 10
- ▩ 1 – 5
- ▢ 0 – 1
- ● Oil and gas production

Can the Arctic be developed sustainably?

Consider these resources:

- Resources 1-2 are about the Arctic's emerging place in the global economy.
- Resources 3-4 assess the challenges of development in cold environments.
- Resources 5-6 are about the activities that have already reached the Arctic.

Resource 1

The Arctic – a new victim of global development?

The vision of the Arctic is a contradictory one. On the one hand, it is the last frontier – a limitless, rich environment that can be exploited for commercial gains – on the other, an unspoiled area of pristine beauty, which can and should be preserved in all its glory.

Resource exploitation is already creating environmental hotspots in the Arctic, which faces the reduction of its wilderness area if it is not managed carefully. Global climate change warms this region at a rate that is twice the world average – melting sea ice, interrupting the food chain, and threatening wildlife on which indigenous populations like the Saami and Inuit depend for food, medicine, and clothing. Long-distance air pollution, which is emanating from the main industrial areas of the world, is also poisoning the entire food chain from micro-organisms to human beings.

The Saami and Inuit, having lived in harmony with their environment for thousands of years, are now seeing their existence threatened by global development. These Arctic peoples are not the causes of the environmental deterioration, but their future is being determined by consumers and decision makers located far beyond the Arctic.

However, nature might yet strike back. The threat to the Arctic is also a threat to the global environment and well-being of the entire planet. Arctic climate change and the melting of the permafrost are accelerating global warming. The reduction of Arctic species and wilderness is also contributing significantly to a reduction in global biodiversity. Over-fishing is putting the global catch at stake.

Can a new kind of sustainable development, based on traditional Arctic societies and cultures, provide an alternative model of development – a challenge to the very basis of the current process of globalisation?

Adapted from an article by Svein Tveitdal in The Environment Times, *September 2004*

▲ Ellesmere Island - Arctic wilderness at risk

Resource 2

The politics of oil

The politics of the Arctic are no longer the politics of the people, but the politics of oil. There is a risk that the costs of development will mostly be borne by the residents of the Arctic, while the benefits are felt mostly outside the Arctic region.
A member of the Arctic Council, which has representatives from the eight Arctic countries, together with members from six indigenous peoples (such as the Inuit)

A challenging environment - taking over nature

Resource 3

Shtokman will be tough!

Gazprom, the Russian energy giant, is currently preparing to develop the world's largest offshore gas field in the Barents Sea – considered one of the most extreme exploration areas in the world. For 110 days each year, when the sea is not covered by a layer of ice, the crews in the gas field will have to battle with 25-metre-high waves. In winter, limited light – and temperatures as low as -60°C – will make their lives even harder.

In order to develop Shtokman, an underwater pipeline 555 km long is to be constructed to transfer the gas to Murmansk, where it will be liquified for export to the USA. An overland pipe across the Kola Peninsula will link up with the Northern European Pipeline to Germany and the European network.

Adapted from BBC News, November 2007

Resource 4

Overcoming hostile climatic conditions

▲ *Industrial development in the Arctic gateway city of Murmansk, which experiences 40 days of total darkness in winter and 40 days of continuous daylight in summer – and a mean temperature range of -20°C to +16°C*

Resource 5

Look what they have done to our world!

From the massive oil developments at Prudhoe Bay in Alaska, to the military installations at Murmansk in northern Russia, the Arctic development record is not great:

● Radioactive leaks from crumbling bunkers at a nuclear waste storage facility in Andreeva Bay, Murmansk, have left the water completely devoid of life. Onshore soils and groundwater are also contaminated.

● In Prudhoe Bay, rusting oil and gas pipelines leak along hundreds of miles of Arctic tundra, alongside the scars of clear-cut logging in the boreal forests. In 1989, the *Exxon Valdez* supertanker ran aground in southern Alaska, spilling crude oil into Prince William Sound (also see pages 94-99).

● Plastic bags, fishing nets, barbed wire, fuel tanks and beer bottles are found washed up on shorelines all round the Arctic, and the ice is also contaminated by windblown pollutants.

Can the Arctic be developed in a sustainable way?

What do you think?

Resource 6

The multinational invasion of Ny Ålesund

The Norwegian Arctic islands of Svalbard are a property hotspot. Ny Ålesund, a tiny former coal-mining settlement, is booming. Scientists from China, India, South Korea and Japan are joining European and American bases there as research into climate change and the riches of the Arctic flourishes. Norway's low-cost airline has reintroduced twice-daily flights to Svalbard, and specialist tourism is taking off. The 'land of the midnight sun' and its wilderness represents another extension of the pleasure periphery (see page 307).

Adapted from The Guardian, August 2007

▲ *Ny Ålesund on the Svalbard archipelago*

Background

The retreat of the ice – lessons from the Arctic

The Arctic region of the twenty-first century (first map) is a fraction of the size of the cold environment that existed 18 000 years ago (second map). Great sheets of ice, which were up to 3.5 km thick, once extended from areas of high latitude and altitude to cover 30% of the Earth's surface. Around the glacial margins were zones of permanently frozen ground (**permafrost**), not dissimilar to those of today's Canadian tundra – known as **periglacial environments**. With the retreat of the ice, new landscapes have been exposed which reveal the actions of moving ice in the past and also the effects of permafrost. Many lessons can be learnt from the Arctic as it thaws, and predicting its future requires only a glance at places previously covered by ice.

Shifting climates and shifting ground

While most parts of the world suffer transport chaos in winter, the roads of the Arctic are generally firm and passable. It is the summer thaw which causes transport problems, because the top layer of permafrost thaws and becomes boggy. The transport of vital commodities to Arctic communities depends on the winter freeze. However, the length of the 'safe' frozen time has been getting shorter and shorter, and the oil, gas and timber industries suffer when they cannot move their products by road or rail. Thawing permafrost is a cause for considerable concern, because it also:

- causes the ground to shift, leading to structural damage to buildings
- deforms vital lines of communication, such as roads and railways
- creates hazardous hollows in the surface
- saturates ground and soils, leading to landslides, rock falls and avalanches
- releases methane (a greenhouse gas) when organic matter thaws and decomposes

The present is the key to the past

Many of today's landscapes owe their characteristics to conditions and processes that existed centuries ago. Landforms sometimes take a very long time to adjust to their prevailing environmental conditions, and exist as **relict features**. Deep, flat-bottomed valleys with misfit streams, like those in the Lake District, and the dry valleys on the Isle of Purbeck, show how glacial and periglacial activity leaves lasting impressions. By studying the present-day processes that are operating in Canada's tundra, or Iceland's glaciers, it is possible to learn how many of Britain's landscapes were created.

The extent of Arctic ice covering today (top) and 18 000 years ago (middle) and permafrost levels today ▶

Uncovering the past

The next two pages consider the effects of glacial and periglacial environments on landscapes in Britain.

- Resources 8-10 consider the ways in which past glacial environments have shaped upland Britain.

- Resources 11-12 consider the ways in which past glacial and periglacial environments have shaped the landscape of Norfolk.

Resource 8

Quaternary news

The Scotsman newspaper rocked Britain's scientific establishment on 7 October 1840, by declaring that 'Glaciers once existed in Scotland'. According to Professor Agassiz, many features of the Scottish landscape matched those now being formed before his eyes by the action of glaciers in the European Alps. On a visit to Britain in 1840, he convinced many British scientists to accept his startling conclusions – but some scientists thought his views were too revolutionary, because there were no glaciers left in Britain in 1840.

Professor Agassiz was explaining the existence of relict landscapes, and was attributing them to processes that can be observed today – where ice caps and glaciers do still exist. During the Quaternary period, large parts of Britain were covered in ice sheets (see Resource 12) and the action of the ice when it retreated left its mark on the British landscape.

Resource 9

Misfits and troughs

▲ *Glen Rosa on the Isle of Arran in western Scotland. The tiny 'misfit' stream flowing through the massive valley, known as a 'trough', cannot have been responsible for such a large feature. The flat-bottomed, steep, parabolic-sloped, 'U-shaped' valley was carved out by glacial ice and illustrates the effects of processes operating under Alpine glaciers today.*

Resource 10

Limestone pavements – ice footprints

The limestone pavements of Britain began with the scouring of the limestone by kilometre-thick glaciers. The weight of the ice removed the soil that lay over the limestone, and also fractured the limestone along existing horizontal surfaces of weakness, known as bedding planes. The fractured rocks were stripped away, leaving level platforms of limestone on which a thick layer of boulder clay (glacial till) was deposited as the glaciers retreated.

Over thousands of years, the characteristic features of limestone pavements were formed on the flat limestone surfaces by water in the glacially deposited soil exploiting cracks and joints in the limestone to form characteristic **grikes** (deep vertical cracks) and **clints** (the blocks of limestone between the grikes). The glacial soil covering the limestone was eventually eroded, exposing the limestone pavements seen today.

Also deposited on the limestone platforms were erratic rocks, ranging in size from pebbles to huge boulders. **Erratics**, being different rock types to limestone, are good indicators of glaciation on limestone pavements.

▲ *Malham in Yorkshire – evidence of glacial scouring*

Ice age legacy on the Norfolk coast

The Norfolk coastline owes much to the frozen wastelands of the past. The melting glaciers and ice sheets at the end of the last ice age – which had lasted for thousands of years – revealed vast changes to the British landscape. The ice which had covered most of Britain (see the map below) transported huge amounts of debris – ranging from massive boulders to fine rock particles – which helped to form much of today's landscape. As the ice melted, the glaciers dumped boulder clay (glacial till), which formed mounds and ridges called **moraines**. Other evidence of the effects of glaciation includes erratics transported hundreds of kilometres from Scandinavia.

Cromer Ridge in Norfolk (see below) was the front line of the ice sheet for some time. The glaciers ground to a halt here. All of the material that was dredged up from the North Sea poured out of the ice to form a ridge 92 metres high and 14 kilometres long.

It was because Norfolk was on the edge of the glacier, that it has such a unique landscape – unmatched anywhere else in East Anglia. The clay soil found in many parts of the region was dumped by the ice and meltwater streams, and provides some of the best land for crops and wildlife in Europe.

Adapted from an article on the BBC Norfolk website

Ice age Norfolk

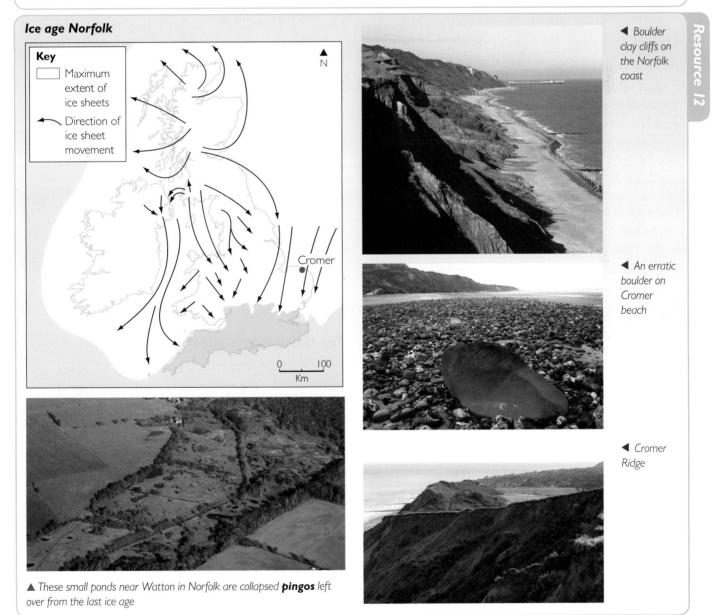

Key
- Maximum extent of ice sheets
- Direction of ice sheet movement

N

Cromer

0   100
Km

◀ Boulder clay cliffs on the Norfolk coast

◀ An erratic boulder on Cromer beach

◀ Cromer Ridge

▲ These small ponds near Watton in Norfolk are collapsed **pingos** left over from the last ice age

● On your own

Look at Resources 1–12 on pages 273–277, and then answer the following questions. Each answer should quote named examples of people, places or incidents as evidence, as you will have to do this later in the exam.

1 What are the consequences of the shrinking Arctic ice sheet on:
   ● the people who work and live there?
   ● the natural environment?
   ● accessibility and exploitation of resources?
2 How far is management of the Arctic area an international issue?
3 How far do natural environmental conditions limit economic activities in the Arctic and similar locations?

Useful websites for research

Remember that this unit is only a taster of what 'Cold environments: landscapes and change' is about. Although it makes a useful case study of cold environments and shows how relict landforms inform us about past conditions, it will not cover the full requirements of Unit 4! The following websites will help you to extend your research further.

News items about exploiting the Arctic and Antarctica can be found at:
● BBC News website – www.bbc.co.uk/news
● United Nations Environment Programme – http://www.environmenttimes.net/index.cfm

In Norway, for more on the challenges of the Arctic:
● Norwegian research – http://npweb.npolar.no/filearchive/Polarnorway.pdf
● Arctic tourism – http://assets.panda.org/downloads/wwfcruisetourismonsvalbard2004_v5p3.pdf

Use key terms to search from this study. For example: tundra, permafrost, periglacial, glacial

Other websites about cold environments include:
● Arctic issues – http://www.arcticpeoples.org/ http://www.arctic-council.org/
● Arctic climate impact – http://www.acia.uaf.edu:80/
● National Geographic Coldscapes – http://ngm.nationalgeographic.com/2007/12/permafrost/barry-lopez-text

Other similar ideas

Building up a file on cold environments will help you to understand the processes and changes that have occurred and still occur in these parts of the world. Useful examples of distinctive landscapes and challenges to human activity include:

● The dry valleys and development of coastal features on the Isle of Purbeck around Lulworth Cove.
● Contemporary periglacial conditions in the Cairngorms and Scottish Highlands.
● The growing threats of modernisation on the Saami reindeer herders of Lapland and the Inuit of northern Canada.

Films, books and music on this theme

Music to listen to
'Glaciation' (2007) by Patrick O'Hearn
'Caught in the Act' (2003) by Pamyua (see http://cdbaby.com/cd/pamyua3)

Books and articles to read
'The big thaw – no-one expected it so fast', *National Geographic*, June 2007
After the Ice Age (1991) by E C Pielou
Iceberg (1996) by Clive Cussler

Films to see
'Warnings from the Ice' (1998) Patrick Gardner
'Arctic Tale' (2007) National Geographic Films

Try these exam questions

When you have completed a programme of research as preparation for the examination, try these questions:

1 Explain why international initiatives are increasingly needed to manage the Polar regions. **(70)**

2 Assess the view that Britain's landscapes owe a great deal to the presence of ice during the Quaternary period. **(70)**

3 Do glacial and periglacial processes produce distinctive landforms? **(70)**

To understand how these questions are marked refer to pages 327–328.

9 Life on the margins: the food supply problem

Introducing this Option

Maintaining an adequate food supply for a rising global population remains a major challenge. For people living in poverty, or on marginal land where crops frequently fail, or those confronted by surges in commodity prices, the realities are stark. This Option focuses on the problems of food insecurity and the attempts being made to ensure food supplies for everyone.

- The intensification of agriculture increased crop yields by developing new high-yield, disease-resistant varieties of crops, and using machinery, fertilizers and pesticides to create an artificial ecosystem to support increasing human needs. These changes became known as the **Green Revolution** and transferred western technology to developing countries in the 1960s and 1970s to help them feed themselves. It had most impact in South and East Asia and South America, but less impact in sub-Saharan Africa.
- Major international businesses now dominate global food supply chains, having acquired premium land for plantations, ranches and crop production at the expense of local smallholders.
- Global talks have tried to establish free trade by opening markets up and reducing unfair tariffs and subsidies that support American and European farmers.

What is this Option about?

This Option will probably appeal to you if you like studying environmental studies, economics, politics, ecology, or have an interest in current social, economic and environmental issues.

The Specification lists four Themes in this Option. If you choose it, you will learn about:

1 Global and local feast or famine: What are the characteristics of food supply and security?

In which you will learn about:
- current issues associated with food supply and security
- the environmental issues resulting from food production
- why food supply varies spatially
- what life on the margins means to differing people, and how food security affects their quality of life

2 The complex causes of food supply inequalities: What has caused global inequalities in food supply and security?

In which you will learn about:
- the short- and long-term direct/indirect causes of famine and food surplus
- the role of population pressure in creating food insecurity
- the impacts of attempts to increase global food supply
- who has been most affected by food insecurity and why

3 Desertification and life on the margins: What is the role of desertification in threatening life at the margins?

In which you will learn about:
- the scale and impacts of desertification
- the characteristics, vulnerability and scale of dry land ecosystems
- why dry lands are vulnerable to over-exploitation and inappropriate land use
- the relationship between food production and supply in desertified regions

4 The role of management in food supply and security: Can management strategies sustain life at the margins?

In which you will learn about:
- attempts to increase global food security
- why greater efforts are increasingly needed
- the most effective initiatives for sustaining life at the margins
- the role of sustainable strategies in food supply and security

How to use this chapter

This chapter contains six pages of resources.
- Page 280 introduces the topic with a case study about Kalahandi in India.
- Pages 281-282 investigate the issue further using resources. You will be guided but you will need to think about each source carefully.
- Page 283 provides background information about the global food crisis and risks.
- Pages 284-285 investigate the issue in other areas of the world, again using resources.
- Page 286 provides activities, useful websites for further research, and examples of the kinds of questions that you will meet in the exam (each worth 70 marks).

The Kalahandi Syndrome

Living on the margins

Kalahandi district, in the western uplands of Orissa state in eastern India (see the map), is home to 1.3 million people, of whom 80% are farmers. However, in Kalahandi it is said that 'agriculture is the gamble of the monsoon', because annual rainfall has become very unreliable and droughts (and therefore famine) are frequent. Since a major famine in 1980, Kalahandi has been home to many of the Indians who go to bed hungry every night.

Orissa remains economically remote from the wealth-generating Indian cities – and Kalahandi is even more remote. Its forested hills are marginal to India's developing economy, and therefore the local people are marginalized from the processes of change – the Green Revolution has not made an impact here. Local livelihoods depend on daily reactions to prevailing circumstances – some within the people's control and some not.

Droughts have persisted for 30 years, and the poor, bare soils on the deforested slopes reflect the people's attempts at survival. The seeds they use are a traditional variety with low yields, which fall every year as the land becomes more degraded; similarly, the livestock breeds used in the villages barely supply enough meat and milk for consumption - and none for sale. Forest products are also gathered in for consumption and sale, but the middlemen who buy these products for sale in the cities exploit the villagers and pay very low prices, so many villagers find themselves having to borrow money or starve. In 2007, starvation played a significant role in 250 cholera-related deaths. BBC journalists visited the district and found people there with no food who were surviving off the bark and leaves of trees. They had been unable to afford rice for a year. However, as you will see below, food supply is not the big problem …

Hunger in spite of plenty

There is a bitter irony in Kalahandi. Between 1998 and 2003, rice production in Kalahandi exceeded local needs and contributed to India's national food reserves. However, most farmers in the district do not own their own land, but instead rent it from absentee feudal landlords. These landless farmers could not afford to buy the rice they had grown for their landlords, and 50 million tonnes rotted in the countryside while the people went hungry.

▼ Villagers trying to cultivate marginal land in Orissa state

This is the 'Kalahandi Syndrome' and it occurs globally. The world today has more food than the hungry need. If the supply of food was evenly distributed in accordance with the World Health Organisation's minimum daily calorie requirement, a surplus big enough to support an additional 800 million people would exist. However, the tragedy is that the WHO estimates that currently 800 million people worldwide cannot afford to buy their recommended daily calorie intake.

The impacts on rural India

Consider these resources:

- Resource 1 is about the impacts of globalisation on rural India.
- Resources 2-4 assess the impacts on many rural Indian communities of the problems they face from spiralling debt.
- Resources 5-6 are about diversification to increase food purchasing power for the poor.

Kalahandi Syndrome reinforced

As the focus of agricultural policies shifts to **agro-processing**, foreign investment and exports, the vital links between agricultural production and access to food have been ignored. In a period when hunger is on the increase, cereals and meat produced by India's most fertile lands are being used to make pet food and whisky for foreign markets. As agriculture becomes industrialised, small farmers are uprooted from their land to become landless labourers, or to join the marginalised urban poor.

Improved agriculture would help India to eliminate hunger and reduce poverty and unemployment, but the World Trade Organisation (see page 156) encourages India to open its doors to international trade. Sadly, cheap food imports will only drive millions of subsistence farmers from their small land holdings, destroy livelihoods and threaten food security. By embracing free trade, India allows foreign companies to take control of its land, seeds and agricultural research – the vital tools which farmers have depended on to produce the nation's food stocks. As farming becomes the target of big business, the fields of India are being switched from food production to flowers and other cash crops. The process of globalisation is accelerating this change and taking food production away from the rural communities.

It is a process that challenges India's ability to provide food for the poor, and it is taking away their jobs. Even cheaper food is too expensive for those without any income at all.

Adapted from essays by Devinder Sharma, a food and trade policy analyst, in 2005

- **Agro-processing** means turning primary agricultural products into other commodities for market, e.g. peanuts into peanut butter.

India for sale

▼ *This sign outside the village of Dorli, northern Maharashtra, in 2006 reads 'Village for sale, along with houses, cattle and fields.' 600 farmers committed suicide in this region in the previous year, because of debt and crop failure.*

Rural indebtedness in India has reached such alarming proportions that village communities are being forced to sell their body organs [their kidneys] and their lands – willing to lose control over their only means of economic security.
Devinder Sharma

Death and despair

25-year-old Betavati Ratan took his own life because he could not pay back debts incurred for drilling a deep tube well on his two-acre farm. The normal wells are now dry, as are the wells in Gujarat and Rajasthan – where more than 50 million people face a water famine.

The drought is not a natural disaster; it is man-made. It is the result of the extraction of scarce groundwater in arid regions to grow thirsty cash crops for export, instead of water-prudent food crops for local needs.

Adapted from a BBC Reith Lecture given by Vandana Shiva in 2000, entitled: 'Poverty and Globalisation'

Hybrids and diversification

An ecological and social disaster

Bhatinda in Punjab has been experiencing an epidemic of farmer suicides. Punjab used to be the most prosperous agricultural region in India. Today, every farmer is in debt and despair. Vast stretches of land have become desert. Even the trees have stopped bearing fruit, because heavy use of pesticides has killed the pollinators – the bees and butterflies.

Farmers in Warangal, Andhra Pradesh, have also been committing suicide. Farmers who traditionally grew pulses and millets and paddy, have been lured by seed companies into buying hybrid cotton seeds – referred to by the seed merchants as 'white gold' – which were supposed to make them millionaires. Instead they became paupers.

The native seeds were replaced with the new hybrids which cannot be saved for reuse the following year, but instead need to be purchased again every year – at high cost. The hybrids are also very vulnerable to pest attacks, so spending on pesticides in Warangal shot up by 2000% from $2.5 million in the 1980s to $50 million in 1997.

Now the farmers are swallowing the same pesticides, as a way of killing themselves so that they can escape permanently from unpayable debt!

Adapted from a BBC Reith Lecture given by Vandana Shiva in 2000, entitled: 'Poverty and globalisation'

The answer is Syngenta!

Laxman Sahu had been growing vegetables alongside his staple crop of rice in Kalahandi District. When the Syngenta Foundation established KARRTABYA – an agricultural extension service – he attended the farmers' workshops to learn about new developments in farming. He acquired details about some promising hybrids of hot chilli peppers, and bought the seeds of 'Roshni', 'HPH 117', 'HPH 404' and 'Flame Hot' from KARRTABYA.

In 2006-07, he cultivated these hybrids on leased land, 2 km from his home. A mountain stream provided the irrigation water. Having followed the advice of experts from the District Agriculture Office, he achieved a bumper crop of red chillies. The net profit came close to 100 000 rupees. On seeing his success, several other local farmers have now followed this pattern.

Adapted from Syngenta's 'Foundation for Sustainable Agriculture' website: http://www.syngentafoundation.com/. Syngenta is a leading biotech agri-business.

Bananas to the rescue

Some of Kalahandi's farmers have now diversified into growing bananas. Around 5000 hectares are under extensive banana cultivation in Kalahandi, as part of Orissa's diversification project. A farmer has to spend about 14 000 rupees ($340) per year for every hectare of banana cultivation, but he can earn around 35 000 rupees ($850) net profit from his crop.

Can cash crops be the answer to localised hunger?

What do you think?

Background

With the world's population reaching 6.7 billion in 2008 – and predicted to reach 9.5 billion by 2050 – pressures are building on food suppliers. The World Bank predicts that global demand for food will double by 2030.

In 1798, Thomas Malthus wrote *An Essay on The Principle of Population*, in which he put forward the theory that the power of global population to increase is greater than that of the Earth to provide sufficient food – in other words, demand for food will always outstrip supply. He also said that war, famine and disease were 'natural checks' which helped to keep population growth under control.

However, Ester Boserup's predictions (in the 1960s) that technology could save the day have largely been

realized – first through the Green Revolution and then by ever-increasing levels of intensification, innovation and capital investment in agriculture. Billions more people than Malthus ever thought possible are now being fed. In terms of overall supply, the world has never been better fed than it is today.

Yet something is wrong. In 2008, over 800 million people worldwide were officially classed as undernourished by the UN's World Food Programme, and there were disturbances and riots over hunger (see Resource 9). Many of the millions of undernourished people are the marginalised who eke out a living on land not suited to cultivation, or who shelter in shanty towns on the fringes of urban areas, having been forced from their land.

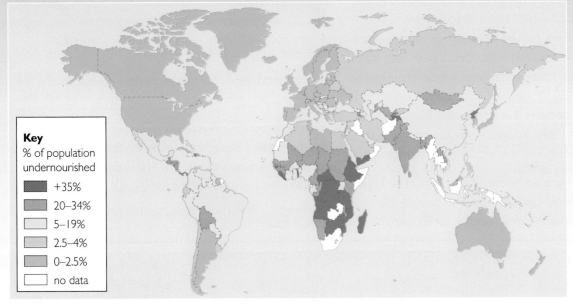

◀ *Global hunger – the percentage of each population considered to be undernourished, according to the UN's World Food Programme*

Key
% of population undernourished
- +35%
- 20–34%
- 5–19%
- 2.5–4%
- 0–2.5%
- no data

Factors influencing the growing global food crisis:
- Overall population growth (the estimated global growth rate is 1.16% a year).
- Changing diets in China and India, due to their economic growth and development. China's meat consumption has gone up by 150% since 1980, and India's by 40% since 1993. Cereal crops are fed to animals to produce meat – it takes 8 kg of grain to produce 1 kg of beef.
- Growing crops for **biofuel** instead of food, which has raised the price of corn and reduced the amount of land available for food crops.

- Declining amounts of productive arable land, because of drought, salinity (see page 65), deforestation and climate change.
- Unstable weather patterns destroying crops – grain harvests in 2007-8 were down 60% in Australia, 10% in China, and 10% in the UK.
- High oil costs forcing up the prices of fertilizers, food processing and transport.
- Altered farming patterns due to The World Bank's free trade/open markets policies, which have removed protective tariffs.
- Wars forcing people into marginal areas where crops often fail.

Global food crisis

The next two pages look at the impacts of food shortages in several countries.

Resources 8-10 are about the emerging global food crisis.

Resources 11-12 suggest that the problems are often in the solutions.

Food riots in Haiti

Haiti is listed by the IMF as the 133rd poorest country on Earth (out of 179). People living here have nothing. Oxfam estimates that the Haitians spend between 50% and 80% of the money they earn on food – just to keep themselves and their families alive. In the last few months, that figure has got closer to 100% (for many here in Haiti well over 100%).

Haiti is facing a most unusual form of famine, caused not by a lack of food (because there is plenty) but by the fact that a whole section of Haitian society can no longer afford enough food to live on.

According to the World Bank, food prices around the world have almost doubled in the last three years, with much of that increase coming in the last six months alone. Wheat prices have doubled in less than a year. Rice is up 70% on a year ago. Maize and beans have been pushed up almost as much. Even those who have been living on food aid are suffering, because the aid agencies buy their food on world markets; their budgets – already struggling to cope with high oil prices – are rapidly being exhausted.

There is real desperation on the streets of Port au Prince. The crisis has already claimed lives, with four people dead in food riots earlier this month – not to mention

▲ Rising food prices sparked violent protests around the world in April 2008. This is Port au Prince, Haiti

the nameless, faceless Haitian children whose deaths are either caused or hastened by malnutrition.

The democratically elected Government of Haiti is another victim, brought down by the food riots. The hard-won democratic reforms in this troubled country may also be a casualty of the food crisis.

James Mates, reporting for ITV News from Port au Prince, Haiti, on 25 April 2008

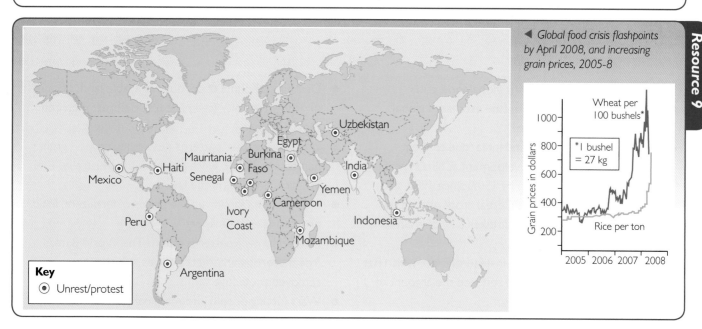

◄ Global food crisis flashpoints by April 2008, and increasing grain prices, 2005-8

Key
◉ Unrest/protest

Resource 10

Life on the edge

The Philippines was a leading exporter of rice, now it is the world's biggest importer.

Rapid urbanisation and a 50% increase in population since 1990 (up from 60 to 90 million in 18 years) now means that sprawling cities have replaced farmland. Hundreds of thousands of landless Filipinos end up living in slums built over garbage dumps – like those on the margins of Manila.

Poverty is the real problem – a day's work scavenging through the piles of waste used to earn the 10p needed to buy a bowl of rice in 2007. By spring 2008, the price of rice had doubled. It is not that there is any shortage of rice, it is simply too expensive for the poor to buy.

Adapted from Anne Walker, Philippines Community Foundation

Technology can provide enough food, but not at the right price.

What do you think?

Resource 11

Don't cry for Argentina

Argentina's population is approaching 40 million, but it produces enough food to feed 300 million. In 1996, Monsanto introduced genetically modified (GM) soya beans into Argentina (also see pages 222 and 242-243). Since 2000, over a million hectares of forest have been destroyed to grow soya. Argentina is now the world's third largest soya producer and its biggest soya exporter. Most of the yield goes to the EU and China as animal feed. Since 1996, GM technology has earned Argentina $20 billion and created 1 million jobs. By 2008, more than 98% of all soya bean, 70% of maize and 60% of cotton produced in Argentina was GM.

GM crops may be providing bumper yields, but there have been consequences: 'Over the last ten years, deforestation has been rapid as areas are literally bulldozed to make way for crops like soya. We lose 250 000 hectares of forest per year in Argentina, which destroys diverse ecosystems and pushes small farmers and indigenous people off the land in the Yungas and the Great Chaco forests.' say Greenpeace Argentina.

There are worries about monoculture too – using so much land for just one crop – because it drains the land of its nutrients. A recent solution involves growing soya in rotation with maize to help restore the soils. Not only is this less damaging, but it also supports Argentina's biofuel boom. It could also boost maize prices, and give farmers the incentive to grow both crops in rotation.

Since 2005, there has been a 10% decline in ranchland as farmers switch from cattle to grain.

Based on extracts from New Agriculturalist in 2007

Resource 12

Argentine Pampas – where have all the cows gone?

◀ *Ever-increasing amounts of grain and soya beans are being turned into biofuels, making many Argentine farmers rich but leaving little left for food. It takes 230 kg of maize to produce enough biofuel to fill a car's tank, which is enough to feed a family for a year.*

Option summary

● *On your own*

Look at Resources 1-12 on pages 281-285, and then answer the following questions. Each answer should quote named examples of people, places or incidents as evidence, as you will have to do this later in the exam.

1 To what extent are food supplies under pressure:
   ● in developed countries?
   ● in developing countries?
2 Does it matter if the distances between growers and consumers become greater?
3 Is population growth really the cause of food insecurity?
4 Is large-scale agri-business the best way to secure food supplies?

Useful websites for research

Remember that this unit is only a taster of what 'Life on the margins: the food supply problem' is about. Although it raises several issues about food security, it will not cover the full requirements of Unit 4! The following websites will help you to extend your research further.

News items about nutrition, obesity and food wastage in the UK can be found at:
● BBC News website – www.bbc.co.uk/news
● Sustainable Development Commission – http://www.sd-commission.org.uk/presslist.php?id=74

International sites about food production and consumption include:
● contemporary articles on food production – http://www.inmotionmagazine.com/npc.html
● technological fixes – http://www.new-agri.co.uk/index.php
● World Resources Institute – http://earthtrends.wri.org/

Use key terms to search from this study. For example: marginal farmers, GM foods, fair trade, free trade, food miles

Other useful websites about food security include:
● for global reports – http://www.fao.org/SOF/sofi/
● for UK government reports – www.research4development.info
● the United Nations – http://www.unep.org/geo/yearbook/yb2008/

Other similar examples

Building up a file illustrating aspects of food supply around the world can strengthen your understanding of this topic. Useful examples include:
● Ethiopia – the focus of the 1980s African food crises, where exports of coffee, nuts and other cash crops continued as millions starved, civil wars raged and droughts made matters worse. After decades of aid, the food problem remains and exports are higher than ever.
● Mexico – where the biotech lobby are fighting off farmers' organisations who fear that the GM crops threaten native corn varieties, livelihoods, and the nation's food sovereignty by increasing dependence on transnational seed companies.
● Post-1982 Structural Adjustment Policies advocated by the IMF and World Bank to help indebted countries 'earn more from exports to pay back their loans' (see pages 186 and 203).

Books to read on this theme

GATT to WTO: *Seeds of Despair and In the Famine Trap* (1997) by Devinder Sharma
Keeping the Other Half Hungry (2008) by Devinder Sharma
How the Other Half Dies (1991) by Susan George
The Debt Boomerang (1992) by Susan George
The Lugano Report (2003) by Susan George
Whose Reality Counts? – Putting the First Last (1997) by Robert Chambers

Try these exam questions

When you have completed a programme of research as preparation for the examination, try these questions:

1 Outline the ways in which the environment influences levels of food security at regional, national and international levels. **(70)**

2 Assess the view that globalisation is forcing new supply lines in the food industry. **(70)**

3 Evaluate the following statement: 'The battle between free and fair traders misses the point – it is feeding people at a price they can afford that matters.' **(70)**

To understand how these questions are marked refer to pages 327-328.

Introducing this Option

Modern communications now allow us to measure geographical distance in time as well as kilometres. Thanks to air travel, e-mail, mobile phones, and social networking websites, people from different countries with different values and experiences come into contact with each other on a regular basis. Information technology spreads ideas faster than ever, and now affects how people consume goods and services. Music, TV programmes and films made by large media TNCs combine with migrations of people to create cultural mixes – or hybrids. This Option focuses on how economic development can lead to hybrid cultures, and examines whether a single 'global culture' could replace cultural diversity.

- Some cultural shifts reflect historical invasions, like those of Central and South America where the Spanish and other European invaders superimposed Christianity and a new economic system on their colonies, in place of their existing ancient cultures (see pages 141-145).
- Other cultural changes are enforced when governments and/or international institutions require local communities to give up their traditional way of life. For example, in the case of rural re-settlement programmes in Brazil – where the IMF urged increased cash crop production instead of traditional subsistence farming.
- And some cultural changes are simply the result of exposure to the global media and instant communications – with fashion, music, films and language reflecting a globalised consumerist society.

What is this Option about?

This Option will probably appeal to you if you like studying economics, politics, sociology, anthropology and ecology, or have an interest in current social, economic and environmental issues.

The Specification lists four Themes in this Option. If you choose it, you will learn about:

1 Defining culture and identifying its value: What is the nature and value of culture in terms of people and places?

In which you will learn about:
- the definitions and origins of the word 'culture'
- the range of human cultures and variety of cultural landscapes
- how some cultures and landscapes are more vulnerable than others
- how the cultural diversity of people and places is valued and protected

2 The geography of culture: How and why does culture vary spatially?

In which you will learn about:
- why some countries and regions are culturally more homogenous than others
- a clear rural/urban cultural divide in terms of human cultural diversity
- how governments and other players can preserve diversity or encourage cultural homogeneity
- how, globally, cultural imperialism affects human cultural diversity and landscapes

3 The impact of globalisation on cultural diversity: How is globalisation impacting on culture?

In which you will learn about:
- varying views on the significance of globalisation on cultural diversity
- how global media corporations can convey dominant cultural values and attitudes
- how cultural globalisation can lead to distinctive hybrid forms of fashion, music and film
- the impact of globalised consumerist society on culture and landscapes

4 Cultural attitudes to the environment: How do cultural values impact on our relationship with the environment?

In which you will learn about:
- how different cultures have different attitudes to the environment
- how different attitudes determine how landscapes are valued, exploited and protected
- how cultural values support and justify consumer cultures
- how the conflict between environmentalism and consumer capitalism has created a 'Green' agenda

How to use this chapter

This chapter contains six pages of resources.
- Page 288 introduces the topic with a case study of the Orang Asli peoples of Malaysia.
- Pages 289-290 investigate the issue further using resources. You will be guided – but you will need to think about each resource carefully.
- Page 291 provides background information on forces of cultural change.
- Pages 292-293 investigate the issue in other areas of the world, again using resources.
- Page 294 provides activities, useful websites for further research, and examples of the kinds of questions that you will meet in the exam (each worth 70 marks).

Malaysia's Orang Asli

Peninsular Malaysia experienced Portuguese, Dutch and British colonization from the sixteenth century onwards, yet the dense, rainforest-covered slopes of the peninsular's interior remained relatively untouched by European influences – as the home of the Orang Asli. The name 'Orang Asli' means 'original or first peoples', and is given to a collection of ethnic groups who live their lives guided by the Malaysian rainforests. In their culture, the land owns their lives and everything they do is tied up with its natural processes. The 105 000 Orang Asli make up just 0.5% of the Malaysian population, and are not a homogenous group; each separate group has its own language and considers itself different.

The Orang Asli group called the Semai live in the forested Cameron Highlands of central Malaysia, grow hill rice and hunt for wildlife and other foods in the forest. They are the descendents of the earliest people to live on the peninsular, and their relative remoteness has, until recently, protected them from the forces of change. In the nineteenth century, wealthy British families built isolated retreats and tea plantations in the cooler highlands, but they remained detached from the indigenous peoples and the two cultures existed side by side with limited contact.

The rights of indigenous peoples have been recognised in Malaysia since *The Aboriginal Peoples Act* of 1954. Schools, clinics and shops have been provided for them, and the various groups have been encouraged to grow cash crops of rubber and oil palm as a way of surviving in a growing commercial economy. The economy of the central highlands area has developed rapidly in recent years – and tea plantations, market gardening/horticulture and tourist resorts now litter the Cameron Highlands. Plans to build hydroelectric power stations and dams there are also putting indigenous cultures at risk.

▲ *The commercial economy of the Cameron Highlands is developing quickly - growing vegetables and strawberries in greenhouses in the valleys*

Some Orang Asli sell forest products like durian fruit and rattan to urban markets and tourists; others now have salaried jobs, with many women working in the new tourist businesses. The land no longer solely dictates their lives, and external forces increasingly determine how the land is to be used. The culture of the Orang Asli is under real pressure as the Malaysian Government tries to integrate the various groups into mainstream Malay society.

So what are the cultural shifts?

Consider these resources:

- Resources 1-3 illustrate traditional Orang Asli culture.
- Resources 4-7 outline the scale of the changes that have occurred in recent years.

This is a way of life but it will not always be like this

In a distant Malaysian valley, wisps of hill fog drift through the tallest trees – their canopies draped in trailing climbers. Butterflies flitter and feed in open patches of shrubs. There is a village of 20 thatched huts on wooden stilts scattered among the trees for shade. Inside these Semai huts the floor is made of split bamboo, the walls of woven rattan, and there is no furniture – just a few shelves for pots and pans. One big room serves as a communal lounge and bedroom where 14 people in an extended family sleep side by side.

The Semai depend on agriculture – slash and burn techniques provide 'gardens' for crops of rice and cassava. After the first year's crops, they plant shrubs and later they allow the forest to take over again for 10-20 years before they farm the same plot.

Women look after goats and chickens and collect fruit. Men hunt and fish. Boys learn how to fish with homemade spears and shoot birds and animals in the forest canopy with carefully crafted blowpipes. These skills are passed on by each generation, and a relative balance is maintained with the rainforest. The gardens are small and satisfy the needs of each village. Soil erosion is prevented by the surrounding forests.

'Spiritual land of their ancestors' is designated between each cultivated plot, to ensure that swathes of trees are never cleared. Life is not easy, but it works. Few people arise before dawn and work then takes 3-4 hours. Talking, making time for each other – and making blowpipes – fills their spare time. Talking used to mean no conflict – they have no word for conflict.

Adapted from Michael Redclift, Development and environmental crisis *(1984)*

▲ *An Orang Asli hunter with his blowpipe in the Malaysian rainforest in 2005*

▲ *A traditional Semai village in the Cameron Highlands*

▲ *Hybrid culture – a Semai man carrying a traditional rattan basket and listening to his MP3 player. Change is all around the Semai.*

Should the Semai and other groups like them be drawn into the global economy?

What do you think?

Modernisation

A few miles away, the tide of capitalism has already engulfed the Cameron Highlands with its crowded hotels, cafes, souvenir shops, golf courses and swimming pools. Shops sell cases of forest butterflies – the magnificent Velvet Black and Rajan Brooke Birdwing. The children are the main suppliers. They all carry butterfly nets.

Market gardens grow vegetables for the urban market, loggers work sections of the forest, and rivers are coffee coloured with erosion.

The Semai are adapting to the modernising economy. They have kerosene lamps, bicycles, radios and televisions running on old car batteries. Men wear boxer shorts and T-shirts, women wear blouses and sarongs made in India. To buy these goods, they have to extract more from the forest than they require for their basic needs. Their working day is longer and precious leisure time is lost.

Some 17 year olds, who should know how to make a blowpipe and catch their prey, have never learned how to use one. They have no interest in learning – preferring to watch TV or visit the nearest disco. They wear flashy trainers which were bought by catching rare butterflies and leaf insects and selling them to Chinese traders. They know the price of every species and the value of none.

The destruction of the culture proceeds hand in hand with the destruction of nature itself.

Colin Nicholas, adapted from the website of The Center for Orang Asli Concerns (http://www.coac.org.my)

Conflict is a new word

Land is now being allocated by the Government on a scale that is too small for the Semai to subsist on. They need to take paid employment in order to survive and, consequently, they become part of the national economy as low-paid peasants. They can no longer determine their own futures – with land being designated for dams, new commercial farms, an airport, and a golf course.

In March 2007, the authorities sent the bulldozers in to clear forests for a botanical garden and dam. Six hundred Semai protested about this and halted the project. Their language gained a new word – conflict.

Colin Nicholas, adapted from the website of The Center for Orang Asli Concerns (http://www.coac.org.my)

▲ *Semai children in western clothes watching television*

Background

Cultural landscapes

Culture changes all the time. Economic change affects how people see the world. Great works of art and literature reflect the times and places of their creation. John Constable's paintings and Thomas Hardy's novels showed rural societies in times of industrial change, while Charles Dickens and L.S. Lowry (see below) explored urban populations and landscapes. Economic development changes landscapes – and therefore people's reactions to them. How have current global changes affected the ways in which people see the world?

▲ *Art and literature are expressions of cultural landscapes*

Globalisation as a threat to cultural diversity

Perhaps globalisation is the twenty-first century term for cultural imperialism, which previous generations called Americanisation, Westernisation or modernisation. As the global economy draws people closer together, the big brand names like Coca Cola, McDonalds, Disney and Nike become part of a global diet. Microsoft and News Corporation are examples of how communications technology and media systems can shape the language we use and how we view the world (also see pages 172-175).

Cultural differences are expressed through language, and phrases do not easily transfer from one region to another. The real 'feel and soul' of a culture is often lost in translation. As control of the global broadcasting, film and music industries becomes more concentrated in the hands of large media TNCs, the use of increasingly common vocabulary begins to erode cultural diversity. This is known as the global homogenisation of culture – with everywhere becoming the same. It is not always a process of dominance and succession; sometimes it's more 'pick and mix', with everything becoming muddled.

Such hybridisation of cultures also means that geographical differences are lost. Cultural differences become blurred as people take their cultures with them when they migrate and then begin to adopt and adapt to what is on offer at their destinations.

So are we all part of a global village?

The global village concept implies close connections between all members of a global society. The reality is far from this – many millions of Indians, Africans and Asians do not have access to television, the Internet, telephones, or cinemas. A 'global culture', if it exists, develops where people are connected – and that means those with wealth and access to modern communications technology. A recent Financial Times advertisement (below) illustrates this point well, by suggesting that it is really an urban world rather than a global village. Millions outside this urban world are more likely to retain their inherited cultural values and attitudes, even if they occasionally wear western-style clothing and drink the odd bottle of Coca Cola or Pepsi.

The attack of the clones

The next two pages look at other cultural shifts, and some attempts at resistance in the name of protecting diversity.

- Resources 9-10 are about the erosion of cultural diversity in wealthy nations.
- Resources 11-12 explore examples where attempts are made to protect cultural diversity.

- Are cloned cultural landscapes inevitable?
- Is it possible for governments to resist the tide of global consumption?

What do you think?

Resource 9

Death of a nation of shopkeepers

Napoleon dismissed Great Britain as 'a nation of shopkeepers', lacking the martial spirit needed to defeat his bid for European dominance. His bid failed, but in recent years the individuality of Britain's shops has also begun to fail – falling victim to the process of cloning, as town and city centres all start to look the same.

In 2004, a survey across Britain by the New Economics Foundation confirmed that 'the individuality of local high street shops has been replaced by a monochrome strip of global and national chains'. It found that many town centres have become 'somewhere that could easily be mistaken for dozens of other bland town centres across the country'. The homogenisation of high streets seems to be an accepted product of progress, but leads to a loss of choice for the consumer as well as the loss of character. Large retail multiples can then take profits out of the local economy and undermine diversity even further.

By making every town centre the same, this identikit culture has a darker side – the death of diversity. By closing small businesses at a local level, the TNCs who develop and build our town centres or malls are making decisions a long way away from the places that they will affect. This, claims Andrew Simms, '… undermines democracy, attacks our sense of place and belonging, and therefore well-being. It hands power to corporate elites.'

Adapted from New Economics Foundation (NEF), Clone Town Britain (2004)

Resource 10

Media manipulation

Four companies – Universal, EMI, Warner Music and SonyBMG own 80% of the world music market. Universal and Sony take a 25% share each, but are themselves part of the much bigger Bertelsmann and Vivendi organisations. The ability of these companies to control the music scene is phenomenal, and - since 2000 - they have been cutting their stables of recording artists to increase their profits. It makes good economic sense to sell 10 million CDs by one artist – such as Madonna – rather than paying to market 10 different artists selling 1 million CDs each.

- The ownership of news broadcasting is also becoming more concentrated. For instance, five corporations control 90% of America's news coverage.
- Rupert Murdoch's News Corporation (see page 175) owns 800 companies around the world, including terrestrial and digital TV channels, news networks, newspapers, magazines, major book publishers, film companies, record companies and sports teams.
- Italy's Mediasat can provide an entire day of leisure for an average Italian, who might spend a Saturday shopping at his local supermarket (owned by Mediasat), then relax in his home – reading a newspaper, flicking through a few TV channels to watch AC Milan play football – all owned by Mediasat.

Composite pop culture, music and cloned programming all lead to mass-marketed, formulaic superstars and programmes like 'Pop Idol' and 'CSI' pumped into living rooms everywhere – regardless of their appropriateness or cultural impact. For instance, 'Who wants to be a Millionaire?' was India's number one TV show in 2008.

▶ From church to supermarket – the shape of things to come? In the 1990s, Sunday opening of supermarkets and other large shops was permitted by law. Religious leaders objected in vain. Have superstores now become the cathedrals of capitalism and has shopping become the new religion?

Resource 11

Iran enforces Barbie ban

Tehran, Iran. The suspect fits the following description: slim, curvaceous, perpetual smile, no head scarf. Goes by the name of Barbie. Iranian police are combing the shelves for the plastic icon of American culture. The doll's uneasy sojourn in the Islamic Republic could be drawing to a close. Agents have been confiscating Barbie from toy stores since a vague proclamation earlier this month denouncing the un-Islamic sensibilities of the idol of girls worldwide, shopkeepers said today. 'They took them all,' said a toy seller whose shop window is plastered with the flower-shaped Barbie logo.

Iran's top prosecutor has called for restrictions in the import of Western toys, saying that they have a destructive effect on the country's youth. He wants measures taken to protect what he called Iran's Islamic culture and revolutionary values, and said that products such as Barbie, Batman, and Harry Potter would have negative social consequences.

But, in Tehran, young people wear T-shirts emblazoned with a tick, watch 'Days of Our Lives' and access 'banned' social and file-sharing networks like Facebook.

Adapted from The Sydney Morning Herald, 23 May 2002

Resource 12

Glocalisation?

Glocal cultures develop where the world exists at a local level. This is not some theme park with 'cultural zones', it is how major cities have been transformed by decades of inward migrations. In 2008, Peterborough reported that 10% of its population was from Eastern Europe and that local service provision reflected that. When well-formed ethnic enclaves evolve, they gain their own identity – like London's East End 'Banglatown', where street furniture, road names, festivals and cuisine add to the city's multicultural character and strengthen cultural diversity.

▲ 'Banglatown' and its festivals adds to the cultural diversity in London's East End.

Option summary

● On your own

Look at Resources 1-12 on pages 289-293 and then answer the following questions. Each answer should quote named examples of people, places or incidents as evidence, as you will have to do this later in the exam.

1 In what ways are global cultures changing:
   ● in remote areas?
   ● in urban areas?
2 Does it matter if a process of cultural homogenisation takes place?
3 Should major companies try to preserve cultural diversity?
4 Should cultures be allowed to evolve to meet people's needs?

Useful websites for research

Remember that this chapter is only a taster of what 'The world of cultural diversity' is about. Although it raises several issues about cultural landscapes, it will not cover the full requirements of Unit 4! The following websites will help you to extend your research further.

News items about cultural changes in the UK can be found at:
● BBC News website – www.bbc.co.uk/news
● Multicultural UK – www.multicultural.co.uk/multiculturallondon.htm

International sites about globalisation and cultural diversity include:
● The Global Policy Forum – http://www.globalpolicy.org/
● National Geographic website – www.nationalgeographic.com/news

Use key terms to search from this study. For example: cultural diversity, hybridisation, homogenisation, disappearing cultures

Other useful websites about cultural changes include:
● for global reports – http://portal.unesco.org/culture
● for defining cultural shifts – www.stephweb.com/capstone/capstone.shtml

Other similar incidents

Building up a file to illustrate aspects of cultural change around the world can strengthen your understanding of this topic. Useful examples include:

● Bhutan – a small Buddhist kingdom in the Himalayas that has long avoided contact with the outside world, but since the opening of an international airport in 1983 – and the arrival of television in 1998 – is facing up to stark choices for its future.
● The Yanomani Indians of the Brazilian rainforest who were only 'discovered' in the 1970s and have been fighting to preserve their way of life since.
● Saipan – a Pacific island where 50% of the population now consists of immigrant workers from China, the Philippines, Sri Lanka and Thailand.

Films books and music on this theme

Music to listen to
'Ghetto Gospel' by 2Pac/Elton John (2004)
'Colours of the Wind' by Vanessa Williams (1995)

Books to read
The City of Falling Angels (2005) by John Berendt
Guns, Germs and Steel (1998) by Jared Diamond
No1 Ladies' Detective Agency (a series of books from 1999-2008) by Alexander McCall Smith

Films to see
'Lost in Translation' (2003)
'On Deadly Ground' (1994)
'Fern Gully: The Last Rainforest' (1992)

Try these exam questions

When you have completed a programme of research as preparation for the examination, try these questions:

1 How might cultural values influence the way in which people see and use the natural environment? **(70)**

2 Assess the view that globalisation is creating one uniform global culture. **(70)**

3 Discuss the role of modern communications systems in the loss of cultural diversity. **(70)**

To understand how these questions are marked, refer to pages 327-328.

Introducing this Option

Pollution incidents vary. Most are linked to economic and human activity – from a factory, traffic, etc. Although media reports about many incidents focus on the effects of pollution on ecosystems, like oil spills from tankers killing wildlife, this Option focuses on pollution and human health. Incidents and their effects on health vary.

- Some are dramatic. In Bhopal, a city in India, toxic gas escaped from a chemical plant in November 1984, killing thousands at the time and leaving a legacy of long-term sickness in the city (see page 262).
- Some are one-off, local incidents. In Camelford, Cornwall, a water filtering plant caused a spill of sulfuric acid into the River Camel. This had serious impacts on ecology and water quality, and the health of hundreds of people who drank the water.
- Some pollution is long-term. Over time, prolonged exposure damages health, such as the example of asbestos on the following pages.

What is this Option about?

This option will probably appeal to you if you like studying Biology, Environmental Studies, Economics, Politics, Ecology, or have an interest in current affairs.

The Specification lists four Themes in this Option. If you choose it, you will learn about:

1 Defining the risks to human health: What are the health risks?

In which you will learn about:
- human health risks
- health risks at different scales (e.g. global, local)
- health risk patterns over time
- how health affects both quality of life and economic development

2 The complex causes of health risk: What are the causes of health risks?

In which you will learn about:
- the complex causes of health risks
- links between socio-economic status and health
- links between diseases and geographical features
- models that may help in the understanding of health risk causes and patterns

3 Pollution and health risk: What is the link between health risk and pollution?

In which you will learn about:
- the link between different pollution types and the health of societies
- the relative health risks associated with incidental and sustained pollution
- the link between pollution, economic development and changing health risks
- the role of pollution fatigue in reducing health risk

4 Managing the health risk : How can the impacts of health risk be managed?

In which you will learn about:
- the socio-economic and environmental impacts of health risk
- how health risk impacts have led to different management strategies and policies
- the different agencies involved in health risk, especially international efforts
- which health risks can be managed effectively and which cannot; and the role of sustainability

How to use this chapter

This chapter contains six pages of resources.

- Page 296 introduces the topic of asbestos, with a case study about Wittenoom in Western Australia.
- Pages 297-298 investigate the issue further using resources. You will be guided but you will need to think about each resource carefully.
- Page 299 provides background information about asbestos and risk.
- Pages 300-301 investigate the issue in other areas of the world, again using resources.
- Page 302 provides activities, useful websites for further research, and examples of the kinds of questions that you will meet in the exam (each worth 70 marks).

Wittenoom – boom town to ghost town

Wittenoom Gorge is in Western Australia, in a region known as the Pilbara. The deep red sandstone gorge is one of many in the Hammersley Ranges part of the region.

The Pilbara is booming. Some of the world's largest reserves of high-grade iron ore are located here, giving the landscape its red colour. The mining industry is flourishing due to increasing demand from China and India (see pages 158-159). So is tourism, with the Hammersley Ranges enjoying a boom as more Gap Year students, career-break adults, and 'sixty-somethings' travel to see Australia's more remote areas.

However, despite the boom in the Pilbara, the Wittenoom Gorge is deserted and the town of Wittenoom has been almost abandoned. Once a boom town, the origin of Wittenoom lay in the thick seams of crocidolite (otherwise known as blue asbestos) that were mined in the Gorge. The town of Wittenoom grew to house the mineworkers. During the 1950s, 20 000 men, women and children lived and worked there. For decades, these workers and their families breathed in asbestos dust. Waste from the mine was used to surface the roads, and for the foundations of buildings. Enormous waste dumps still cling to the side of the Gorge (see the photo). One report described school trips going to the area and pupils 'scree running' down the waste heaps.

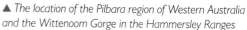

▲ The location of the Pilbara region of Western Australia and the Wittenoom Gorge in the Hammersley Ranges

Nowadays we know that exposure to asbestos can cause asbestosis and the cancer mesothelioma. Workers at the Wittenoom mines started to die. The resulting court cases, in the 1970s, found that the dangers of exposure to asbestos were known well before the Wittenoom mine closed in 1966. In the late 1970s, the State Government of Western Australia tried to close the town by withdrawing services and declaring it a health risk. A few residents refused to move. A clean-up operation was attempted in the early 1980s, but it proved impossible. Now, only a few determined residents remain – there is a basic caravan park but no power supply and no fuel. During 2006, three residents took up the Government's offer to buy their houses for A$40 000 (about £18 000), plus a grant of 10% and A$10 000 relocation costs. The town's population is now down to eight.

The Pilbara has several gorges like this one at Wittenoom. However, not all of them are as dangerous as Wittenoom. The cause of the danger is Wittenoom mine, abandoned and left since 1966. There were three mines in total: Wittenoom, Colonial and Yampire. ▼

So … how safe is Wittenoom now?

Consider these resources:

- Resource 1 is about Wittenoom in 2007.
- Resource 2 assesses the risks of being in Wittenoom for different groups of people.
- Resources 3-4 are about the companies who mined and processed asbestos products.

> *Should Wittenoom be sealed off for good and further access prevented?*
>
> **What do you think?**

Resource 1

Wittenoom gone – but eight stay

By **Nick Taylor**, 20 June 2007

The infamous asbestos mining town of Wittenoom has finally been wiped off the map by the Western Australia government. Regional Development Minister Jon Ford said that the risks for most types of land users in and around Wittenoom were in a medium-to-high risk. Mr Ford said the removal of town status would help to progress the closure of the town, including closing all roads into the area.

But the eight residents who still live in the Pilbara ghost town have vowed to stay. 'We're not going anywhere,' said Lorraine Thomas, who owns the Wittenoom gem shop and has lived there for 23 years. 'This is our home and whether it's on the map or not doesn't mean a thing to us.' Mrs Thomas had a solar-power system installed in the town on Monday to supply four homes owned by her and her husband, Les. The State Government turned off the electricity a year ago.

(An extract from an article from *Australian Associated Press*)

Resource 2

Descriptor	Indicative response
Extreme	The land is not suitable for the proposed/current use. **Remediation** is required, or conditions applied, to control activities at the site for other use. There is imminent risk of exposure to harmful levels.
High	The land is not suitable for use unless remediation is carried out, or conditions are applied to control the activities at the site.
Medium	The land may be suitable for use. Give consideration to whether remediation may be required or conditions applied to control the activities at the site.
Low	The land is suitable for use without conditions.

▲ *An assessment of risk carried out by the Western Australia Government when it researched the Wittenoom area in 2006. The table assesses risk in different places, and the different people who are at risk.*

USERS	STUDY AREA			
	Wittenoom mine	**Wittenoom Gorge**	**Flood plain**	**Town site**
Aboriginal people	High	Extreme	Extreme	High
Tourists	High	Extreme	Medium	Medium
Remediation workers	Medium	Medium	Low	Medium
The Press	Medium	Medium		Medium
Mining industry		High	Medium	Medium
Pastoralists			High	High
Residents				Extreme
Transitory road users				Low
Road users (pedestrian)				Medium
Construction contractors				High

- Environmental **remediation** means removing pollution from the environment (such as the soil, air, ground or surface water), for the general protection of human health and the environment.

The invisible hazard – 2

Asbestos compensation – who is responsible?

CSR Limited

CSR Limited is a major Australian company, which produces aluminium, sugar and construction products. Between 1948 and 1966, it operated the asbestos mines at Wittenoom. During this time, workers, their families, visitors, tourists, and officials were exposed to lethal levels of blue asbestos. Despite warnings from the Western Australia Health Department, CSR continued operations - and was later charged with negligence for this.

That charge opened the way for court cases for compensation from the people affected by exposure to asbestos. The first court victory for a Wittenoom victim was in 1988, when a judge ruled that CSR acted with 'continuing, conscious' disregard for its workers' safety, and awarded the victim A\$676 000 in compensation and punitive damages.

It is estimated that, by 2020, a third of the people who lived in or passed through Wittenoom while the mines were operating will be diagnosed with a fatal disease caused by exposure to the blue asbestos. This would total 2000 cases and cost CSR payments for damages of around A\$500 million.

CSR initially resisted claims that it was also liable for harm suffered by those who worked for its **subsidiary companies**. Eventually it agreed to compensate those victims as well. CSR pays for the damages out of its current earnings.

James Hardie

James Hardie describes itself as a 'leading international building materials company. Australian-owned, it manufactured building products containing asbestos through smaller subsidiary companies. Now, its legal liabilities for compensation claims are being tested in the Australian courts.

- In 1999, the company moved its assets and headquarters to the Netherlands – claiming 'tax reasons' for the move. With this move, it could no longer be prosecuted under Australian law to compensate Australian victims of asbestos.
- However, in 2001, it set up a charitable trust with A\$293 million to fund compensation claims (instead of paying out of its profits).

- James Hardie also handed over the ownership of its subsidiary companies to the trust – meaning that the trust, and not James Hardie, was liable for any compensation claims.
- In 2004, lawyers investigating this trust found that there was a A\$1.3 billion shortfall in what was needed to compensate victims of asbestos. They asserted that more would be needed to meet claims for the next 20 years.
- James Hardie refused to top up its compensation funding, because – as a parent company – it was not liable for its subsidiaries.
- By 2006, arguments raged. The Australian Government accused James Hardie of leaving Australia to avoid paying compensation claims.

Background

What is asbestos?

Asbestos is a naturally occurring mineral. There are two types – white and blue. Each consists of densely packed fibres, which can be milled into powder. However, the milling process creates huge quantities of dust. Blue asbestos (known as crocidolite) is more lethal, with fine fibres that are more easily inhaled. During the period between 1950 and 1980, asbestos was widely used in:

- building materials – as a fire retardant
- industrial processes involving high temperatures – as an insulator
- brake linings – as a damper where friction caused high temperatures

What is asbestosis?

Asbestos inhalation can lead to two diseases - asbestosis and mesothelioma.

- Asbestosis is an incurable lung disease. Dust fibres inflame the wind tract and affect breathing. It can leave patients breathless and needing oxygen masks, and can develop into lung cancer.
- Mesothelioma is a cancer of the chest lining. It is almost always fatal and most patients die within a year of diagnosis.

Because asbestos fibres are so tiny, they lodge in the lungs and any ill effects develop slowly. Though negligible for up to 10 years following exposure, the ill effects are substantial after 15-20 years.

Who is at risk?

Construction and vehicle booms in Europe, Australia and North America during the 1960s and 1970s led to widespread use of asbestos in those industries. The greatest exposure for workers there was in about 1970. As former workers age, the number of cases of asbestosis has increased. Those most at risk worked in mining and milling asbestos, and in the building trades. Men born between 1943 and 1948, who worked in the high-risk industries, are most at risk.

Peak deaths from mesothelioma are likely to be in about 2020 (see the graph below). Asbestos removal has reduced the risk, so death rates will eventually fall.

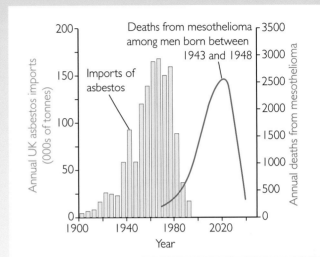

▲ The rise in imports of asbestos into the UK, and predictions of deaths from mesothelioma. From an article in the medical journal, The Lancet.

- The bars show annual UK imports of asbestos in thousands of tonnes.
- The line shows the annual number of deaths from mesothelioma in men born between 1943 and 1948 – the generation most affected by working with asbestos in the UK.
- Notice the 'time lag' between asbestos usage, and the number of cases.

Asbestos is now known to be responsible for more occupationally related deaths than any other cause. Although the numbers affected are nowhere near the numbers of deaths caused by heart disease, they do compare with other important causes of death (see the table). By 2025, there could be 20 000 deaths annually. In the USA, asbestos may eventually kill 5.4 million people.

Cause of death	Number of deaths in 2005	Trend
Leukaemia	3910	Rising
Transport accidents	2740	Falling/stable
Melanoma	2075	Rising
Asbestos-related conditions	1843	Rising
AIDS-related illnesses	500	Stable

Deaths in the UK linked to asbestos, compared with other causes. Source: The Office for National Statistics, 2006 ▶

Asbestos closer to home

The next two pages take the story of asbestos away from Australia, and look at the implications of asbestos use in other countries.

- Resources 6 and 7 are about cases of asbestosis from Leeds.
- Resources 8-10 explore what has happened to the risks from asbestos, and how people try to cope with it.

Resource 6

June Hancock's story from Leeds

In Armley (an inner suburb of Leeds), lies an old mill that was owned and operated by JW Roberts as an asbestos mill. The mill's history is a lesson in globalisation. JW Roberts was taken over by Turner and Newall (T&N), an industrial conglomerate, and, in 1998, T&N was taken over by Federal Mogul, an American company. When it bought the mill, each company became responsible for the legal liabilities of JW Roberts.

Under JW Roberts' ownership, asbestos dust from the mill used to accumulate and blow around the streets of Armley. At that time, the white fibrous dust is said to have filled the air and built up in drifts in corners and against walls. When the children played in the streets and park, some are said to have thrown asbestos 'snowballs' – as the deadly particles steadily accumulated in their lungs. Mill workers used to come home with dust on their clothing.

In March 1996, June Hancock, who grew up in Armley, was awarded £65 000 compensation from T&N for damage done to her health by asbestos from the JW Roberts mill. She had mesothelioma, diagnosed in January 1994 when she was 59. Her compensation was the first awarded in British courts, and opened up thousands of other claims. She died in 1997.

The victims of the JW Roberts factory are not the only ones in Leeds to suffer. Joyce Ives lost her husband Trevor in 2002. The couple had run the Cardigan Arms pub on Kirkstall Road, Leeds. Asbestos lagging in the beer cellar had broken up and was hanging from pipework, releasing asbestos dust into the air.

Resource 7

The claims build up

By 2020, worldwide insurance claims from asbestos victims are expected to total over US$50 billion. Over 200 000 asbestos claims have been made. By the late 1990s, T&N had paid out over £250 million globally. The new owners, Federal Mogul, would have been unable to survive financially if all claims had to be paid at once. So they voluntarily filed for Chapter 11 Bankruptcy in the USA. This is a legal process which allows companies under financial stress to continue to trade and make profit, but puts all claims (e.g. court judgments) on hold.

- *For how long are companies responsible for past errors?*
- *Should polluters pay, wherever they are? If 'yes', under whose law could they be prosecuted?*

What do you think ?

Exporting the problem elsewhere

As health and safety restrictions have regulated the use of asbestos in the USA and the EU (including the UK), the asbestos industry has declined. But world production hasn't! Building and industrial processes still rely on the almost unique properties of asbestos – for instance in heat absorption from tension cables in lifts. The production and processing of asbestos has now shifted to developing countries where health and safety regulations are less strict.

Research in the 1990s showed that:

- exposure to asbestos was not monitored in South Korea, and there was no protection for workers. The exposure of workers to asbestos dust was higher there than either American or German standards would allow.
- in Taiwan, asbestos particles were found in the lungs of asbestos workers.
- loss of lung capacity was recorded among workers in asbestos mines and mills in Zimbabwe.

▲ Breaking up asbestos with no protection in an Indian ship-breaking yard. In this shipyard, 1 in 6 workers has asbestos poisoning.

Who is to blame?

In the USA, a Government directive required that all schools built in the 1940s and 1950s had to use asbestos-based materials in order to reduce any fire risk. However, research gradually built up evidence during the 1950s and 1960s that exposure to asbestos might prove harmful. When the link between asbestos and mesothelioma became clear, the door opened for thousands of compensation claims. However, like most governments, the American Government was exempt from such legal action.

In cases like this, the only way open for compensation claims is to sue the building companies who used the asbestos in their construction. Those who sue have to prove a case to show that companies knew of the risk of exposure to asbestos. This is not difficult. A study of British asbestos workers found that 80% of those working in the industry had asbestosis ... in 1928! However, the link between lung cancer and asbestos was not proved until 1955, and between mesothelioma and asbestos until the mid-1960s. Workers were not the only ones affected; a woman died in Cardiff in 1995, almost certainly from the asbestos that she shook from her husband's clothes every evening before she washed them.

▲ How workers now protect themselves against asbestos dust in European countries

● On your own

Read and look at Resources 1 to 10 on pages 297-301 and then answer the following questions. Each answer should quote named examples of people, places or incidents as evidence – because you will have to do this later in the exam.

1 What impacts does asbestos have on:
   • people who work with it?
   • people who live close to it?
   • people who manage areas where there is risk?

2 a What are the arguments in favour of making polluters pay?
   b What difficulties stand in the way, based on these case studies?
   c Could these difficulties be resolved?

Useful websites for research

Remember that this chapter is only a taster of what 'Pollution and human health at risk' is about. Although the asbestos example makes a useful case study of pollution and the risks posed to human health, it will not cover the full requirements of Unit 4! The following websites will help you to extend your research further.

News items about asbestosis can be found in the UK at:
● BBC News website – www.bbc.co.uk/news
● Sky News website – http://news.sky.com

In Australia, for more news on Wittenoom try:
● The Australian newspaper: www.theaustralian.news.com.au/
● The Sydney Morning Herald: www.smh.com.au

Use key terms to search from this study. For example: asbestosis, mesothelioma, Wittenoom, deaths from asbestosis

Other useful websites about pollution include:
● For air pollution in the UK: www.airquality.co.uk
● For water pollution: www.water-pollution.org.uk (basic but sound!)
● Friends of the Earth website: www.foe.co.uk
● The Environment Agency website: www.environment-agency.gov.uk

Other similar incidents

Building up a file of pollution incidents will help you to understand the issues about pollution and human health. Other useful incidents include:

● The London smog in 1952, where a lethal combination of weather conditions and smoke pollution caused deaths from bronchitis and other lung infections.
● Bhopal in India in 1984, where a faulty pipe in a chemical plant ruptured, causing toxic gas to escape, poisoning thousands of people.
● Chernobyl in 1986, where human error caused an explosion in a nuclear power station in Chernobyl, Ukraine, that spread pollution across Europe.

Films, books and music on this theme

Music to listen to
'Blue Sky Mine' (1994) by Midnight Oil

Books to read
The Appeal (2008) by John Grisham

Films to see
'Erin Brockovich' (2000), which is based on a true story

Try these exam questions

When you have completed a programme of research as preparation for the examination, try these questions:

1 Explain why international initiatives are increasingly needed to cope with the risks of disease and pollution. **(70)**

2 Assess the view that the social costs of pollution are usually greater than the economic costs. **(70)**

3 How far can pressure groups influence the management of pollution issues that affect people's health? **(70)**

To understand how these questions are marked, refer to pages 327-328.

Introducing this Option

Rural areas are being consumed by a wave of leisure and tourism – challenging local economies once devoted to food production or dependent on other primary activities. This Option focuses on the challenges posed by this ever-expanding sector of the global economy. Rural areas, which previously produced for consumption, are now being consumed themselves. The challenges vary according to the location.

- In the rural-urban fringes, short spells of leisure time are spent on a wide range of formal and informal activities that consume the landscape, such as golf courses, sports arenas, country parks, theme parks, walking and 'horsiculture', which are forcing people to reconsider how the countryside appears and is used.
- The Spanish Costas and Australian Gold Coast went from boom to almost bust as their carrying capacities were exceeded by over-development – degrading the natural appeal which first attracted the tourists and putting a strain on natural resources such as water supplies.
- Some remote wilderness areas, such as Antarctica, are now being developed for tourism. The litter trails of the Himalayas and Machu Picchu, show what could happen here if uncontrolled tourism is allowed.

What is the Option about?

This Option will probably appeal to you if you like studying leisure and tourism, economics, politics, ecology and sociology, or have an interest in current social, economic, land management and environmental issues.

The Specification lists four Themes in this Option. If you choose it, you will learn about:

I The growth of leisure and tourism landscapes: What is the relationship between the growth of leisure and tourism and rural landscape use?

In which you will learn about:
- the rise of leisure and tourism and the spread of the pleasure periphery
- the range of rural landscapes, from urban fringe to wilderness, used for leisure and tourism
- the attitudes of different groups of people involved in this relationship, e.g. governments, intergovernmental agencies, businesses, pressure groups, communities and individuals
- how different leisure and tourism activities in rural landscapes may lead to conflicts

2 The significance and fragility of rural landscapes: What is the significance of some rural landscapes used for leisure and tourism?

In which you will learn about:
- the physical significance and ecological value of some rural landscapes
- how rural settlements may be classed as fragile landscapes
- the degree of threat to rural landscapes, by using models such as carrying capacity and resilience
- the use of qualitative and quantitative environmental quality measures, and their usefulness in designating protected areas

3 Impact on rural landscapes: What impact does leisure and tourism have on rural landscapes?

In which you will learn about:
- the range of negative impacts which leisure and tourism has on rural landscapes, e.g. trampling, pollution, erosion
- the range of positive impacts which leisure and tourism has on rural landscapes, e.g. wildlife conservation, river restoration, conservation of heritage sites
- how impacts can change over time as the nature and level of use varies
- the threats and opportunities posed in areas of differing economic development

4 Rural landscape management issues: How can rural landscapes used for leisure and tourism be managed?

In which you will learn about:
- whether rural landscapes should be managed or not
- the range of different management strategies, including preservation, conservation, stewardship, sustainable management and the growth of ecotourism
- the attitudes and strategies of different groups and the conflicts that can arise
- the effectiveness of different approaches to managing rural environments

How to use this chapter

This chapter contains six pages of resources.
- Page 304 introduces the topic with a case study of ecotourism in Peru.
- Pages 305-306 investigate the issue further using resources. You will be guided but you will need to think about each resource carefully.
- Page 307 provides background information on the pleasure periphery and models of tourism development.
- Pages 308-309 investigate the issue in other areas of the world, again using resources.
- Page 310 provides activities, useful websites for further research, and examples of the kinds of questions that you will meet in the exam (each worth 70 marks).

12.1 Save the rainforests!

The Posada Amazonas eco-lodge is located deep within the pristine Amazon rainforest of south-eastern Peru – much of it now protected by the Peruvian Government. The Bahuaja Sonene National Park was established by the Government in 1990, and expanded in 1996, to completely protect 1 million hectares of unique rainforest from any development whatsoever. Tambopata National Reserve of 275 000 hectares (also established in 1990) is located next to the National Park. Some limited, small-scale sustainable development is permitted within the Tambopata Reserve, and Posada Amazonas was opened in 1998 on its border – within land owned by the local indigenous community.

Posada Amazonas is a 30-room eco-lodge owned by the local Ese'eja community of Infierno, and managed in partnership with Rainforest Expeditions, a local Peruvian company. The project offers tourists a brief insight into the rainforest ecosystem, and also boosts the economy of the local community.

Successful ecotourism can create new forms of income for local communities, while also encouraging care of the environment. Instead of cutting down trees for timber, the forest can be marketed to tourists as a resource for adventure and education. Also, keeping the number of tourists down to just 40-50 arriving by small boat helps to limit their environmental impact. In this way, tourism can begin to develop an image of 'consumption with a conscience', and valuable landscapes can be marketed many times over without being destroyed.

However, by making 'commodities of culture and nature', it is clear that the indigenous people do have to alter their values. They are – after all – selling their identity, culture and way of life. At Posada Amazonas, there is currently a profit-sharing agreement – with the Infierno tribal community receiving 60% for providing the knowledge, labour, culture and access to 10 000 hectares of tribal lands, and Rainforest Expeditions receiving 40% for managing the operations and staff. However, under the terms of the original contract, the Infierno community will take over total responsibility for Posada Amazonas in 2016.

Ecotourism offers an alternative route to development without destroying the very things that visitors go to see. In Infierno, television and radio had already given the community a view of the outside world, and the Posada Amazonas project does not set out to deny anyone the chance of an improved life – levels of literacy, healthcare and nutrition have all improved in Infierno as a result of the economic benefits brought by the project (see Resource 2).

▲ *Posada Amazonas eco-lodge*

Posada Amazonas eco-lodge:
- was constructed from local, natural materials
- avoids the use of non-renewable materials
- uses recycled materials where possible
- is designed to harmonise with its surroundings
- offers educational, conservation and research facilities
- is a small-scale development

What has Posada Amazonas achieved?

Consider these resources:

- Resources 1-2 are about the effects of Posada Amazonas.
- Resource 3 defines ecotourism.
- Resources 4-6 consider the risks of opening up remote rainforest areas to development.

Resource 1

Posada Amazonas, 2000

As income from ecotourism conflicts with earning a living from more traditional activities, it is not surprising that, four years after the project was begun, normal subsistence activities do seem to be reducing. Equally, modern tools are now being used to cut trees and cultivate the land, and motorboats are replacing canoes.

Handicrafts are now sold as souvenirs, and the children of the forest perceive the economic value of local fauna and flora. Tourism places a value on natural species and, as more tourists arrive, so that value increases – along with the significance of protecting them. Ecotourism adds to sustainability and is reinforced by the role of the indigenous community as stakeholders.

Local farmers gain as their market now extends to both tourists and those locals who no longer have time to farm, although there is a need to designate land for domestic production. A proposed new road designed to boost development was rejected by the community, because of fears about its impact on the wildlife; the very attractions that tourists come to see.

Extracts from an academic report in 2000

Resource 3

Defining ecotourism

Ecotourism is: 'responsible travel to natural areas that conserves the environment and improves the well-being of local people.' So ecotourism activities should:

- minimise impact
- build environmental and cultural awareness and respect
- provide positive experiences for both visitors and hosts
- provide direct financial benefits for conservation
- provide financial benefits and empowerment for local people
- raise sensitivity to host countries' political, environmental, and social climate
- support international human rights and labour agreements

Resource 2

Posada Amazonas, 2008

Rainforest Expeditions shares the management of Posada Amazonas with the elected 'control committee' from Infierno. Eduardo Nycander, one of the founding members of Rainforest Expeditions, has listed the following benefits of Posada Amazonas to the Infierno community and the area by 2008:

- profits of $130 000 in 2007, plus $140 000 in wages – because most of the staff are from the local community
- the provision of training programmes in readiness for full handover of control to the community in 2016.
- improved literacy, healthcare and nutrition levels in Infierno.
- reduced levels of hunting in the rainforest because of the income received from tourists and the value of the wildlife for the success of the project.
- keeping the rainforest unspoilt and undegraded.
- benefits to conservation and social development due to the profits being retained locally.

Nycander believes that, by protecting his own interests, he is helping conservation and making money at the same time. But he warns of potential problems ahead. The success of the 70+ eco-lodge projects across Peru's Madre de Dios region is leading to improvements in local infrastructure.

The road that was previously rejected has become a reality. Road crews are completing an upgrade of the old dirt track and, by 2010, the last 700 km will have been paved to form the Interoceanic Highway to Brazil. This upgraded link between Peru's Pacific coast and Brazil will reduce journey times from 3 days to 1 and open the area up to more visitors. It will run just 15 km away from Posada Amazonas.

Traditionally, new roads through the rainforest tend to lead to development and rainforest destruction up to 50 km deep on either side of the road – through deforestation for logging, mining and agriculture. Nycander hopes that the creation and promotion of an ecotourism corridor alongside the new road will lead to the preservation of up to 150 000 hectares of rainforest which would otherwise be under threat.

Threats to the rainforest

New roads across Brazil provoked waves of uncontrolled development and deforestation extending 50 km either side of the road

▲ *Over use of mud roads in the rainforest can lead to mudslides*

> *Can ecotourism sustain the rainforests?*
>
> **What do you think?**

What can go wrong? – lessons from RINCANCIE

In Ecuador, the Quichua are involved in 20 ecotourism projects, coordinated by RINCANCIE (Indigenous Network of Communities of the Upper Napo for Intercultural Co-Living and Ecotourism). In the 1990s, these projects boosted local incomes when oil and maize prices fell. The tourists are led through the rainforest and then make presentations about their own culture in exchange for the Quichua's traditional singing and dancing. Tours also include educational elements about biodiversity and local family life.

After a few years, the community's enthusiasm was waning and it was clear that western values were influencing local habits. Drugs, alcohol and earrings were becoming common amongst people who previously did not even have words for those things. Female tourists were also known to be arranging late-night rendezvous with Quichua men.

Background

Pushing the boundaries – 'the pleasure periphery'

Increasing leisure time, higher disposable incomes, greater personal mobility, and cheaper air travel, mean that more people have more time to travel further for pleasure. In effect, the boundaries created by travel costs and journey times have been broken, and tourists now reach areas that were once considered remote and inaccessible. The rural-urban fringe and coastal areas little more than an hour from home used to be on the boundary. Those peripheral zones are now being leap-frogged as tourists demand new experiences further and further away. The furthest distance which tourists travel is known as the *pleasure periphery.*

Changing destinations – changing impacts

The development of the tourist industry inevitably changes the character of a destination. Benidorm's high-rise skylines – which represented success in the 1960s-80s – have since been associated with 'tackiness and excess'. Butler's life-cycle model on the right explains how Benidorm was once 'the place' – but success reached saturation point, the **carrying capacity** was exceeded, and a lack of investment resulted in deterioration. Meanwhile, other locations were catching up and exceeding Benidorm as an experience, and its decline became inevitable.

Destinations and their host populations can become 'commodities for sale', and traditional crafts can be distorted to meet the demand for tourist souvenirs – it is common for craftsmen to alter designs to suit tourists' tastes. For instance, woodcarvers in Bali's Ubud district carve wooden giraffes with slotted necks to store CDs, yet the Balinese live thousands of miles away from the nearest giraffe and few of them own CD players!

Too much tourism?

Misunderstandings and misinterpretations can occur when people from different cultures meet, and the risk of conflict is high. Doxey's Irritation Index suggests that a euphoric phase associated with the early prizes

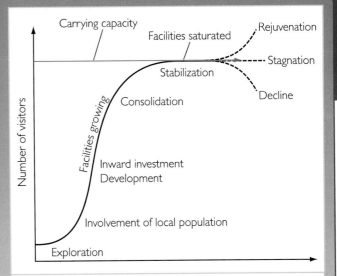

- The exploration and early growth phases involve small numbers of visitors who enjoy real contact with the local people. Few changes are experienced and the local community has full control. The visitor is exploring and opening up a new territory.
- The involvement phase sees benefits from investment, and then – from this point on – a growth in numbers inevitably draws in non-local interests and a significant change to local characteristics.
- Further growth heralds the arrival of the masses.

▲ Butler's tourist destination life cycle

of tourism is transformed into one of antagonism. The arrival of tourists may overshadow the needs of their hosts. Eventual confrontation reinforces Butler's view that tourist numbers will decline and the local economy will be at stake. This situation shows that tourists are not necessarily welcomed – and have therefore exceeded their 'social carrying capacity'.

Places to visit before you die

Designating places as National Parks/Heritage Sites may not protect sensitive landscapes from the pressures associated with leisure and tourism. The Grand Canyon, the Great Barrier Reef, Machu Picchu, Uluru, The Maldives and the Pyramids all feature in the top twenty 'World's Treasures To See Before You Die', and such publicity makes them endangered sites.

12.2 ## Postcards from Paradise?

The next two pages look at other ways in which leisure and tourism is consuming the rural landscape.

- Resources 8-10 are about the wider impacts of leisure and tourism.
- Resources 11-12 explore examples of managing leisure landscapes.

Resource 8

Call some place Paradise and you kiss it goodbye!

On the Hawaiian island of Kaua'i, the cliffs of Nā Pali rise straight from the green Pacific like giant palisades that keep the modern world at bay. It is a lovely illusion, of course. Nā Pali is emblazoned on every tourist map of Kaua'i and a single photo in a 1960 *National Geographic* article on Hawaii unveiled a lush valley amid the ridges offering an unspoiled world to a generation hungry for just such a place. Nā Pali has come to represent paradise on Earth; starring in the movies *Jurassic Park*, *King Kong*, and *South Pacific*.

Despite the difficulty of the trail, 500 000 visitors from all over the world flock there each year. Dozens of campers, some apparently long-term, are scattered among the trees behind the beach. A group of college kids have a boombox blaring, and a woman with bright red hair is shaving her legs in the valley's famous waterfall. Bags of garbage, old coolers, and discarded tents are strewn about the campsites and sea caves – waiting for work crews to haul them out by helicopter. Dan Quinn, state parks administrator, says: 'The challenge of managing Nā Pali is its isolation, which is also its attraction. If more people carried out what they carry in, it would be a better experience for everyone.'

Adapted from Joel K. Bourne, National Geographic *'Fortress Paradise'*

Resource 9

Cultural prostitution

In 2007, there were 7.5 million visitors to Hawaii; by 2010 they expect 12 million. The dramatic development of tourism has affected the relationship between the local people and the land and sea which lies at the heart of their culture.

Tourism has destroyed ancient burial grounds, archaeological sites and sacred places, as new hotels and amenities have been built over them. Many Hawaiians are offended by the exploitation and packaging of their culture for economic benefit. The Rev. Kaleo Patterson says: 'The culture is romanticised to appeal to the exotic fantasies of tourists. This reinforces racist and sexist stereotypes that are culturally inappropriate and demeaning. The issue becomes one of cultural prostitution.'

Adapted from Jonathan Croall, Preserve or Destroy *(1997)*

Resource 10

Wearing out our welcome

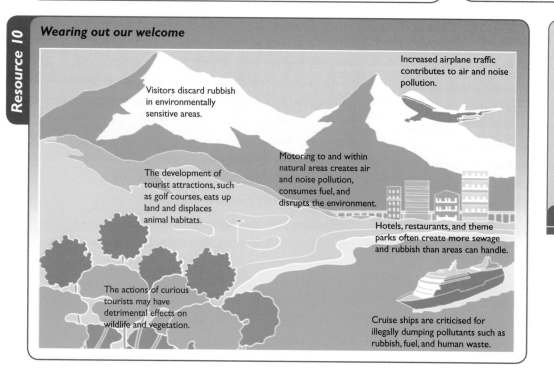

Increased airplane traffic contributes to air and noise pollution.

Visitors discard rubbish in environmentally sensitive areas.

The development of tourist attractions, such as golf courses, eats up land and displaces animal habitats.

Motoring to and within natural areas creates air and noise pollution, consumes fuel, and disrupts the environment.

The actions of curious tourists may have detrimental effects on wildlife and vegetation.

Hotels, restaurants, and theme parks often create more sewage and rubbish than areas can handle.

Cruise ships are criticised for illegally dumping pollutants such as rubbish, fuel, and human waste.

- Is it possible for tourists to adopt the motto: 'Kill only time, take only photographs and leave only footprints'?
- Who should be responsible for managing leisure landscapes?

What do you think?

Whose landscape is it?

Skiing on the edge

> From the ski tourism induced traffic pollution and increasing urban sprawl of hotels and holiday homes in former Alpine villages, to the visually intrusive and habitat-wrecking ski lifts, the ski industry's environmental record is in all honesty more a contender for the wooden spoon than any Olympic medal.
> The Independent, 6 February 2006

In common with many Alpine valleys, the Glemmtal, above Zell am See, suffers from a number of hydrological problems. Steep slopes, thin soils, impermeable rocks and heavy summer storms cause rapid runoff and flooding of the river Glemm. Devastating floods and avalanches were always common.

These natural sensitivities have been made worse by changes in land use associated with tourism. The ski industry flourishes here and Saalbach has grown dramatically with its hotels, bars and shops catering for the winter sports brigade. Farmers have switched to chalet owning, slopes have been deforested as new pistes scar the landscape. Compacted soils and reduced vegetation cause rapid surface runoff and increase the flood risks.

To protect the interests of the tourist economy, the valley is now littered with concrete avalanche traps, raised riverbanks, canalised channels and stepped sediment barriers along the valley side tributaries. The natural landscape has been tamed and the hotel industry is safe.

Walking on the edge

Located 4000 feet (1220 metres) above the Grand Canyon's floor, and sticking out 70 feet (20 metres) beyond its rim, the Skywalk is being described as an engineering first. David Jin, a Las Vegas businessman, funded the Skywalk and gave it to the Hualapai Indians, who own the site, in exchange for a percentage of the profits. They are hoping to attract more visitors to an area which suffers from poverty and high unemployment.

However, some Hualapai don't like the project because they think it is a desecration of sacred ground. 'When that Skywalk came about, it hit me like I was being stabbed,' said Dolores Honga, a tribal elder.

Since March 2007, and for $25 each, up to 120 people at a time have been able to 'walk the sky'. To reach the transparent deck, tourists must drive on twisty, unpaved roads through rugged terrain. The Hualapai hope that the Skywalk will double tourist traffic to the Reservation this year, from about 300 000 visitors to about 600 000.

This project is part of management strategies for the Grand Canyon which are designed to concentrate visitors at selective sites and leave everywhere else relatively free from visitor pressure.

▶ The Skywalk juts out 1220 metres above the floor of the Grand Canyon and has a reinforced glass walkway floor for those with a head for heights!

Option summary

● On your own

Look at Resources 1-12 on pages 305-309 and then answer the following questions. Each answer should quote named examples of people, places or incidents as evidence, as you will have to do this later in the exam.

1 To what extent is the countryside under pressure from the leisure and tourism industry:
   ● in remote areas?
   ● in areas closer to towns and cities?
2 Does it matter if rural economies shift away from primary production to service provision?
3 What are the potential conflicts arising from such changes?
4 Who should be responsible for maintaining the countryside?

Useful websites for research

Remember that this unit is only a taster of what 'Consuming the rural landscape – leisure and tourism' is about. Although it raises several issues about this topic, it will not cover the full requirements of Unit 4! The following websites will help you to extend your research further.

News items about leisure and tourism in the UK can be found at:
● BBC News website – www.bbc.co.uk/news
● Guardian Unlimited – http://www.guardian.co.uk/

International sites on ecotourism include:
● Global Development Research Centre – http://www.gdrc.org/uem/eco-tour/eco-tour.html
● Global Journal of Practical Ecotourism – http://www.planeta.com/ecotravel/tour/year.html
● United Nations Environment Programme – http://www.uneptie.org/pc/tourism/sust-tourism/about.htm

Use key terms to search from this study. For example: commodification of culture, staged authenticity, pleasure periphery, sustainability

Other useful websites about international tourism include:
● For global reports – World Tourism Organisation http://www.world-tourism.org/
● Tourism Concern – http://www.tourismconcern.org.uk/

Other similar incidents

Building up a file to illustrate the impacts of the leisure and tourism industry around the world can strengthen your understanding of this topic. Other useful examples include:

● the development of specialist leisure activities – like safaris, wildlife, adventure and active pursuits – in Kenya, Zimbabwe and Botswana
● opening up cold environments like Antarctica, Iceland and Svalbard
● the increasing numbers of second homes in British National Parks
● the impacts of 'All-inclusive packages' on the Caribbean islands
● the relocation of weekly leisure activities to the urban-rural fringes of American and British cities

Films books and music on this theme

Music to listen to
'The Last Resort'/'Hotel California' by The Eagles (1976)

Books to read
Black Swan Green (2006) by David Mitchell
The City of Falling Angels (2005) by John Berendt
The Lost Village (2008) by Richard Askwith

Films to see
'The Beach' (2001) reveals the attractions of Thailand
'Lord of the Rings' (2001) was filmed in New Zealand

Try these exam questions

When you have completed a programme of research as preparation for the examination, try these questions.

1 To what extent do the impacts of tourism increase as the pleasure periphery expands? **(70)**

2 How far is it possible to predict and then manage the demands placed on rural landscapes by leisure and tourism? **(70)**

3 Can the needs of the leisure and tourism industry ever have neutral impacts on the natural environment? **(70)**

To understand how these questions are marked refer to pages 327-328.

How to use this chapter

In this chapter, you will develop an overview of Unit 3. Although it focuses on Dubai, it also links the six Core topics of the A2 Specification and addresses the question of 'managing the contested planet.' It contains seven pages of resources:

- Pages 311-312 outline the scale of change taking place in Dubai.
- Page 313 provides background information about the United Arab Emirates, of which Dubai is a major part.
- Pages 314-315 use resources to investigate the ways in which Dubai is changing. You will need to think about each resource carefully.
- Pages 316-317 question the sustainability of Dubai's development strategies, again using resources.
- Page 318 provides an additional resource, useful websites for further research, and an example of the kind of question that you will meet in the exam.

Dubai – the world in one city

Dubai, on the northern coast of the Arabian Peninsula (see the map on page 313), is planning for a future without oil – and is now in the process of re-inventing itself as a 'global city'. Converting arid lands into the world's largest building site has attracted thousands of overseas workers, put pressure on resources, and altered the natural environment beyond recognition. Dubai is redefining the concept of sustainable development; development for the future with wealth creation the driving force.

Dubai's oil and natural gas reserves are expected to be exhausted within 20 years, and Sheikh Mohammed bin Rashid Al Maktoum, the ruler of Dubai, has embarked on a period of rapid reconstruction and re-invention to prepare for this. Oil revenues fell to just 3% of national GDP in 2006, so something else has to take oil's place. Dubai is remodeling itself as:

- the world's prestige tourism location. In 2007, tourist receipts exceeded £1.6 billion as 7 million visitors chose Dubai.
- a centre of international finance. It announced its arrival on the international stage by hosting the 2003 World Bank/IMF talks.
- a media, knowledge and information hub at the heart of the world's new economy.
- … and it also plans to become the sports capital of the Arab world, with the construction of Dubai Sports City (see Resource 7).

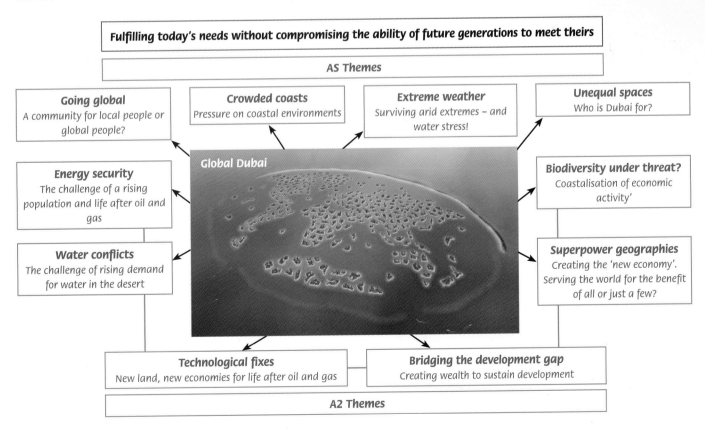

Fulfilling today's needs without compromising the ability of future generations to meet theirs

AS Themes

Going global
A community for local people or global people?

Crowded coasts
Pressure on coastal environments

Extreme weather
Surviving arid extremes – and water stress!

Unequal spaces
Who is Dubai for?

Global Dubai

Energy security
The challenge of a rising population and life after oil and gas

Water conflicts
The challenge of rising demand for water in the desert

Biodiversity under threat?
Coastalisation of economic activity'

Superpower geographies
Creating the 'new economy'. Serving the world for the benefit of all or just a few?

Technological fixes
New land, new economies for life after oil and gas

Bridging the development gap
Creating wealth to sustain development

A2 Themes

The Vision – economic success without oil

Vision for Dubai (2010)

Dubai's development strategy part 1, 2000–2010

Dubai's primary objectives are:

- to attain the status of a developed economy by pursuing planned economic growth through modernisation and diversification
- to build the first non-oil economy in the region

> We aim to create an infrastructure, environment and attitude that will enable Information and Communications Technology (ICT) enterprises to operate locally, regionally and globally – from Dubai – with significant competitive advantage.
> **Sheikh Mohammed bin Rashid Al Maktoum, ruler of Dubai**

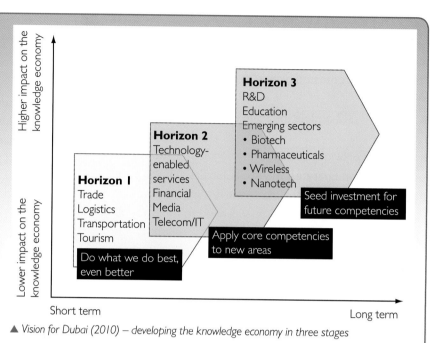

▲ Vision for Dubai (2010) – developing the knowledge economy in three stages

The speed of change – 1990s to today

These photos of Sheikh Zayed Road illustrate the speed of change in Dubai. However, some critics are suggesting that Dubai is moving too fast – construction firms lack capacity, power is in short supply, water is scarce, and workers are demanding higher wages.

Dubai Strategic Plan (2015)

Dubai's development strategy part 2 ... where the future begins

The ruling sheikhs determine the development strategies of the United Arab Emirates (UAE), because the emirates are still controlled by family dynasties. Sheikh Mohammed bin Rashid Al Maktoum, Vice President and Prime Minister of the UAE – and ruler of Dubai – unveiled his revised vision for Dubai in a speech on 3 February 2007:

'We have come a long way towards achieving the objectives of an economy independent of oil. Indeed we have exceeded all expectations and predictions. When I announced my Vision for Dubai, in the year 2000, I spoke of economic aims for the year 2010. In fact, not only have these aims already been realized, but they were exceeded in half the time.

In 2000, the plan was to increase GDP to $30 billion by 2010. In 2005 that figure was exceeded, with GDP reaching $37 billion. The plan also included an increase in income per capita to $23 000 by the year 2010. In 2005 the average income per capita reached $31 000. In other words, in five years we exceeded the economic targets that were originally planned for a ten-year period.

Based on the economic performance of recent years, and on expected future global trends, the economic objectives for Dubai for the year 2015 are to sustain real economic growth at a rate of 11% per annum, to reach a GDP of $108 billion in 2015, and to increase income per capita to $44 000 dollars.'

Background

The United Arab Emirates

◀ *The United Arab Emirates is a constitutional federation of seven emirates: Abu Dhabi, Dubai, Sharjah, Ajman, Umm al-Qaiwain, Ras al-Khaimah and Fujairah. The federation was formally established on 2 December 1971. The seven emirates make up the world's fourth largest oil producer, and are considered the new commercial hub of the Middle East.*

▶ *A data file for the United Arab Emirates in 2006*

GDP (US$ billion)	129.7
HDI ranking	39
Annual population growth (%)	3.5
Life expectancy (years)	79
Adult literacy rate (% age 15+)	88.7

▼ *Dubai's economic growth, 2000-2005, compared with other Arab countries. Dubai's economic growth has been over 11% per year since 2000 (16% in 2007), which far exceeds that of the other Arab countries, as well as China, India, and the USA.*

Dubai's advantages

Geographical advantages

Dubai is a gateway location – midway between Europe, the Far East, Africa and Russia It provides access to a market covering over 1.5 billion people, and serves as a time-zone bridge between Europe and South-East Asia. It is accessible – served by over 120 shipping lines and linked, via 100 airlines, to over 140 global destinations. It also sits in the middle of one of the world's richest regions, with abundant resources of oil and aluminum – as well as benefiting from its own reserves.

Political and economic stability

Since the formation of the UAE in 1971, Dubai's governing family has provided a stable environment in which business investments can flourish. An open economy, with no exchange controls, quotas or trade barriers, has allowed massive direct foreign investment. Foreign companies operating there take advantage of its minimal regulations, pay no direct taxes on their profits or personal income, pay low customs duties, and are permitted repatriation of 100% of their profits. Only oil companies and international banks pay tax on their profits – at rates of 55% and 20% respectively. Liberal migration policies have also encouraged overseas workers to flock to Dubai – creating a large available workforce.

Economic performance

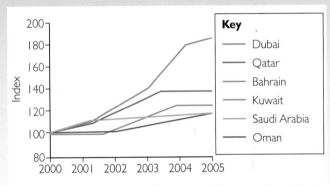

Infrastructure and services

Investment in transport and telecommunications has enabled Dubai to connect itself to the global economy. Designated industrial areas, business parks, free-trade zones, two international seaports – and a new major international airport – all serve as magnets to the emirate (see page 315).

The manufacture of metal products, aluminum ingots, textiles, clothing, gold and jewelry, and consumer electronics – as well as food processing and the oil industry with its associated by-product industries – has been part of Dubai's diversification strategy since the 1990s. Hotels, residential and commercial properties, recreational and leisure facilities have also mushroomed as part of this strategy as multi-national companies have moved in.

13.2 What has been going on in Dubai?

Consider these resources:

- Resources 4-5 are about Dubai's population surge.
- Resource 6 assesses the way in which the future is being built.
- Resource 7 shows the scale of current developments in Dubai.

Why are overseas workers vital to Dubai's success?

What do you think ?

Dubai's population surge

Resource 4

75% of Dubai's population is male!

The Dubai Statistics Centre issued the following statistics in March 2007:

- An extra 292 000 people became Dubai residents in 2006 (the emirate's population increased from 1 130 000 in 2005 to 1 422 000 in 2006).
- Remarkably, 75% of Dubai's total population was male in 2006, and only 25% female.
- The census also revealed that 39% were between the ages of 25 and 34, having nearly doubled from 276 000 in 2000 to over 513 000 in 2006.
- Only 20% of Dubai's population were born there.

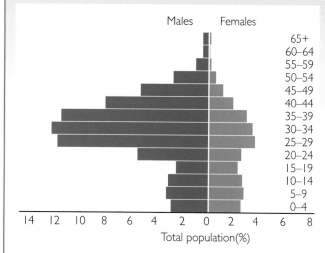

▲ This 2005 population pyramid for the United Arab Emirates typifies the demographic trends outlined above

Resource 5

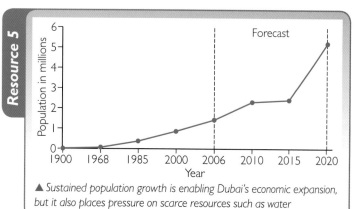

▲ Sustained population growth is enabling Dubai's economic expansion, but it also places pressure on scarce resources such as water

Resource 6

Expatriate workers – welcome to the other side of Dubai

It is a landscape of building sites, full of workers feverishly constructing the highest, largest and deepest in the world. Half the world's cranes now operate in Dubai, on projects worth $100 billion – twice the World Bank's estimated cost of reconstructing Iraq and double the total foreign investment in China (the world's third-largest economy).

But there is a downside. More than 2500 workers at the site of the world's tallest building, the Burj Dubai, went on strike in a country where striking – and unions – are illegal. Indian workers, complaining of unpaid wages and demanding better conditions, marched out of the cramped, stifling dormitories where they live 25 to a room, in protests which cost $1m.

80% of Dubai's population comes from 160 different countries, and this proportion is growing. Entry visas are tied in with jobs and there is always the risk of being thrown out of the country when a job contract ends.

The average pay for an unskilled labourer is $4 a day, which is enough to lure impoverished workers from India, Pakistan, Sri Lanka and Bangladesh. The jobs are arranged through contractors. Those who get offered a job then have to take out loans, often at exorbitant rates of interest, to pay for their transport to Dubai. On arrival, their passports are confiscated to prevent them leaving before the end of the contract. 39 workers died in building-site accidents last year, some due to inadequate safety provisions. Another 84 committed suicide – up from 70 in 2004.

Adapted from The Independent, March 2006

▶ Manual workers' living conditions in Dubai. Labour has to be imported to sustain Dubai's growth.

Creating a new global city

Artificial islands, artificial world

Dubai's natural beachfront is 45 kilometres long. To increase the opportunities for tourism, the coast is being extended. Artificial islands in the shape of huge palm trees, and even a world map, are now adding an extra 1500 kilometres of beachfront to Dubai's coastline (see below).

- The World (**1**) – 300 artificial islands made from sand dredged from the sea floor and either dumped or pumped into forms that mimic the shape of the world's continents (also see the photo on page 311).
- Palm Jumeirah (**2**) – using 100 million m³ of sand and rock to create 60 km of artificial beaches in 17 'fronds'. When finished, it will house 2000 luxury villas, 2500 apartments and more than 50 hotels – creating an instant community of up to 50 000 people.
- Palm Jebel Ali (**3**) – with 2000 homes and entertainment facilities such as an aquatic theme park with whale stadium, a sea village and an amusement park.
- Palm Deira (**4**) – the biggest 'palm island', which will cover an area greater than Manhattan Island in New York when completed by 2015.
- The Universe (**5**), announced in January 2008, and Waterfront (**6**) are both in their early stages. Waterfront will be twice the size of Hong Kong when it is finished.

Economic diversification

Dubai's economic diversification programme includes major new developments:

- Dubailand (**A**) – Dubai's answer to Disneyland, which is costing $5 billion. It will employ 300 000 people, servicing 15 million visitors (7 times larger than Disneyland Paris).
- Dubai Silicon Oasis (**B**) – a centre of advanced electronic innovation, design and development, creating the world's most integrated microelectronics technology park.
- The Dubai International Financial Centre (**C**) – a catalyst for regional economic growth, development and diversification, functioning as a global financial centre like Wall Street in the USA, the City of London in the UK and Hong Kong in Asia.
- International Media Production Zone (**D**) – a Free Zone dedicated to media production activities.
- Business Bay (**E**) – a global commercial and business centre, with fast-track start-up facilities.
- The Dubai Mall (**F**) – intended to be the largest shopping mall in the world, with a world-class aquarium, gold souk, fashion outlets and leisure facilities.
- Dubai Sports City (**G**) – the world's first purpose-built sports city, featuring a 60 000 seat multi-purpose outdoor stadium, a 25 000 capacity cricket stadium, a 10 000 seat multi-purpose indoor arena, and a field hockey venue for 5000 spectators.

Oil revenues are helping to fund Dubai's new developments; oil was $35 a barrel in 2000 and hit $147 in 2008.

Port Rashid Container Port

Jebel Ali Container Port

Jumeirah Beach

Dubai International Airport

Free Zones – designated tax and custom free areas

Consequences of Dubai's development

The next two pages consider some of the consequences of rapid development in the desert.

- Resources 8-9 are about increasing pressures on Dubai's water supplies.
- Resources 10-13 question Dubai's sustainability credentials.

Resource 8

Dubai has only two day's water supply

All of the recent economic developments in Dubai have brought about a massive increase in water demand. Changing lifestyles and an extra million people – with modern properties and services – require huge amounts of water (see the graph). Dubai's prosperity would suddenly end if the taps were to run dry. The city's freshwater reserves are held in two giant tanks at sea and contain just two day's supply. An oil spill in the Persian Gulf would contaminate Dubai's desalination plants and put everything at risk.

The UAE is a water-scarce country – rainfall is slight, evaporation is high and groundwater levels are low. But money buys water in the desert! In March 2008, construction began on the world's largest concrete drinking water reservoirs. With a total capacity of 180 million gallons, the Mushrif scheme will help to meet the exponential increases in demand for water.

◀ *Growth in water consumption in Dubai, 1999-2004*

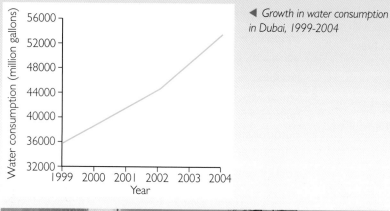

▲ *A plant in Dubai for recycling effluent (sewage)*

Dubai golf drive upsets greens

Leading environmentalists have raised concerns about the rapid construction of golf courses in Dubai, after the emirate announced plans to establish itself as one of the world's leading golfing destinations. Dubai's 11 new courses will be designed and endorsed by Ernie Els and Tiger Woods. A golf course needs about 1 million m³ of water per hectare per year – equivalent to the water consumption of a town of 12 000 inhabitants. Dubai's golf courses are irrigated with desalinated water, treated effluent or a combination of the two.

Last year, the World Wide Fund for Nature reported that: 'They do very little to manage water in Dubai. If a supply shortage looms, they look for ways to increase the amount of water available, rather than conserve it or use it efficiently.'

Caroline Shearing, Daily Telegraph, April 2008

Resource 10

Creating better environments?

▲ *The new wetland at International City (see Resource 11)*

Dubai goes blue!

Sheikh Mohammed launched the Blue Communities Initiative with Dubai's development company, Nakheel, promoting sustainable coastal projects like those used on the Jumeirah Palm and artificial reef. The aims include:

- raising awareness of and leadership in the development of sustainable coastal communities
- involving NGOs, governments and private companies
- incorporating environmentally friendly and reliable transport systems, e.g. a monorail
- using natural materials to encourage coral growth on the breakwater, increasing fish stocks
- creating openings in the Palm breakwater to allow water circulation and maintain water quality
- creating a Ramsar-recognized wetland (see page 129) next to International City, by pumping treated effluent into a disused quarry (see Resource 10)

Dubai goes green!

Air conditioning is everywhere in Dubai, because winters are hot and summers are blazing! Chilled water swimming pools and an indoor ski slope that produces snow when it is 48 °C outside, push up the electricity bill. There is little public transport in most of Dubai.

According to WWF, the UAE's ecological footprint is 11.9 global hectares per person. A global hectare is a unit of the amount of productive land and water a person requires to produce their annual resource consumption and absorb their annual waste. The American footprint is 9.6, and the global average is 2.2.

Dubai also has a heavy per capita carbon footprint – it takes 9 hectares of land to absorb each person's carbon emissions in a year. The American footprint is 5.7, and the world average 1.7.

In 2007, the UAE became the first government to sign an agreement with WWF to study the country's ecological footprint and reduce it to a sustainable level by 2010.

Resource 13

Gridlock!

▶ According to official statistics, the number of registered vehicles in Dubai increased from 415 242 in 2002 to 853 827 in 2006. Dubai has a car ownership rate of 541 cars per 1000 people – more than New York City (444), London (345) and Singapore (111). If this trend continues, there will be 5.3 million registered cars in Dubai by 2020.

Synoptic summary

City searching for its soul

A complex social experiment is taking place in the desert heat to create a global megalopolis and communications hub. But is it anything more than concrete and money?

Pico Iyer's book *Global Soul* describes a new transnational super-class – people with compound identities and no fixed country of residence – who live 'between categories', such as international sportsmen Roger Federer and Ernie Els, both part-time Dubai residents and always passing through the international airport there on their way to somewhere else. Nearly everyone you meet in Dubai is from somewhere else – from the Indian subcontinent, the Philippines, Iran, Europe, the Americas. They live and work in the emirate but are not part of any larger cohesive community.

There is, in Dubai, no such thing as society. Instead, everyone lives in his or her own discrete micro-community. If there is wider social interaction, it takes place only in the cavernous malls – those temples of ostentatious consumerism – and in the lobbies of the vast hotels.

Adapted from The Guardian, *May 2008*

On your own

This chapter gives an indication of what it means to be synoptic – bringing several geographical issues and threads together in order to understand the challenges of managing a contested planet. Although it provides useful case-study material, it will not tell you everything! Consider the links between each of the themes and find evidence elsewhere in this book, and from your own studies, to illustrate contrasting and similar situations.

Useful websites for research

The following websites will help you to extend your research further.

News items about Dubai can be found at:
- BBC News website – www.bbc.co.uk/news
- UAE News – http://www.uaeinteract.com/
- Gulf News – http://www.gulfnews.com/home/

Specific details about the developments in Dubai can be found at:
- Weekly updates – http://www.tutztutz.com/2008/02/dubai-what-the-hell-is-going-on-over-there/
- Map of new developments – http://www.propertyportal.ae/downloads/Dubai-Map.pdf
- Critical review of developments – http://yaleglobal.yale.edu/display.article?id=5992 and http://blogs.abcnews.com/theblotter/2006/11/dark_side_of_du.html

- Dubai's Strategic Plan 2015 – http://www.metro-press.com/dubai/dubai_07.pdf and http://www.aerlines.nl/issue_38/38_Knorr_Eisenkopf_Emirates_Business_Model.pdf
- The UAE Yearbook – http://viewer.zmags.co.uk/showmag.php?mid=stfrs&preview=1&_x=1

Try this synoptic question

You may use examples from any part of the A Level course to support your answers to the synoptic investigation in the exam (Section B). When you have completed a programme of revision and research as preparation for the exam, try this question:

Synoptic question

a Explain the factors that have helped Dubai to move away from being an oil-dependent economy. **(10)**

b Assess the human and environmental impacts of the development strategies adopted by Dubai. **(18)**

c To what extent are Dubai's strategies in conflict with the basic principles of sustainable development? **(12)**

EXAMS: HOW TO BE SUCCESSFUL

No matter how much you enjoy geography, preparation for the exam is essential. To be successful, you not only need to know and understand the geography that you have been studying, but you also need to know how you will be examined, what kinds of questions you will come up against, how to use what you know, and what you will get marks for. That is where this chapter can help.

What is the Edexcel A2 Geography Specification all about?

It is a new specification written for the twenty-first century. It includes new ideas, e.g. 'The Technological Fix?' and offers a new twist to old geographical favourites, e.g. 'Bridging the Development Gap'. Like the AS, it provides a balance between your own physical, human and/or environmental interests, and key geographical topics that provide you with the knowledge, understanding and skills to ready you for further study in higher education – or for employment.

What does the A2 Specification include?

The A2 Specification consists of two Units, broken down into Topics and Options.

Unit 3 Contested Planet investigates the distribution of resources, plus the physical factors resulting in this distribution. The use and management of resources is a key issue for geographers in today's world. Many resources are finite, and increasing consumption means that difficult decisions need to be made about their future use and management. You will consider how resources are used, and also the problems and costs involved in obtaining them.

Three types of resource are considered in three topic areas:
- *Topic 1: Energy Security*
- *Topic 2: Water Conflicts*
- *Topic 3: Biodiversity under Threat*

Inequalities in global wealth, power and influence are investigated in:
- *Topic 4: Superpower Geographies*
- *Topic 5: Bridging the Development Gap*

The role of technology in overcoming resource scarcity, income inequality and environmental management is considered by investigating:
- *Topic 6: The Technological Fix?*

Part of your learning will involve linking the content and concepts from Unit 3 above with Unit 1 (Global Challenges) and Unit 2 (Geographical Investigations) in the AS course.

Unit 4 Geographical Research offers six Options for research, of which you choose ONE. These range from those with a physical geography focus, to those exploring environmental, social and cultural geographies. They are designed to allow you not just to learn about new geographical ideas, but also to learn how to research independently.

The Options on offer are:
- *Option 1: Tectonic Activity and Hazards*
- *Option 2: Cold Environments – Landscapes and Change*
- *Option 3: Life on the Margins – the Food Supply Problem*
- *Option 4: The World of Cultural Diversity*
- *Option 5: Pollution and Human Health at Risk*
- *Option 6: Consuming the Rural Landscape – Leisure and Tourism*

The more you study the themes in the two A2 Units, the more you should spot links between them – and also between them and the two AS Units. For example:

▶ The Daintree Rainforest in northern Queensland, Australia, is under threat from development (explored in Unit 3 'Biodiversity under Threat'), as more and more tourists visit the area (explored in Unit 4 'Consuming the Rural Landscape – Leisure and Tourism').

▶ GM crop protests in France. GM crops are supposed to offer 'fixes' to the food supply problem (explored in Unit 3 'The Technological Fix?'), but bring social, economic and environmental impacts for those who use them (explored in Unit 4 'Life on the Margins – the Food Supply Problem').

How will you be assessed?

There are two exams for A2 Geography (and no coursework). The table below shows what the exams are like.

Unit	Assessment information	Marks and method of marking
Unit 3 Contested Planet	This exam is 2 hours and 30 minutes long, and consists of two sections (A and B). It includes a pre-release resource booklet (about 4 pages long), which is designed to draw all the themes of the course together – known as the *synoptic* element. The synoptic resources will be pre-released as advance information four weeks before the exam. They will be about ONE of the Topics in Unit 3, and will form the basis of the question(s) in Section B of the exam. Please note that this Topic will NOT be examined in Section A as well. There is no restriction on the prior use of the pre-release resource booklet, but you must not take it into the exam with you. The synoptic resources will be reproduced in the exam resource booklet. **Section A** of the exam consists of five questions about the Topics in Unit 3. You have to select and answer TWO of them. Each question has two parts (a and b), worth 25 marks in total – giving a combined mark for Section A of 50. Part (a) will normally be about a resource, such as a map, diagram or data. Part (b) will normally require you to answer based on your own knowledge and understanding. **Section B** is the synoptic investigation, based on the resource booklet. It consists of one compulsory question in three parts – worth 40 marks in total.	There are 90 marks in total for this exam, which are then converted to a UMS mark out of 120. All questions are level marked.
Unit 4 Geographical Research	This exam is 1 hour and 30 minutes long. Whichever Option you choose for Unit 4, there are four Themes within it. Four weeks before the exam, you will be notified about which two of these Themes will be assessed in the exam. It will be a single essay question drawn from the two Themes. In the exam, there will be six questions in total – one for each of the six Options. You must select and answer the question relating to the Option that you have studied. You will need to write a single long essay, in which you use and apply the results of your research. As with the Unit 3 exam, you cannot take any research materials into the exam room with you.	There are 70 marks for this exam, which are then converted to a UMS mark out of 80. Questions are level marked.

What will the exam questions be like?

The table above explains what type of questions you will get in the exams.
Here are three examples.

Sample question for Unit 3 Section A

2 Study Figure 2.

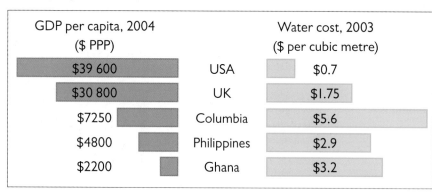

GDP per capita, 2004 ($ PPP)		Water cost, 2003 ($ per cubic metre)	
$39 600	USA	$0.7	
$30 800	UK	$1.75	
$7250	Columbia	$5.6	
$4800	Philippines	$2.9	
$2200	Ghana	$3.2	

▲ Figure 2: Per capita GDP compared to the cost of water for five countries.

a Suggest how water resources and human well-being might be affected by the data in Figure 2. **(10)**

b Using named examples, assess the role of different players and decision-makers in trying to secure a sustainable 'water future'. **(15)**

(Total 25 marks)

Note:
Costs are via piped household connections in the USA and UK
Costs are via informal water sellers in Columbia, Philippines and Ghana
$ PPP (Purchasing Power Parity) is adjusted to reflect the cost of living in each country

Sample question for Unit 3 Section B

6 a Explain the factors that have led to Latin America's rapid adoption of GM farming technology. **(10)**

b Assess the human and environmental impacts of GM farming in Latin America. **(18)**

c To what extent does GM technology provide a technological fix that is economically sustainable? **(12)**

(Total 40 marks)

Sample question for Unit 4

3 To what extent do food security issues vary spatially and temporally?

(Total 70 marks)

Making the step to A2

The biggest step up between AS and A2 is the requirement for **extended writing**. Whereas some questions at AS carry a few marks and require short answers, at A2 the questions range between 10 marks – in part (a) of Section A in the Unit 3 exam – and a full-blown 70-mark essay in Unit 4. Therefore, it is important that you develop your writing skills so that you can earn the full marks.

Developing extended writing

Most candidates can write at length. The difference is whether they can:

● understand the question
● plan their answer, so that it stays focused
● keep track of time
● learn and use case-study examples to support their argument(s)
● organise their essay

The following steps will help you to prepare for this.

Understanding the question

When reading an exam question, check the command words. They tell you what the examiner wants you to do. The table on page 322 gives you some of the most commonly used A2 command words. The A2 command words are different from those at AS. For instance, you are unlikely to find 'describe' at A2, so refresh your knowledge of likely command words using this list.

The key rule is to read the exam question carefully and answer the question that has been set – not the one that you hoped would be set! Therefore, read and interpret the question to work out exactly what is being asked. Review the question through its key words (see the yellow box on the right).

Command words – these have distinct meanings, which are listed overleaf.

Theme or topic – this is what the question is about. The examiner who wrote the question will have tried to narrow the theme down, so that you do not write everything you know about the theme.

Focus – this shows how the theme has been narrowed down – e.g. the environmental impacts of development in China.

Case studies – look to see if you are asked for specific examples.

Here is an example of a question that has been interpreted using the key words above:

Command words
● 'Explain' You must give reasons for the pattern.
● 'Suggest possible impacts' You must give more than one possible impact.
● 'Evaluate' Examine and weigh up the relative importance of threats to the ecosystem.

Theme or topic
This question is from 'Biodiversity under Threat', and looks specifically at the factors and processes that threaten biodiversity.

Study Figure 3.

a Explain the pattern of alien species invasions, and suggest the possible impacts of alien species on ecosystems. **(10)**
b Evaluate the relative importance of global and local threats to one named global ecosystem. **(15)**
(Total 25 marks)

Focus
● Part (a) asks you to 'explain the pattern' of alien species invasions and 'suggest the possible impacts' that this can have on ecosystems generally.
● Part (b) asks you to give a named example of an ecosystem and 'evaluate the relative importance' of threats to it.

Case studies
Choose one named global ecosystem only.

Command word	What it means	Example
Account for	Explain the reasons for. Marks are given for explanation, rather than description.	Account for the loss of biodiversity in coastal wetlands.
Analyse	Identify the main characteristics and rate the factors with respect to importance.	Analyse the social, environmental and economic impacts of exploiting tar sands.
Assess	Examine closely and 'weigh up' a particular situation, e.g. strengths and weaknesses, for and against.	Using examples, assess the view that the relationship between the developed and the developing world is a neo-colonial one.
Comment on	This is asking you to assess a statement. You need to put both sides of the argument.	Comment on the view that the USA should end its dependence on oil.
Compare	Identify similarities *and* differences between two or more things.	Compare the strengths and weaknesses of communism and capitalism as theories.
Contrast	Identify the differences between two or more things.	Contrast the threats to biodiversity in two biodiversity hotspots.
Discuss	Similar to assess.	'Aid donors should retain greater power than the recipients.' Discuss.
Evaluate	The same as assess.	Evaluate this statement. 'The battle between free and fair traders misses the point – it is feeding people at a price they can afford that matters.'
Examine	You need to describe *and* explain.	Examine the attempts to manage the threats to the Daintree rainforest.
Explain	Give reasons why something happens.	Explain how the desire to develop ecosystems threatens them.
How far?	You need to put both sides of an argument.	How far are the conflicts in the Daintree a case of economic versus environmental interests?
Illustrate	Use specific examples to support a statement.	Illustrate the ways in which superpowers use resources as economic weapons.
Justify	Give evidence to support your statements.	Rank the factors affecting access to energy and justify your rankings
Outline	You need to describe *and* explain, but more description than explanation.	Outline the ways in which the environment influences levels of food security at regional, national and international levels.
To what extent?	The same as 'How far?'	To what extent do you agree with the view of the World Bank that 'many countries are poor because they do not use their water potential'?

Planning an answer

All the research into students who plan shows that they get more marks than those who do not. This is because:

- they stay focused – having a plan stops them from going off-track
- they do not suffer from 'memory blanks', in which they forget what to say

So, even if you never have before, learn to plan now! It does not have to be complex, or take long. The plan gets marked – so do not cross it out! Take roughly 5-10% of the exam time to plan your answers. A 35-minute answer will require 2-3 minutes of planning. People plan differently – some make a list, some use spider diagrams, some make notes in the margin. Do what works for you.

Keeping to time

Like planning, keeping to time not only makes sense but will help you earn marks. Bear in mind that you earn most marks in the first half of an answer. Prolonging the answer beyond its time slot, will progressively earn you fewer and fewer marks. Therefore, it is better to draw a slightly incomplete question to a close and start a new one – you'll earn more marks that way.

Generally, use these rules:

- Work out how long a complete question should take to answer, and then how long each sub-question should be given.
- Work on a basis of 5-10% planning time, 80-85% writing time, and 10% checking time.

Learning and using case study examples

Case studies are in-depth examples, of which there are a number in this book. Some are brief (2 pages), while others are much longer (4-6 pages). Apply them to help you answer the exam questions.

However, in comparison with AS, two main differences apply when using case studies at A2:

- One example is occasionally enough at AS (e.g. the study of one coastal stretch under threat from erosion). But, at A2, you need to use *a range* of examples. For instance, one example of a development project in an answer about 'Bridging the development gap' will not normally be enough.
- Make sure that you are not just describing or reeling off points you have learned. At A2, you will need to *apply* the material to the question.

Here is an example where you would need to use – and apply – a case study to answer a question on 'Biodiversity under Threat':

> **b** Evaluate the relative importance of global and local threats to **one named** global ecosystem.

For this question, knowing examples is not enough. You have to:

- understand and be able to write about the **range of threats**, which might mean one set of threats in one location (e.g. increasing tourism in the Daintree), to which are added threats from another (e.g. mangrove depletion).
- be able to say – with evidence – which threats are greater, and which are lesser. Ideally, you would not just say this at the end, but would have built up an answer, progressing from those which are lesser to those which are greater, or vice-versa.

So, knowledge by itself is not enough. Here is another example from a question on 'Water Conflicts':

> **b** Using **named examples**, assess the role of different players and decision-makers in trying to secure a sustainable 'water future'.

This question involves:

- using **more than one** example – you will limit your marks if you restrict yourself to just one
- saying who the different players and decision-makers are for each example, and what part they actually play
- assessing the role that each plays (i.e. how important each player is) in securing a sustainable 'water future'

There are some things you can do to help yourself learn and use case studies.

A portfolio of case studies

Build up a 'portfolio' of case studies, so that you know which part of the Specification and content the case studies fit. A grid like this might help:

Topic name …

Subject content	Relevant theory	Case studies

The grid works like an index – or pigeon hole. If you fill it in, you will know exactly which case study and theory is appropriate for each part of a topic. It also helps to make links between topics.

Remembering case studies

Look again at the question on page 323 from 'Biodiversity under Threat'. There are two case studies in this book that you could use to answer this question, the Daintree rainforest and mangrove ecosystems. One way of remembering these is, *in each case*, to:

● draw a spider diagram with each threat and details about it
● extend the spider with ways in which threats affect the ecosystem (with examples), and identify the threats as global, national or local
● use a large or small '+' around the outside to show how big the threat is – that would help you to assess each threat.

The spider diagram below has been started as an example.

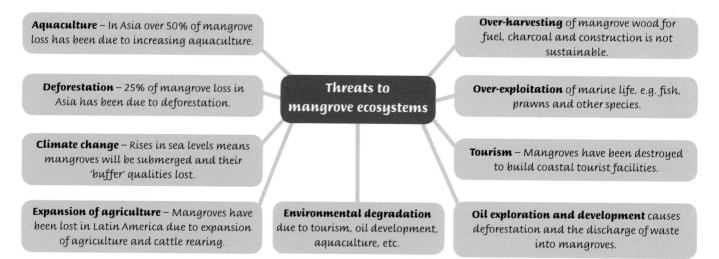

Aquaculture – In Asia over 50% of mangrove loss has been due to increasing aquaculture.

Over-harvesting of mangrove wood for fuel, charcoal and construction is not sustainable.

Deforestation – 25% of mangrove loss in Asia has been due to deforestation.

Over-exploitation of marine life, e.g. fish, prawns and other species.

Climate change – Rises in sea levels means mangroves will be submerged and their 'buffer' qualities lost.

Threats to mangrove ecosystems

Tourism – Mangroves have been destroyed to build coastal tourist facilities.

Expansion of agriculture – Mangroves have been lost in Latin America due to expansion of agriculture and cattle rearing.

Environmental degradation due to tourism, oil development, aquaculture, etc.

Oil exploration and development causes deforestation and the discharge of waste into mangroves.

Learn it!

Finally, make sure you learn the detail. Get basic details right – learn key facts and figures. Accuracy and detail will gain marks. Be careful with dates and places; there is little worse than starting out with *'The Asian tsunami in the Pacific Ocean in 2002 …'*. Common inaccuracies include: incorrect location, poor knowledge of physical processes, lack of terminology (e.g. 'soaks into the ground', instead of 'infiltration'), lack of factual detail (e.g. 'lots of' instead of actual data), poor spelling and quality of grammar and punctuation.

Organising an essay

Beware! Very good students can underachieve on essays. Most commonly, they show excellent knowledge and understanding – in amazing depth – but run out of time, or fail to answer the question set. A weaker student who actually answers the question can do well.

Essay practice – including timed essays – is essential. Essays are assessed in Unit 4, where you need to be able to write a formal essay. These points apply less to Unit 3, where writing needs to be more focused and to the point. Here are some handy tips to add to those about extended writing:

1 **Introduce the essay**

   Define the key terms, set out the context to which the question is referring (e.g. how far biodiversity is under threat), and outline your argument.

2 Arrange the **main body** of the essay. Argue or show knowledge of both sides. Do not just write 'whatever comes next'. Organise your examples so that you:
   - classify those on one side of an argument from those on the other – or into social, economic and environmental
   - develop from those that are major threats to those that are minor
   - progress towards an answer, e.g. from factors that strongly support the argument, to those that support it less so

   In each case, keep coming back to the title – 'This example shows how big a threat …' etc.

3 Summarise your answer to the question in a **conclusion**. Here you should answer the question fully – showing where the balance of the argument lies, or evaluating which are the major threats, for example. Wait until then – do not give it all away in the introduction!

Preparing for the synoptic resources in Unit 3

Four weeks before the Unit 3 exam, you will receive the pre-release resource booklet. This is likely to contain 4-5 pages of resources about ONE of the six Topics in Unit 3. It should be easy to identify which one. For example, the sample materials on Edexcel's website are about GM foods, and therefore link to the Unit 'The Technological Fix?'. Once you have established this, you know that this topic will ONLY be examined in Section B, and NOT in Section A.

Note the following:
- The resource booklet will form the basis of lesson work in the weeks leading up to the exam. Lessons will be like normal, except that you will be going through the pre-release resource materials instead of learning new taught material. Four weeks is plenty of time to grasp the issue and absorb the detail.
- The booklet itself will give you enough detail to answer the exam questions, but will always provide extra website references for further research. Clearly, you will benefit if you follow these up.
- In the exam, Question 6 in Section B will relate to the pre-release resource materials. You are recommended to spend 70 minutes on this question – longer than the others that you will answer on this paper. It carries 40 marks, which will normally be spread over three sub-questions – a, b and c.

To succeed with the synoptic resource questions:

- read the pre-release materials thoroughly. You will not be allowed to take your own copy into the exam, so be sure to keep notes of all the points made in class so that you can prepare.
- use the evidence available in the resource materials (photos, diagrams, tables of statistics, etc.) to support your answers.
- plan what you want to say at the start of every question. You do not need to do much, just make a few notes.
- keep coming back to the question and make sure that you answer it. If the question is:

> Assess the human and environmental impacts of GM farming in Latin America.

  simply listing the impacts will only get you so far. If you classify them into social, economic and environmental, you will do better.
- keep to a rigid time schedule – if you have 70 minutes for 40 marks, then take about 10 minutes for planning, 50 minutes for writing and 10 minutes for reading, checking and adding any other points that come to mind.

How to be synoptic!

The questions in Section B are **synoptic**, i.e. they are intended to draw out what you have learnt across the whole of the AS and A2 course. Using the resource materials, you will be expected to make links between Topics. For instance, the example of GM crops links in with many topics in Unit 3 and elsewhere in the AS and A2 course:

- Biodiversity and its links to economic development
- The development gap – how technology in the developing world may bring benefits or problems
- Water and energy – the resource demands of GM crops

To make sure that you obtain high marks, you need to use examples in your writing which demonstrate what you have learnt. For instance, with the question:

> Assess the human and environmental impacts of GM farming in Latin America.

you should refer to examples where you may have come across similar instances.

You can refer to examples from the AS course (e.g. how you might have studied examples in 'Rebranding Places' where investment in rural areas has helped to develop local or organic foods without GM technology; or where globalisation is linked to the spread of technology), or from Unit 3 in the A2 course (e.g. in 'Bridging the Development Gap' where you found out that some activities of TNCs and international organisations do not always benefit developing countries). These examples need only be brief – a couple of sentences, with names or data – but they will help you to earn maximum marks. You will not achieve maximum marks unless you do this.

How to research for Unit 4

The purpose of Unit 4 is to allow you to investigate areas of geographical study in depth. Although background concepts and theories might be taught in classroom style, your ability to develop independent research skills is as important as the content and concepts. By working independently, you will be able to develop your own level of study, find up-to-date materials, and use your teacher as a tutor rather than someone on whom you rely for everything in your file. In this way, you will get a range of examples to use in the exam. Some will be common to all students, but it is in your interest to find your own materials to prove independent research.

Use a wide range of sources

Until now, you will probably have relied on one or two textbooks, plus teacher materials (e.g. video), to help you through the course. Fieldwork is also likely to have been organised by your teachers. However, this unit now gives you the chance to research for yourself. Make sure that you do. In your essay, evidence of your research will come from how you reference authors and organisations, together with a small bibliography. Make sure that you refer to more than Wikipedia, the BBC or Google! What you need to include is the original sources of data, or named authors with their publications.

When quoting material that you have studied, use the authors' names as proof of your reading and research. For example:

Smith (2007) shows how tectonic hazards are no more frequent now than they were in the early twentieth century. However, Jones (2008) shows that the actual reporting of tsunami now occurs far more frequently since the Asian tsunami of 2004. So hazards aren't more frequent – but we are hearing about them more often.

Where to go for research?

- Use your Internet skills. If you can use Facebook or MySpace, you can research!
- Wikipedia is really useful for basics, such as definitions or basic understanding. However, it is much more useful for follow-up references to other articles and websites, which will provide further detail and also confirmation of information.
- Get to know the websites of the best research organisations. Universities and government organisations publish widely now, so you are more likely to find good sources if you Google 'research in tectonic hazards' rather than just 'tectonic hazards'. Be specific,

e.g. Googling 'data on tectonic hazards' will bring up the CRED database for tectonic hazards.
- Use your school or college library! In particular, ask the librarian to guide you to any resources which are available. Let them know about any books, journals or magazines which you would find useful.
- Have a look at any geography department library that your teachers might have built up. This could include a filing cabinet containing relevant news articles. However, do not expect your teachers to save all of these for you – after all, it is your research!
- Visit a local city or university library, if one is within travelling distance. Many universities will let school students use their libraries. You will not be able to take items home, but many university libraries will let you have a temporary readers' card, especially if your teacher has telephoned in advance to introduce you. Telephone them before visiting and explain what you are looking for, so that you arrive with a clear focus and do not spend time just wandering aimlessly around the library shelves.
- Save newspapers taken by your school/college library or staff room and file away relevant articles. Read newspapers at home and save good articles.

Preparing for the Unit 4 exam

Four weeks before the exam a pre-release statement will be issued by Edexcel, to tell you which part of each Unit 4 Option the exam will assess. Each of the six Options has four Themes (see the first page of Chapters 7-12 in this book), and the exam essay title will be set from a combination of two of them – which will be rotated in each exam cycle. The pre-release statement will tell you which two Themes are going to be assessed this time. You will not be told the title – just the Themes.

Once you know this, you can use your research to help you prepare. Now is the time to intensify your essay practice. The exam will require a formal essay style, with paragraphs, a clear introduction and conclusion, clear syntax, and accurate grammar and spelling. Careful reading of the mark scheme will help you to achieve the best possible result. Regular use of this mark scheme by your teacher when feeding back on essays will also be invaluable.

How are the exam papers marked?

An examiner will mark your exam papers, and will have clear guidance about how to do it. You will be rewarded for what you know and can do, rather than being penalised for what you have left out. All questions at A2 are level marked. If your answer matches the best qualities in the top level of the mark scheme, you will get full marks.

Level marking

Unit 3 Section A questions come in two parts:

- Part (a) is worth 10 marks and has 3 levels in the mark scheme.
- Part (b) is worth 15 marks and has 4 levels in the mark scheme.

Look at this question:

Study Figure 3.

a Explain the pattern of alien species invasions, and suggest the possible impacts of alien species on ecosystems. **(10)**

There are 10 marks for this question and the examiner will award them using the mark scheme on the right.

Level	Mark	Descriptor
Level 1	1-4	The structure is poor or absent. There are one or two basic ideas explaining the pattern. It lacks an understanding of ecosystem processes. It is likely to describe the map. Any explanations are over-simplified and lack clarity. Geographical terminology is rarely used with accuracy. There are frequent grammar, punctuation and spelling errors.
Level 2	5-7	The structure is satisfactory. It explains the pattern with some clarity. There is some understanding about the impact on ecosystems. It uses some incomplete geographical terminology. It makes reference to the map. Some explanations are clear, but there are areas with less clarity. It lacks a full range. There are some grammar, punctuation and spelling errors.
Level 3	8-10	The structure is good. There is a sound explanation of the pattern, an understanding of several processes, and a thorough use of geographical terminology. The answer is likely to illustrate the impact on ecosystems. Descriptive language is precise. Explanations are always clear. Grammar, punctuation and spelling errors are rare.

So, to gain the most marks, your answer needs to meet the criteria for level 3.

Unit 3 Section B is the synoptic investigation, and you will need to show strong synoptic links with other parts of the Specification in your answer in order to gain maximum marks. For example, in this part of a question from Section B:

a Assess the human and environmental impacts of GM farming in Latin America. **(18)**

the synoptic links given in the mark scheme are:

1.2 Migration of people to urban areas

1.3 Questions over the costs and benefits of globalised trade

2.3 Rural inequality – social polarisation and marginalisation of some farmers

3.3 Biodiversity and the threat of economic development

3.6 The neutrality of technological innovation in terms of impacts

Unit 4 – you answer one question and have to write a long essay for which 70 marks are available. Marks are awarded for the following:

- Introducing, defining and focusing on the question
- Researching and methodology
- Analysis, application and understanding
- Conclusions and evaluation
- Quality of written communication and sourcing

Each of the above is divided into four levels, so to gain the most marks your answer needs to meet the criteria for level 4.

How to gain marks and not lose them

Gaining marks instead of losing them is just a matter of technique. You might get a lower mark than you really ought to, not because you do not understand the material but because you do not know how to use it. Here are some helpful hints based on choosing the right question, case studies, the type of question you are answering and communication.

Choose the right question

Do not be seduced by an attractive cartoon or photo – how difficult is the whole question?

Read all the questions carefully and then choose. At this stage, you might want to sketch out a quick few points in pencil to make sure you include examples you think you might forget.

Always choose questions from the part of the specification you have studied – the other questions only look easy because you have not studied them.

If you choose the wrong question have the guts to change in the first 5-6 minutes. If you find yourself running out of time, bullet points are better than no points.

Case studies

Choose the right case study – don't put down something just because you have revised it

Get the length right – look at the mark allocation.

Get a balance between breadth and depth. There is a clear differentiation here between AS and A2. Reference to several examples at this level is better than focusing just on one. A2 requires breadth and understanding of ideas as much as detailed knowledge.

Answer the question

Read the whole question and underline the command words.

Structure your answer into paragraphs.

Include some key words from the question in every paragraph.

Write brief, punchy introductions and conclusions.

Answers must contain evidence, examples or data. Detailed knowledge and understanding will keep you from waffling!

Refer to any resources provided.

Be prepared to use geographical skills, e.g. looking for evidence in photos, reading tables of data to discover trends, interpreting map evidence.

Be accurate when describing trends on graphs or distributions on maps, e.g. 'Biodiversity is greatest in the tropics and declines towards the poles'.

Sketch out a quick plan for your answer. Make a list, a spider diagram or notes – whatever works for you. This will keep you on track and stop you going off the point.

Use maps and diagrams – a well-labelled diagram takes less time than writing a paragraph of text.

Communication

Make sure your writing is legible.

Use correct spelling, punctuation and grammar. Only use abbreviations where they are commonly understood, e.g. UK, USA, UN.

Use a suitable style of writing – you are writing for an intelligent adult.

Write in logical well-planned paragraphs.

Use precise geographical terminology.

Avoid sweeping generalisations.

Use diagrams or sketch maps – these are a simple, effective and impressive way of showing information. They should be used when a sketch is quicker than a text explanation.

GLOSSARY

A

altruism – benefiting other persons or groups of people. Caring about the needs and happiness of other people more than your own

anaerobic – any organism or process which can or must exist without free oxygen from the air

antiretroviral drugs – drugs which inhibit the reproduction of retroviruses (viruses composed of RNA rather than DNA). The best known of this group is HIV, human immunodeficiency virus, which causes AIDS

aquifer – a rock, such as chalk, which will hold water and let it through

areal – relating to a roughly bounded part of the space on a surface; a region

B

bilateral aid – foreign aid (in the shape of money, expertise, education or technology) from a single donor to a recipient country

biofuel – solid, liquid or gas fuel derived from relatively recently dead biological material. This includes ethanol, diesel or other liquid fuels made from processing plant material or waste oil. But it also includes wood, or liquids made from wood and biogas (methane) from animals' excrement

biomanufacturing – concerned with the manufacture of biopharmaceuticals, which are medical drugs produced using biotechnology – the modification of biological organisms, largely through genetic engineering

biome – a naturally occurring community, characterised by distinctive life forms which are adapted to the broad climatic type

British Overseas Territory – fourteen territories that are under the sovereignty of the United Kingdom, but which do not form part of the United Kingdom itself. They are mostly small islands, mainly in the Caribbean and the South Atlantic

C

carrying capacity – the maximum population that a particular environment can sustain without incurring environmental damage

Cold War – this term is used to describe the relationship between America and the Soviet Union from 1945 to the late 1980s. Neither side ever fought the other – both countries had nuclear weapons and the consequences would have been appalling – but they did 'fight' for their beliefs using 'client states' who fought on their behalf, e.g. South Vietnam was anti-communist and was supported by America, while North Vietnam was pro-communist and fought the south (and the Americans) using weapons from communist Russia and communist China

command government – where macro-economic policy and entrepreneurial activity is controlled by the State. The Government decides what to produce, how to produce it and who to produce it for – but with some freedom for individual decisions

commodity – a product or a raw material that can be bought and sold, especially between countries

commodity trading exchanges – these are where various commodities (i.e. raw or primary products) are traded. Most commodity markets across the world trade in agricultural products and other raw materials (like wheat, barley, sugar, maize, cotton, cocoa, coffee, milk products, pork bellies, oil, metals, etc.) and contracts based on them

conservation – the protection of natural or man-made resources for later use

coral bleaching – the loss of colour of corals. Under stress, such as a change in water temperature or a bacterial infection, coral will expel the symbiotic unicellular algae which give it its colour

D

desalination – the conversion of salt water into fresh water by the partial or complete extraction of dissolved solids

devaluation – the reduction in the value of the currency of one country when it is exchanged for the money of another country

digital blackout – where people are without some or all of the following: e-mail, the Internet, television and telephone connections. This may be because of a malfunction or because of a switch to new technology

disparity – a difference, especially one connected with unfair treatment

DNA – the chemical in the cells of animals and plants that carries genetic information. It is a type of nucleic acid

E

ecosystem diversity – the concept that biodiversity (i.e. the varied range of flora and fauna) is essential for the functioning and/or sustainability of an ecosystem

El Niño – a southerly warm ocean current, which develops off the coast of Ecuador about fourteen times a century. It is associated with major variations in tropical climates

electronic colonialism – this is a theory that electronic mass media (film, TV, the Internet and commercials) are having an impact on the minds of audiences around the world, making them think along the same lines and receptive to the same influences. This 'electronic empire' is not based on military power or land acquisition, but on controlling the mind – and the English language is the means by which it achieves its ends

energy poverty – when a country or region has insufficient access to reliable sources of power

energy surplus – when a country or region has more than enough sources of power for its needs and is able to export its surplus power to other countries

environmental determinism – the view that human activities are governed by the physical environment. People and peoples are what they are because they have been shaped by their physical surroundings – climate, vegetation, etc.

erratic – a large boulder which has been transported by a glacier so that it has come to rest on country rock of a different composition and structure

Euro-centric – seeing things from a European point of view

evangelism – persuading people to become Christians by building churches, preaching and setting up schools

F

formal economy – the economy that is regulated by the State. It is economic activity that is taxed and monitored by the Government; and is included in the Government's Gross National Product (GNP)

futures market – this is where futures are traded. A future is a financial contract obligating the buyer to purchase an asset (or the seller to sell an asset), such as a physical commodity, at a predetermined future date and price

G

genetic diversity – the theory that genetic diversity and biodiversity are dependent on each other – that diversity within a species is necessary to maintain diversity among species, and vice versa.

genetic modification – the manipulation of DNA by splitting the DNA molecule and then rejoining it to form a hybrid molecule. This technique bypasses biological restraints to genetic exchange and is used in agriculture to produce 'transgenic' plants, which have greater resistance to pests, herbicides or harsh environmental conditions

Gini co-efficient – a measure of the difference between a given distribution of some variable, like population or income, and a perfectly even distribution

glasnost – a Russian word meaning 'openness'. Policies based on this idea were introduced into the USSR by Mikhail Gorbachev between 1985 and 1990, as part of his wider policy of perestroika. By cultivating a spirit of intellectual and cultural openness, which encouraged public debate and participation, Gorbachev hoped to increase the Soviet people's support for and participation in perestroika.

Green Revolution – the transformation of agriculture that began in 1945 when agricultural research led to the development of more productive varieties of wheat in order to feed the rapidly growing populations of third world countries. The consensus among some agronomists is that the Green Revolution has allowed food production to keep pace with worldwide population growth

groundwater – all water found under the surface of the ground which is not chemically combined with any minerals present, but not including underground streams

H

hegemon – the controlling country or organization within a particular grouping or confederacy

high-pressure system – a region of high atmospheric pressure, otherwise known as an 'anticyclone'. In Britain the term is generally applied to pressures of over 1000 mb

high-technology – the most modern methods and machines, especially electronic ones

I

ideology – a set of ideas that an economic or political system is based on, or a set of beliefs that influences the way people behave

infiltration – the process of water entering rocks or soil

informal economy – all economic activities that fall outside the formal economy regulated by the State. It is economic activity that is neither taxed nor monitored by the Government; and is not included in the Government's Gross National Product (GNP)

information and communications technology (ICT) – this is a blanket term to cover all technologies involved in the manipulation and communication of information

infrastructure – the basic systems and services that are necessary for a country or an organization

intermediate technology – tools and technology for developing countries that are significantly more effective and expensive than traditional methods, but still much cheaper than developed world technology. Such items can be easily bought and used by poor people, and can lead to greater productivity without creating social dislocation. Much intermediate technology can also be built and serviced using locally available materials and knowledge

inter-tidal area – otherwise known as the 'inter-tidal zone' or the 'foreshore', is the area that is exposed to the air at low tide and submerged at high tide, for example, the area between tide marks

irrigation – the supply of water to the land by means of channels, streams and sprinklers in order to permit the growth of crops

L

La Niña – an extensive cooling of the central and eastern Pacific. Globally La Niña means that parts of the world that normally experience dry weather will be drier and those with wet weather will be wetter. Typically La Niña will last for up to 12 months and will be a less-damaging event than the stronger El Niño

lenticels – spongy areas in the corky surface of plant parts, such as twigs and stems, that allow gas exchange between the atmosphere and the internal tissues of the plant. In some plant species, lenticels are raised round dots; they may also be seen as vertical or horizontal slits

low-carbon standard – an initiative first introduced in California in 2007, which is aimed at reducing the carbon intensity of transportation fuel by 10% by 2020. Fuel providers can choose how to achieve this target by various means, including blending low-carbon ethanol into petrol, buying credits from utilities supplying electricity to electric cars, and diversifying into low-carbon hydrogen as a fuel for motor vehicles

M

moraine – any landform directly deposited by a glacier or ice sheet

multilateral aid – foreign aid (in the shape of money, expertise, education or technology) from a group of countries or donors to a single country

multiplier effect – an effect in economics in which an increase in spending produces an increase in national income and consumption greater than the initial amount spent

N

nanoparticle – a small object, used in nanotechnology, that behaves as a whole unit in terms of its transport and properties. It is sized between 1 and 100 nanometres

nationalist – a person (or a political party) who feels that their country should be independent and has a great love and pride for their country. But it can also mean people who think their country is better than any other

neo-liberalism – the doctrine that market exchange is an ethic in itself, capable of acting as a guide for all human action. Under neo-liberalism, State interventions in the economy are minimized, while the obligations of the State to provide for the welfare of its citizens are diminished

net primary productivity – primary production is the production of chemical energy in organic compounds by living organisms. Net primary production is the rate at which all the plants in an ecosystem produce net useful chemical energy. Some net primary production will go towards growth and reproduction of primary producers, while some will be consumed by herbivores

Non-Governmental Organisation (NGO) – a legal organisation created by private organisations or people with no participation or representation by any government. Where NGOs are funded by governments, the NGO maintains its non-governmental status by excluding Government representatives from membership in the organisation

P

pandemic – a disease that spreads over a whole country or over the whole world

pasteurisation – where a liquid, especially milk, is heated to a particular temperature and then cooled in order to kill harmful bacteria

percolation – the filtering of water downwards through soil and through the bedding planes, joints and pores of a permeable rock

perestroika – the policy of economic and governmental reform instituted by Mikhail Gorbachev in the Soviet Union during the mid-1980s. It is a Russian word meaning 'restructuring'

periglacial environments – arid areas with a tundra climate where temperatures are below 0 °C for at least six months, and summers are warm enough to allow surface melting to a depth of approximately one metre

permafrost – areas of rock and soil where temperatures have been below freezing point for at least two years. Permafrost does not have to contain ice – a sub-zero temperature is the sole qualification

pervasive – existing in all parts, or spreading gradually to affect all parts. Technology is spreading everywhere pervasively

photosynthesis – the chemical process by which green plants make organic compounds from atmospheric carbon dioxide and water, in the presence of sunlight

phreatic eruption – a volcanic eruption where meteoric water (water precipitated from the atmosphere) is mixed with lava

pingo – a large ice mound formed under periglacial conditions from an unfrozen pocket confined by approaching permafrost, possibly on a former lake. When the inner ice lens melts, the pingo collapses leaving a depression surrounded by ramparts

pneumatophores – specialized aerial roots which enable plants to breathe air in habitats that have waterlogged soil. The roots may grow down from the stem, or up from typical roots

poverty line – the official level of income that is necessary to be able to buy the basic things you need such as food and clothes and to pay for somewhere to live

precipitation – the deposition of moisture from the atmosphere onto the Earth's surface. This may be in the form of rain, hail, frost, sleet or snow

prevailing – most frequent, or most common

privatisation – the sale of a business or industry so that it is no longer owned by the Government

prop roots – modified roots that grow from the lower part of a stem or trunk down to the ground, providing a plant with extra support

pyroclastic flows – the result of the bursting of gas bubbles within the magma during a volcanic eruption. Lava is fragmented and a dense cloud of fragments is thrown out (a mixture of hot gases, volcanic fragments, ash, and pumice). Pyroclastic flows are also known as 'nuees ardentes'

R

rain shadow – an area of relatively low rainfall to the lee (sheltered from the wind) side of uplands. The incoming air has been forced to rise over the highland, causing precipitation on the windward side

reciprocity – a situation in which two countries or people provide the same help or advantages to each other

relict features – a geomorphological feature which existed under past climatic regimes but still exists as an anomaly in the changed, present-day conditions

relief rainfall – this forms when moisture-laden air masses are forced to rise over high ground. The air is cooled, the water vapour condenses, and precipitation occurs

riparian – relating to a river bank. Owners of land crossed or bounded by a river have 'riparian rights' to use the river

S

security premium – the extra cost built into the price of oil to allow for any disruption in supply

segregation – the act or policy of separating people of different races, religions or sexes and treating them differently

seismic – of an earthquake

spatial imbalance – the uneven distribution or location across a landscape or surface of, for example, population

species diversity – a measure of the diversity within an ecological community that incorporates both species richness (the number of species in a community) and the evenness of species' abundance

strategic – something that is done as part of a plan that is meant to achieve a particular purpose or to gain an advantage

streamflow – the flow of water in streams, rivers, and other channels. It is a major element of the water cycle and the main mechanism by which water moves from the land to the oceans

subsidiary companies – companies that are controlled by a bigger and more powerful company

surface runoff – the movement over ground of rainwater. It occurs when the rainfall is very heavy and when the rocks and soil can absorb no more

sustainable use – a strategy of reducing the environmental impacts associated with resource use and to do so in a growing economy. This would require changes in consumption patterns, better education, new technologies and higher prices for exploiting ecosystems

T

tariff – a tax that is paid on goods coming into or going out of a country

technology poor – places and people who lack access to a regular and reliable source of electricity are technology poor. Where there is electricity, there may not be access to the Internet. The gap between digital 'haves' and 'have-nots' is sometimes referred to as the 'digital divide'

technology rich – places and people who have access to reliable electricity and to a good communications infrastructure are technology rich

tied aid – where foreign aid benefits the donor in the shape of interest payments, access to new markets or by political allegiance. Tied aid sometimes has to be spent in the country providing the aid (the donor country), or in a group of selected countries

U

urbanisation – the migration of rural populations into towns and cities. It indicates a change of employment structure from agriculture and cottage industries to mass production and service industries

V

value-added – the additional value of a good over the cost of commodities used to produce it from the previous stage of production. It refers to the contribution of the factors of production, i.e. land, labour, and capital goods, to raising the value of a product

W

water rights – the legal right of a user to use water from a water source, e.g. a river, stream, pond or source of groundwater

water wars – international conflict as a result of pressure on water supplies. This is an increasing possibility as a growing and increasingly urbanised global population will increase demand for food and water, at the same time as climate change and other trends put greater pressure on their supply

world water gap – the difference between those people, nearly two-thirds of the world's population, who live in water poverty (with either a physical or economic scarcity of water, or both) and those who have ready and reliable access to water for drinking and sanitation

Z

zonation – The distribution of organisms in biogeographic zones

(Page numbers in italics denote illustrations.)

A

Afghanistan, mobile phones 238, *238*
Africa,
    European colonial control *143*
    Internet access 224
    water crisis 76-81, *77*
        rainfall *76*
        River Nile Basin 77, *77*
        Toshka Project 79, *79*
        WaterAid projects 80-1
aid or investment 206-12
Alaska *see* Arctic National Wildlife Refuge (ANWR)
Amazon rainforest, eco-tourism 304-5
Aral Sea, decreasing size *62*
Arctic National Wildlife Refuge (ANWR) 94-9, *95*
    environmental groups 97, 99
    oil exploration 96-9, 273
Arctic peoples 97, 273
Arctic region,
    climate change 271, 272-5, *272*
    control of 28-9, *29*, 272
    energy resources 28-9, *28*, 272-3
    retreat of the ice 275, *275*
    sustainable development 273-4
Argentina, GM crops 285
asbestos 296-301
    cause of death 299-301
    compensation claims 298, 300, 301
    risk assessment 297, 299
Australia,
    Aboriginal people 142
    biodiversity under threat 105
    colonisation 142
    Pilbara
        asbestos 296-8
        iron ore boom 158-9, *158*
    tourism 109, 110-11, 296
    Wet Tropics 106, *106*
    see also Daintree Forest; Murray-Darling Basin

B

Bangladesh,
    development goals 215-16
    Dhaka/flood protection 230-1, *230*
    poverty 215, 230
biodiversity 95, 130-1
    Convention on 131
    global variations 100-1, *100*
    hotspots 101, 104, *105*
    influences on 102
    Millennium Development Goals 130-1
    Millennium Ecosystem Assessment 132-3
    threats to 103-5, *103*, 122, *122*
biomes 101, *101, 102*
    world distribution *101*
Bolivia, water riots 90
British Empire 142, *142*, 172

C

California,
    energy security 8-10, *8*
    low-carbon standard 10
    water supply 52-9
        drought years 53
        environmental consequences 58-9
        natural supplies 54-5, *54*
        Sacramento-San Joaquin River Delta 58, *58*
        Salton Sea *58*, 59
        spatial imbalance 52, *52*
        supply system 53, *53*
        see also Colorado River
Canada,
    Arctic Region 28, *29*
    tar sands 10, 30-1, *30*
capitalism 148
capitalist world system 192-3
    core and periphery model 192
carbon capture technology 254-5
carbon sequestration 109, 124
China,
    coal 14, 19, 20
    earthquakes 268
    economic development 28, *36*, 88-9
    economic superpower 17, 137
    electricity 19, *19*
    energy 15, 17-21, *20*
        consumption/demand 17-18
    natural gas 20
    nuclear power 46
    oil 20, 21, *23*, 159
    raw materials 158-9
    Russian oil 23-4
    as superpower 160-3
        environmental costs 161-2
        military strength 163
        social concerns 162
    water supply 88-9, *88*
        increasing demand 88, *88*
        pollution 89
        Three Gorges Dam 67, 89
climate change 40-1, 124
    the Arctic region 271-5
    CO_2 emissions 40, *40*
CO_2 emissions 244-5, *244*
    India 247, *247*
    Western Europe 245
Cold War 146
colonialism 142, 144-5
    development of 143, *143*
    see also Ghana, neo-colonialism
Colorado River 53, 55-8
    California's water 53
    demand management 57
    river basin 55, *55*, 56-7
    river delta 58
    sharing the water 56-7
communism 148, 164-5
cultural imperialism 174
culture 287
    Americanisation 172-3
    change and diversity 288-93
    diversity under threat 290-2
    global 172-5

D

Daintree rainforest, Australia 106-17
    buy-back land 117
    Douglas Shire Council 115, 116
    environmental threats 112
    logging 109
    management/preservation 115-17, *117*
    tourism 109, 110-12
    tropical rainforest 106-9
debt,
    developing countries 186-9, *188*
    Highly Indebted Poor Countries initiative (HIPC) 187, *187*
deforestation 113, *113*
Dependency theory 153, 154
development,
    aid or investment 206-13
    higher education 236, *236*, 237
    innovation/technology 236-7
    see also Millennium Development Goals
development gap 179, 180-3
    see also Uganda
Development theory 154
Digital Access Index 221, *221*
Dubai 311-18
    development strategy 312, *312*
    economic growth 313
    geographical advantages 313
    new global city 315, *315*, 318
    oil 311, 315
    population surge 314, *314*
    sustainability 317
    water supplies 316, *316*

E

earthquakes 268-9
eco-tourism 304-6
economic man theory 193
ecosystem,
    processes 126
    services 108, *109*
ecosystems,
    tropical rainforest 108-9, *109*
    tundra 95
Ecuador, eco-tourism 306
Egypt, Toshka Project 79, *79*
electricity, world consumption 6
endemism 102, 106
energy
    Arctic Region 28-9
    biofuels 285
    China 17-21
    classification and costs 11
    conservation/recycling 11, 48-9
    demand and supply *18*, 36-9, *37*
    dependency 21
    distribution of sources 12-13, *12*
    environmental costs *11*
    fossil fuel consumption 40-1
    Gazprom 25-6, 27, 274
    hydro-electric power (HEP) 19, 47
    low-carbon standard 10
    micro-hydro generation 16
    pathways 22, 24
    poverty 16

renewable 13, *13*, 15, 37, *37*, 47, *47*
    Slovakia 258, 259
resources 28-31, 36-9
Russia 26
solar 13, 15, 240, 241
UK 14
world consumption 7, 15, *37*
energy security 5
    California 8-10
    Eastern Europe 256-9
    Europe 26-7
    USA 6-10
Ethiopia, water supply 248-51, *248*
    small dams 251
    Tekeze Dam 250
    Tigray 249-51, *249*
EU (European Union),
    Economic Partnership agreements 205
    expansion 168, *168*, 169
    influence of 168-9
Euphrates River, disputed water 69, 70, 71
Euro-centric world view 190-3
Europe 168
    colonial control of Africa *143*
    energy security 26-7
    see also EU
evangelical Christianity 145
exam marking 328-9

F
farming,
    agro-processing 281
    marginal agriculture 280-2
    technology 222-3
food,
    prices 178
    security 179
    supply problems 279
    see also global food crisis; global hunger
fossil fuels, consumption 40-1

G
gas,
    peak gas 39
    pipelines *27*, 45
    reserves 37
    South Caucasus 171, *171*
Gazprom 25-6, *27*, 274
genetic modification see GM
geopolitics 23, 70-1, 72
Germany, Berlin Wall 136
Gersmehl's nutrient cycle 114, *114*
Ghana,
    Akosombo Dam 207
    aluminium smelter project 208
    cocoa trade 153-7, *153*
        farmer cooperatives 157
        World Trade Organisation 155
    neo-colonialism 152-3
    water provision 209
Gini Co-efficient index 196
glaciation and the aftermath 275-7
    Britain 276-7, *277*
global,
    capitalism 141, 148

culture 172-5, 291
    Americanisation 172-3
    News Corporation 175
food crisis 284, *284*, 285
hunger 280, 283, *283*
income levels 183, *183*
globalisation 291
glocalisation 293
GM (genetically modified) crops 222, *222*,
    242-3, 285
Green taxation 49
Gross Domestic Product (GDP) 180, *181*
Gross National Income (GNI) 180
Growth Report 236-7

H
Haiti, food riots 284
HIV/AIDS pandemic 232-3, *232, 233*
    combat with technology 234-5
Human Development Index (HDI) 181-2,
    182

I
ICT (information & communications
    technology) 221
India,
    Bangalore, new economy 198-9, *198*
        future challenges 201
    car ownership 246-7, *246, 247*
    caste system 199, 200
    economic development 28, 36, 137
    film industry 174
    food supply problems 280-2
    Narmada Project 67
    solar energy 241
International Monetary Fund (IMF) 150, 151
    debt cancellation 187
    Structural Adjustment Packages (SAPs)
        186, 203
Internet,
    access 224, *224*
    international communications 220, *220*
Iran,
    culture protected 293
    oil and gas 45
Iraq, US invasion 44
iron ore 159
    Australia and China 158
    superpowers *139*
Israel, water insecurity 69, 72-5
    disputes 72, 73, 74
    Litani River 74, *74*
    management strategies 74-5
    sources 73, *73*

J
Japan,
    earthquakes 269
    economic development 151
    energy insecurity 21
    oil imports *23*
    Russian oil 23-4

K
Kalahandi Syndrome 280-1
Kenya, water use 78

L
Lomonosov Ridge 28, 29, 272
London, flood risk 228, 229, *229*
low-carbon standard 10

M
Mackinder's heartland theory 144, *145*
Malaysia,
    cultural diversity 288-90
    Pergau Dam, funding 210, *210*
mangroves 118-29
    aquaculture, impact of 122, 123
    climate change 124
    ecosystems 120-1
    global distribution 119, *119*
    loss of 122
    significance of 118, 121, *121*
    Sunda Shelf mangroves 118, 119, *119*
    threats to 125, 127
    wetland management 127-9
map projections 190-1
MDGs see Millennium Development Goals
media influence/manipulation 141, 292
Mercator, Gerardus, map projection 190,
    190
Middle East 42-3, *42, 43*
    oil 44-5
        world dependence on *41*, 42
    regional geopolitics 72
    security/stability 42
    water conflict 69
    water stress 68-9, *68, 69*
    see also Egypt; Israel, water insecurity;
        Turkey
Millennium Development Goals 130, 214-17
mobile telephone use 225, *225*, 238-9, *238*
modernism 144
Moldova, UK charity project 212
Montserrat, volcanic activity 264-7
    risk management 266, 267
Murray-Darling Basin 82-7, *82*
    agriculture 84, *84*, 85
    environmental degradation 85
    management of 86-7
    rainfall 83, *83*
    regulation of river flow *84*

N
NGOs (Non-Governmental Organisations)
    206
Nile River Basin 77, *77*
North Africa, water stress 68, *69*
North Atlantic Treaty Organisation (NATO)
    169, *169*
North-South divide 180, *180*, 182
nuclear electricity producers *46*
nuclear power 14, 46, 258
    Slovakia 256, 257, 258
nutrient cycles 114, *114*

O
OECD (Organisation for Economic Co-
    operation and Development) 7
oil 45
    Alaska 96, 99
    Arctic Region 28-9, 272, 274

China 20, 21, 23
consumption 36, *36*
Dubai 311, 315
East Siberia-Pacific Ocean pipeline 22-4, *22*
Kuwait 24, *24*
peak oil 39
pipelines, Central Asia 45, *45*
price 32, *32*, 33, 43, 186, 315
production 38, *38*
reserves 12, 36, *37*
Russia 22-4
USA 6, 7
world producers *17*, *42*
see also Middle East
OPEC (Organisation of the Petroleum Exporting Countries) 32-3, *33*, 43, 186

P
Pakistan, solar energy 240, *240*
Palestine, water management 72-5
Peters, Arno, map projection 190-1, *191*
Philippines 179, 285
Polluter Pays Principle (PPP) 244
pollution 244-5, 295
remediation 297
see also asbestos; CO_2 emissions
Purchasing Power Parity (PPP) 181

R
rail transport, high speed rail 226, *226*, 227
Rainforest Co-operative Research Council, Australia 116
Ramsar Sites 129
religions, world picture 141, *141*
Russia 138
Arctic Region 28, 29
energy strategy 26
oil pipeline/supply 22-4, 256
resurge of power/influence 137, 167, *167*, 170
see also Gazprom

S
sanitation, International Year of Sanitation 63
shrimp farming in Asia 123, *123*
Singapore, economic development 151, 172
Slovakia, energy security 256-9
social Darwinism 145
solar energy 15
India 241, *241*
Pakistan 240, *240*
South Africa 195
income inequalities 194, 196-7, *197*
South Caucasus/gas pipeline 171, *171*
species diversity 95, *95*
superpowers 138
economic *137*, 140, *140*
military force 140, *140*
physical size *138*
population 139, *139*
resources 139

T
tar sands 10, 30-1
technology 221-5
and development 236-7
farming 222-3
global warming 252-5
leapfrogging 239
Thames Barrier 228-9
transport 226ff
see also GM crops
tectonic hazards 263-70
risks/disasters 267, *267*
Thailand,
Had Chao Mai National Park 129
shrimp farming 123, *123*
Thames Barrier 228-9
Tigris River, disputed water 69, 70, 71
tourism and leisure 303-9
eco-tourism 304-6
Grand Canyon 309
Hawaiian islands 308
South American rainforests 304-6
trade and development 202-5
tropical rainforest ecosystem 108-9, *109*
Turkey,
Southeastern Anatolia Project (GAP) 67, 70-1, *70*
water management strategies 69, 70-1

U
Uganda,
bottom-up development 211
coffee 204
debt cancellation 188-9, *189*
development goals 217
Economic Partnership agreements (EU) 205
education 185, 189
IMF/World Bank 203
life expectation 184-5
national debt 186
trade 202-5
traditional values 203
World Trade Organisation 204
UK energy 14, 46, 47
Ukraine, and Russia 170, 256
United Arab Emirates (UAE) 312, 313
United Nations,
Development Programme (UNDP) 214, 215
World Food Programme 283
USA,
energy consumption 6-7
invasion of Iraq 44
modernisation theory 149
oil companies 35
oil consumption 34
political funding 34-5, *34*
superpower 138, 146, 147, *147*
see also California
USSR 146, *146*
collapse of communism 136, 137, 164-5
economic/social impacts 165-6
influence in Africa 148, *149*
superpower 147, *147*
see also Russia

V
virtual water 75, 78, 91
volcanic activity, Montserrat 264-7

W
water, global picture
access and poverty 64, *64*
conflict hotspots 66-7
demand/supply 91
distribution *60*
health and well-being 63
hydrological system *60*
management 66, *66*
pollution 64, *65*
privatisation of supply 67, 90
quality crisis 64-5
rainwater harvesting 91, *91*, 249
sanitation 63
storage of resources *60*
stress and scarcity 61, *61*, 69
use 61, *61*
water wars 66, 67
water, regional issues 62, *62*
Bolivia 90
Ghana 209
Kenya 78
North Africa 68, 69
political threats 63
salinity 65
UK 82
see also Africa, water crisis; California, water supply; China, water supply; Israel, water insecurity; Middle East; Murray-Darling Basin
Wet Tropics Management Authority 115
Wet Tropics World Heritage Area *115*
Wetlands International 128
World Bank 150, 151, 183, 187
World Trade Organisation (WTO) 155, 156, 280
World Water Assessment Programme (WWAP) UN 67
world's largest companies *160*

Y
Yangtze River 88, 89

Z
Zimbabwe, land reform 213